Introduction to Analysis

Introduction to Analysis

by

MAXWELL ROSENLICHT
University of California at Berkeley

DOVER PUBLICATIONS, INC.
New York

This Dover edition, first published in 1986, is an unabridged and unaltered republication of the work first published by Scott, Foresman and Company, Glenview, Illinois, in 1968.

Library of Congress Cataloging-in-Publication Data

Rosenlicht, Maxwell.
Introduction to analysis.

Reprint. Originally published: Glenview, Ill. : Scott, Foresman, and Co., 1968.
Bibliography: p.
Includes index.
1. Mathematical analysis. I. Title.
QA300.R63 1986 515 85-25300
ISBN-13: 978-0-486-65038-8 (pbk.)
ISBN-10: 0-486-65038-3 (pbk.)

Manufactured in the United States by LSC Communications
65038319 2017
www.doverpublications.com

Preface

This text is the outgrowth of a course given at Berkeley since 1960. The object is to redo calculus correctly in a setting of sufficient generality to provide a reasonable foundation for advanced work in various branches of analysis. The emphasis is on abstraction, concreteness, and simplicity. A few abstract ideas are introduced, almost minimal in number. Such important concepts as metric space, compactness, and uniform convergence are discussed in such a manner that they will not need to be redone later. They are given concrete illustration and their worth is demonstrated by using them to prove the results of calculus, generalized in ways that are obviously meaningful and practical.

The background recommended is any first course in calculus, through partial differentiation and multiple integrals (although, as a matter of fact, nothing is assumed except for the axioms of the real number system). A person completing most of the material in this book should not only have a respectable comprehension of basic real analysis but should also be ready to take serious courses in such subjects as integration theory, complex variable, differential equations, other topics in analysis, and general topology. Experience indicates that this material is accessible to a wide range of students, including many with primary interests outside mathematics, provided there is a stress on the easier problems.

The quest for simplicity has resulted in the elimination of a host of mathematical synonyms and the omission or relegation to the problem sets of a number of important ideas. Some problems, easily recognizable, assume a familiarity with linear algebra that was considered unwise to presume of all students. Indeed only a few simple facts on determinants are needed for the chapters on multivariable calculus, but things are so arranged that the instructor who wishes to avail himself of the conveniences of linear algebra may easily do so with no break in continuity. Differential forms were excluded, regrettably, to avoid exorbitant algebraic detours.

This text can be used for courses ranging in length from one to two quarters. The original semester course at Berkeley covered the first eight chapters, with the omission of most of the third section of the eighth chapter. At present Berkeley has a two-quarter sequence with the first quarter, somewhat sped up, covering the first six chapters and parts of the seventh.

Here are some comments on the individual chapters. Chapter I, which discusses material on basic set theory that is familiar to many students, can be covered very rapidly. Chapter II gives a brief account of how all the properties of the real numbers can be deduced from a few axioms. This material can also be covered rapidly. It should not be bypassed, however, for contrary to a widespread faith in modern pedagogy, my experience has been that time spent here is not wasted. Chapters III and IV, on metric spaces and continuous functions, are the meat of the book. They must be done with great care. After this, Chapters V, VI, and VII flow along smoothly, for their substance (elementary calculus) is familiar and the proofs now make sense. Chapter VIII, which is about existence theorems, is of a slightly greater order of difficulty. One may save some time by going very lightly over the implicit function theorem if it is intended to do the general case later, in the following chapter. The last section of the chapter treats ordinary differential equations, and the classroom discussion of these may with relative impunity be restricted to the very first theorem. Chapters IX and X, on multivariable calculus, conclude the book. They should cause no difficulty for anyone who has come this far. However, omitting them entirely would be preferable to an attempt to rush through.

It is impossible to write a text such as this without an obvious indebtedness to J. Dieudonné's classic *Foundations of Modern Analysis*. My gratitude is also due my colleagues of the curricular reform committee at Berkeley which instituted the new Mathematics 104 course, to Mrs. Sandra Cleveland for writing up my original lectures, to Adam Koranyi, whose revised notes have long been in use, for many conversations, to numerous other colleagues and students for their comments, to my family for its patience, and to Scott, Foresman and Company for its affable efficiency.

Berkeley, California — Maxwell Rosenlicht

Contents

Introduction to Analysis

CHAPTER I

Notions from Set Theory

Set theory is the language of mathematics. The most complicated ideas in modern mathematics are developed in terms of the basic notions of set theory. Fortunately the grammar and vocabulary of set theory are extremely simple, at least in the sense that it is possible to go very far in mathematics with only a small amount of set theory. It so happens that the subject of set theory not only underlies mathematics but has become itself an extensive branch of study; however we do not enter deeply into this study because there is no need to. All we must do here is familiarize ourselves with some of the basic ideas so that the language may be used with precision. A first reading of this chapter can be very rapid since it is mainly a matter of getting used to a few words. There is occasional verbosity, directed toward the clarification of certain simple ideas which are really somewhat more subtle than they appear.

§ 1. SETS AND ELEMENTS. SUBSETS.

We do not attempt to define the word *set*. Intuitively a set is a *collection*, or *aggregate*, or *family*, or *ensemble* (all of which words are used synonymously with set) of objects which are called the *elements*, or *members* of the set, and the set is completely determined by the knowledge of which objects are elements of it. We may speak, for example, of the set of students at a certain university; the elements of this set are the individual students there. Similarly we may speak of the set of all real numbers (to be discussed in some detail in the next chapter), or the set of all straight lines in a given plane, etc. It should be noted that the elements of a set may themselves be sets; for example each element of the set of all straight lines in a given plane is a set of points, and we may also consider such less mathematical examples as the set of married couples in a given town, or the set of regiments in an army.

We shall generally use capital letters to denote sets and lower-case letters to denote their elements. The symbol $\in$ is used to denote membership in a set, so that

$$x \in S$$

means that x is an element of the set S. The statement "x is not an element of S" is abbreviated

$$x \notin S.$$

Instead of writing $a \in S$, $b \in S$, $c \in S$ (the commas having the same meaning as "and") we often write $a, b, c \in S$.

The statement that a set is completely determined by its elements may be written as follows: If X and Y are sets then $X = Y$ if and only if, for all x, $x \in X$ if and only if $x \in Y$. Equality here and elsewhere in this book (denoted by $=$) means identity; X and Y happen to be different symbols, but they may very well be different names for the same set, in which case the equation $X = Y$ means that the sets indicated by the symbols X and Y are the same. $X \neq Y$ of course means that the sets indicated by the symbols X and Y are not the same.

Thus we imagine ourselves in a world peopled by certain "objects" (certain of which are called "sets"), and for some pairs of objects x, X, where X is a set, we write $x \in X$, the symbol $\in$ having the property that two sets X and Y are equal if and only if for each object x we have $x \in X$ if and only if $x \in Y$. The symbol $\in$ must also have other properties (which we don't specify here) that enable us, given certain sets, to construct others. The important thing is that everything which follows is expressible in terms of the fundamental relation $x \in X$.

Sets are sometimes indicated by listing all their members between braces. For example, the set

$$\{\text{Jane, Jim}\}$$

has Jane and Jim as its members,

$$\{1, 2, 3, \dots\}$$

(with the three dots read "and so on") is the set of positive integers, and

$$\{a\}$$

is the set having one element, the object named a. (Note that $\{a\}$ is not the same as a. In the same way there is a difference between a university class consisting of one student and the student himself, or between a committee consisting of one person and that person.) The above notation however is not always feasible. A more frequently used notation is

$$\{x : (\text{statement involving } x)\},$$

which means the set of all x for which the statement involving x is true. Thus

$$\{x : x \text{ is a positive integer}\}$$

is the set of positive integers, and for any set S we have

$$S = \{x : x \in S\}.$$

For any set S, the symbol

$$\{x \in S : (\text{statement involving } x)\}$$

denotes the same set as

$$\{x : x \in S \text{ and } (\text{statement involving } x)\},$$

which is the set of all elements of S for which the statement is true. Thus if $\mathbf{R}$ is the set of real numbers,

$$\{x \in \mathbf{R} : x^2 = 1\} = \{1, -1\}.$$

If X and Y are sets and every element of X is also an element of Y, we say that X is a *subset* of Y; this is written

$$X \subset Y, \quad \text{or} \quad Y \supset X.$$

Thus $X \subset Y$ is shorthand for the statement "if $x \in X$ then $x \in Y$". $X = Y$ is equivalent to the two statements $X \subset Y$ and $Y \subset X$. If $X \subset Y$ and $Y \subset Z$ then clearly $X \subset Z$; the two first statements are sometimes written more succinctly

$$X \subset Y \subset Z.$$

For any object x and set X the relation $x \in X$ can now be written in another (less convenient!) way, namely $\{x\} \subset X$. The negation of $X \subset Y$ is written

$$X \not\subset Y \quad \text{or} \quad Y \not\supset X.$$

X is called a *proper subset of* Y if $X \subset Y$ but $X \neq Y$.

The *empty set* is the set with no elements. It is denoted by the symbol $\varnothing$. A source of confusion to beginners is that although the empty set contains *nothing*, it itself is *something* (namely some particular set, the one characterized by the fact that nothing is in it). The set $\{\varnothing\}$ is a set containing exactly one element, namely the empty set. (In a similar way, when dealing with numbers, say with ordinary integers, we must be careful not to regard the number zero as nothing: zero is *something*, a particular number, which represents the number of things in "nothing". Thus zero and $\varnothing$ are quite different, but there is a connection between them in that the set $\varnothing$ has zero elements.) Note that for any set X we have

$$\varnothing \subset X \quad \text{and} \quad X \subset X.$$

A special case of both of these statements is the statement

$$\varnothing \subset \varnothing,$$

which occasions difficulty if, as is often improperly done, one reads "is contained in" for both of the symbols $\subset$ and $\in$. The statement $\varnothing \subset \varnothing$ is true because the statement "for each $x \in \varnothing$ we have $x \in \varnothing$" is obviously true, and also because it is "vacuously true", that is there is no $x \in \varnothing$ for which the statement must be verified, just as the statement "all pigs with wings speak Chinese" is vacuously true.

§ 2. OPERATIONS ON SETS.

If X and Y are sets, the *intersection of* X *and* Y, denoted by $X \cap Y$, is defined to be the set of all objects which are both elements of X and elements of Y. In symbols,

$$X \cap Y = \{x : x \in X \quad \text{and} \quad x \in Y\}.$$

The *union of* X *and* Y, denoted $X \cup Y$, is the set of all objects which are elements of at least one of the sets X and Y. That is. $X \cup Y$ is the set of all objects which are either elements of X or elements of Y (or of both), in symbols

$$X \cup Y = \{x : x \in X \quad \text{or} \quad x \in Y\}.$$

The word "or" is used here in the manner that is standard in mathematics. In ordinary language the word "or" is often exclusive, that is, if A and B are statements, then "A or B" is understood to mean "A or B but not both", whereas in mathematics it *always* means "A or B, or both A and B".

If X is a subset of a set S, then the *complement of X in S* is the set of all elements of S which are not elements of X. If it is explicitly stated, or clear from the context, exactly what the set S is, we often omit the words "in S" and use the notation $\mathcal{C}X$ for the complement of X. That is

$$\mathcal{C}X = \{x \in S : x \notin X\}.$$

These operations are illustrated in Figure 1, where the sets in question are sets of points in plane regions bounded by curves.

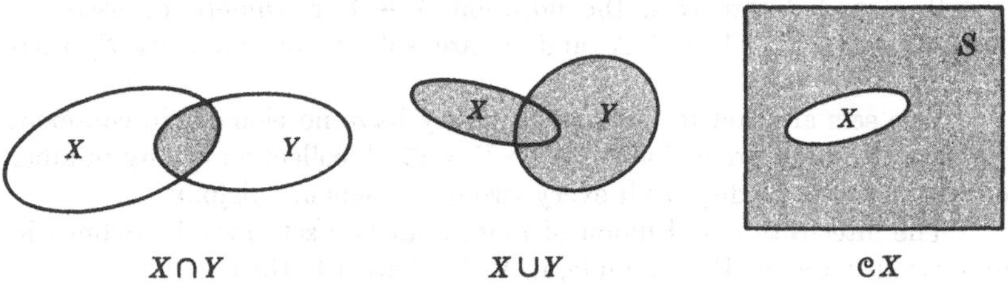

FIGURE 1. Intersection, union, and complement.

For another example, let S be the set of real numbers, and let

$$X = \{x \in S : x \geq 1\}, \quad Y = \{x \in S : 0 \leq x \leq 1\}.$$

(The symbols $\geq$ and $\leq$ will be defined later.) Then

$$\begin{aligned} X \cap Y &= \{1\} \\ X \cup Y &= \{x \in S : x \geq 0\} \\ \mathcal{C}X &= \{x \in S : x < 1\}. \end{aligned}$$

Certain relations hold among the symbols $\cap$, $\cup$, $\mathcal{C}$. For example, if X and Y are subsets of a set S, then

$$\mathcal{C}X \cap \mathcal{C}Y = \mathcal{C}(X \cup Y).$$

This is illustrated in Figure 2. A proof of this formula is given below.

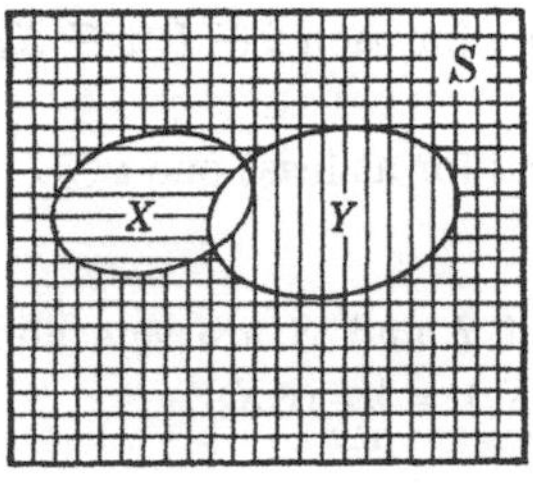

FIGURE 2. $\mathcal{C}X$ is shaded ||||, $\mathcal{C}Y$ is shaded ☰, $\mathcal{C}(X \cup Y)$ is shaded ⊞, illustrating $\mathcal{C}X \cap \mathcal{C}Y = \mathcal{C}(X \cup Y)$.

EXERCISE. Prove that if $X \subset S$, $Y \subset S$, then $\mathcal{C}X \cap \mathcal{C}Y = \mathcal{C}(X \cup Y)$.

It must be shown that the two sets have the same elements, in other words that each element of the set on the left is an element of the set on the right and vice versa.

If $x \in \mathcal{C}X \cap \mathcal{C}Y$, then $x \in \mathcal{C}X$ and $x \in \mathcal{C}Y$. This means that $x \in S$, $x \notin X$, $x \notin Y$. Since $x \notin X$, $x \notin Y$, we know that $x \notin X \cup Y$. Hence $x \in \mathcal{C}(X \cup Y)$.

Conversely, if $x \in \mathcal{C}(X \cup Y)$, then $x \in S$ and $x \notin X \cup Y$. Therefore $x \notin X$ and $x \notin Y$. Thus $x \in \mathcal{C}X$ and $x \in \mathcal{C}Y$, so that $x \in \mathcal{C}X \cap \mathcal{C}Y$. This completes the proof.

If X and Y are sets, the notation $X - Y$ is sometimes used for $\{x \in X : x \notin Y\}$. Thus if X and Y are subsets of some set S, then $X - Y = X \cap \mathcal{C}Y$.

Two sets are said to be *disjoint* if they have no element in common. That is, X and Y are disjoint if $X \cap Y = \emptyset$. A collection of any number of sets is said to be disjoint if every two of the sets are disjoint.

The intersection and union of more than two sets may be defined in an obvious manner. For example, if X, Y, Z are sets then

$$X \cap Y \cap Z = \{x : x \in X,\ x \in Y,\ x \in Z\},$$

and

$$X \cup Y \cup Z = \{x : x \in X \quad \text{or} \quad x \in Y \quad \text{or} \quad x \in Z\}.$$

Clearly $X \cap Y \cap Z = (X \cap Y) \cap Z = X \cap (Y \cap Z)$, and similarly for the union of three sets. More generally the intersection and union of arbitrary families of sets may be defined, and in an obvious way. The only problem is finding an adequate notation for an arbitrary family of sets, and this is done as follows. Let I be any set and for each $i \in I$ let X_i be another set (so that we may speak of I as being an *indexing family*, whose elements are *indices* used to specify the sets at which we direct our main attention). The set of all sets X_i as i ranges over I is denoted

$$\{X_i : i \in I\} \quad \text{or} \quad \{X_i\}_{i \in I}$$

and the *intersection* and *union* of this family of sets, together with their respective conventional symbols, are defined by

$$\bigcap_{i \in I} X_i = \{x : \text{for each } i \in I, x \in X_i\},$$

$$\bigcup_{i \in I} X_i = \{x : \text{for at least one } i \in I, x \in X_i\}.$$

EXERCISE. Prove that if I and S are sets and if for each $i \in I$ we have $X_i \subset S$, then $\mathcal{C}(\bigcap_{i \in I} X_i) = \bigcup_{i \in I} (\mathcal{C}X_i)$.

It must be shown that each element of the set on the left is an element of the set on the right, and vice versa.

If $x \in \mathcal{C}(\bigcap_{i \in I} X_i)$ then $x \in S$ and $x \notin \bigcap_{i \in I} X_i$. Therefore $x \notin X_j$, for at least one $j \in I$. Thus $x \in \mathcal{C}X_j$, so that $x \in \bigcup_{i \in I} (\mathcal{C}X_i)$.

Conversely, if $x \in \bigcup_{i \in I} (\mathcal{C}X_i)$, then for some $j \in I$ we have $x \in \mathcal{C}X_j$. Thus $x \in S$ and $x \notin X_j$. Since $x \notin X_j$ we have $x \notin \bigcap_{i \in I} X_i$. Therefore $x \in \mathcal{C}(\bigcap_{i \in I} X_i)$. This completes the proof.

If a and b are objects, by the *ordered pair* (a, b) we mean the two objects a and b in a definite order, a first, b second. Thus if a, b, c, d are objects then $(a, b) = (c, d)$ if and only if $a = c$ and $b = d$. Note the distinction between (a, b) and $\{a, b\}$; the latter is a set with two elements (unless, of course, a happens to equal b, in which case $\{a, b\} = \{a\}$, a set with one element), and $\{a, b\}$ can equally well be written $\{b, a\}$, spoiling the order. We remark that instead of introducing the new concept "ordered pair" into set theory, we can actually define the ordered pair (a, b) in terms of the primitive notions about sets that we already have: we set $(a, b) = \{\{a\}, \{a, b\}\}$. This definition does precisely what we want: to any two objects a, b (distinct or not) it assigns an object (a, b), and it does this in such a fashion that $(a, b) = (c, d)$ if and only if $a = c$ and $b = d$.

Given two sets X and Y we define the *cartesian product* (or *product*) *of* X *and* Y, denoted $X \times Y$, to be the set of all ordered pairs the first member of which is in X, the second in Y, that is

$$X \times Y = \{(x, y) : x \in X,\ y \in Y\}.$$

Ordinary rectangular coordinates in the plane give the usual pictorial representation of the cartesian product: the whole plane can be identified with the product of the two coordinate axes. In Figure 3 there is a more complicated picture in which X, Y are subsets of the two coordinate axes and the cartesian product is a subset of the first quadrant.

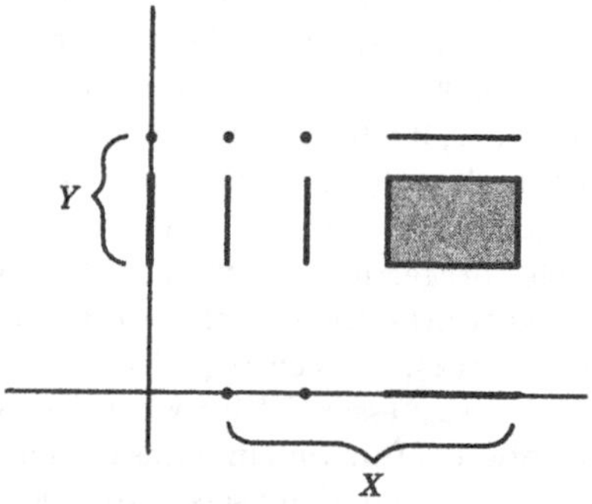

FIGURE 3. Cartesian product.

§ 3. FUNCTIONS.

If X and Y are sets, by a *function from X to Y* (or a *function from X into Y*, or a *function on X with values in Y*) is meant a rule which associates with each element of X a definite element of Y. (The word *mapping*, or *map*, is often used instead of *function*.) The "rule" can be given in many ways, some of which are discussed below, but the essential thing is that given any element of X there is associated, somehow, some definite element of Y. Two functions from X to Y are considered equal if and only if both functions associate with each specific element of X the same element of Y.

Functions are usually denoted by small letters, such as f. The statement "f is a function from X to Y" is often written

$$f\colon X \to Y.$$

For any $x \in X$, the element of Y that the function f associates with x (the *value of f at x*) is denoted $f(x)$. Thus if $f\colon X \to Y$ and $g\colon X \to Y$ then $f = g$ if and only if $f(x) = g(x)$ for all $x \in X$. We say that f *sends x into $f(x)$*, or that f *maps x into $f(x)$*, or that *x and $f(x)$ correspond under f*.

The rule defining a given function $f\colon X \to Y$ may be given in various ways. One way, which is usually not very practical, is to list all the elements of X, listing with each one the corresponding one of Y. Or the rule may be given by a mathematical formula. For example, if X and Y are both taken to be the set $\mathbf{R}$ of real numbers, an equation like

$$f(x) = x^3 + 3x - 2$$

defines a real-valued function f on $\mathbf{R}$; in such a case one often speaks (imprecisely!) of the function $x^3 + 3x - 2$. Again, if X is a subset of $\mathbf{R}$, a real-valued function on X may be given geometrically by its *graph*, that is the set of points $\{(x, f(x)) : x \in X\}$ in the plane; note that this method may or may not be practical, depending on what f is like, for it may not be possible to "draw" the graph. In fact any subset of the plane defines a real-valued function on a subset of the real numbers, provided that any vertical line (x = constant) intersects the subset of the plane in at most one point. Finally we remark that the "rule" defining a function need not be practically computable. For example, for x any real number, let $f(x)$ denote that integer $0, 1, \ldots, 9$ which is in the billionth decimal place of x (to be precise, since a real number x may have more than one decimal representation, as in $1.0000\ldots = .9999\ldots$, we might better take $f(x)$ to be the smallest possible integer in the billionth decimal place of x); this rule gives an honest-to-goodness function $f\colon \mathbf{R} \to \mathbf{R}$, but who would hazard a guess as to the value of $f(\pi)$, or even $f(\sqrt{2})$?

Though this is in no way essential for what follows, we remark that it is easy to define the notion of function in terms of more primitive concepts of set theory, as follows: If X and Y are sets, a function from X into Y is a subset of $X \times Y$ with the property that for any $x \in X$ there is one and only one $y \in Y$ such that (x, y) is in the subset. If the function is denoted $f\colon X \to Y$ then the unique y referred to above is, of course, $f(x)$. In analogy with the case of real-valued functions on the real numbers, this subset $\{(x, f(x)) : x \in X\}$ of $X \times Y$ is called the *graph* of the function, so that we are again saying that the graph determines the function.

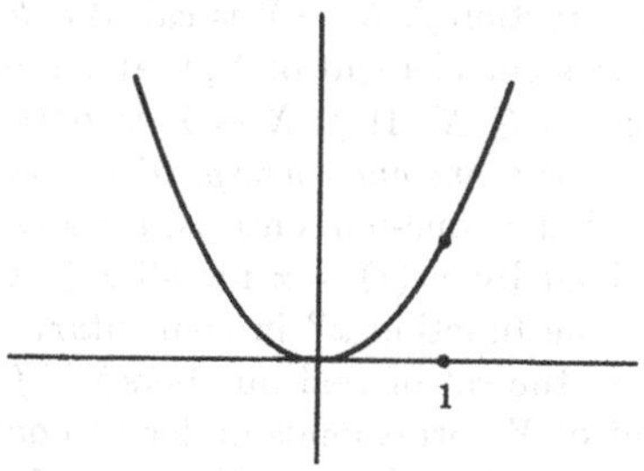

FIGURE 4. Graph of the function $f\colon \mathbf{R} \to \mathbf{R}$ given by $f(x) = x^2$ for all $x \in \mathbf{R}$.

It is useful to note that the word "function" alone can be defined in primitive terms, not only the more complete concept "function from X into Y": a function is an ordered pair whose first member is an ordered pair of sets, say (X, Y), and whose second member is a function from X into Y. That is, a function is something of the type $((X, Y)$, (a certain kind of subset of $X \times Y$)). This emphasizes that the sets X and Y are to be considered as essential parts of the function $f\colon X \to Y$. For many purposes it is important to bear this fact in mind, but most often we do not make any explicit mental note of it. For example, if $f\colon X \to Y$ is a function and Y is a subset of another set Y', then we get a function $f'\colon X \to Y'$ by setting $f'(x) = f(x)$ for all $x \in X$, that is, giving f' the same graph as f, which is possible since $X \times Y \subset X \times Y'$. Although f and f' are really different functions, we usually denote them by the same symbol, even writing down the technically incorrect expression $f\colon X \to Y'$. In the same way, given any function $f\colon X \to Y$ and any subset X'' of X, we can define a function $f''\colon X'' \to Y$ by $f''(x) = f(x)$ for all $x \in X''$; f'' is called the *restriction of f to X''*. Of course, f'' is not the same as f, though we often denote them by the same symbol, for example by writing the technically incorrect expression $f\colon X'' \to Y$ when there is no danger of confusion. The graph of the latter function is of course a subset of the graph of the original function $f\colon X \to Y$, since $X'' \times Y \subset X \times Y$.

If $f\colon X \to Y$ and $g\colon Y \to Z$ are functions, one can define the *composition of f and g*, or *composed function*, a function from X into Z, by associating to each element of X an element of Z in the obvious way: given an element of X, one first uses f to get an element of Y, then one uses g to get from this last element an element of Z. The composed function is usually denoted $g \circ f$, so that we have

$$g \circ f: X \to Z,$$

with $(g \circ f)(x) = g(f(x))$ for each $x \in X$.

A function $f: X \to Y$ is called *one-to-one*, or *one-one*, if different elements of X correspond under f to different elements of Y, that is if $f(x_1) = f(x_2)$ only if $x_1 = x_2$. A function $f: X \to Y$ is called *onto* if each element of Y corresponds under f to some element of X, that is if each $y \in Y$ is of the form $y = f(x)$, for some $x \in X$. If $f: X \to Y$ is both one-one and onto it is called *one-one onto*, or a *one-one correspondence between X and Y*. An example of a function that is one-one onto is, for any set X, the *identity function* $i_X: X \to X$, given by $i_X(x) = x$ for all $x \in X$. (Note that what goes under the name "the function x" in elementary calculus is actually the identity function on the set of real numbers.) If $f: X \to Y$ is one-one onto then each element of Y corresponds under f to one and only one element of X, so we can define a function $f^{-1}: Y \to X$ by $f^{-1}(y) = x$ if $y = f(x)$. f^{-1} is called the *inverse function of f*, and is also one-one onto. Clearly $(f^{-1})^{-1} = f$, and $f^{-1} \circ f = i_X$, $f \circ f^{-1} = i_Y$.

If $f: X \to Y$ is a function and $X' \subset X$, then the subset of Y given by

$$f(X') = \{f(x) : x \in X'\}$$

(where the last symbol is shorthand for $\{y :$ there exists $x \in X'$ such that $y = f(x)\}$) is called the *image of X' under f*, or simply the *image of X'*, if there is no danger of confusion. The two uses we have made for the symbol $f(\ )$ are related by the equation

$$f(\{x\}) = \{f(x)\} \text{ for each } x \in X.$$

If $f: X \to Y$ is a function and $Y' \subset Y$, then the subset of X given by

$$f^{-1}(Y') = \{x \in X : f(x) \in Y'\}$$

is called the *inverse image of Y' under f*. It consists of those elements of X which correspond under f to elements of Y'. If $f: X \to Y$ happens to be one-one onto we have another use for the symbol f^{-1}, namely the inverse function $f^{-1}: Y \to X$. These two uses of f^{-1} must be carefully distinguished, though confusion rarely arises. If $f: X \to Y$ is one-one onto then the two uses of f^{-1} are related by $\{f^{-1}(y)\} = f^{-1}(\{y\})$ for each $y \in Y$.

§ 4. FINITE AND INFINITE SETS.

We are familiar with the set of *positive integers*, or *natural numbers* $\{1, 2, 3, \ldots\}$. This set, together with the various ideas associated with it, such as its ordering (the fact that its elements can be written down in a definite order), or such as the fact that two of its elements may be added to obtain a third with certain general rules holding for this addition, can be obtained from the primitive principles of set theory. In this text we shall instead assume the basic properties of the real number system and from those derive all the properties of the set $\{1, 2, 3, \ldots\}$. In this section we shall for convenience assume a few simple facts about the natural numbers in order to get as quickly as possible to certain other easy matters of set theory. However all the facts about the set of natural numbers that are used here will be proved explicitly in the next chapter. The notions developed in this section will not be applied until later, so no circular reasoning occurs.

Let us therefore assume knowledge of the set $\{1, 2, 3, \ldots\}$. A set X is called *finite* if it is empty or there is a positive integer n such that X can be put into one-one correspondence with the set $\{1, 2, \ldots, n\}$, that is there is a one-one function from $\{1, 2, \ldots, n\}$ onto X. Thus a set is finite if we can count its elements and run out of elements after we count a certain number, say n, of them. The number n depends only on the set X, not on the order in which its elements are counted off; n is called the *number of elements in* X. For completeness we say that the number of elements in the empty set is zero. Any subset of a finite set is itself finite, and if it is a proper subset it has a smaller number of elements.

A set is *infinite* if it is not finite. This means that we can pick an element $x_1 \in X$, then an element x_2 from the complement of $\{x_1\}$, then an element x_3 from the complement of $\{x_1, x_2\}$, etc., and we never run out of elements of X. Thus there exist distinct elements $x_1, x_2, x_3, \ldots$ in X.

It is easy to show that a set X is infinite if and only if it may be put into one-one correspondence with a proper subset of itself. To do this, note first that if X is finite then any proper subset has a smaller number of elements, whereas two finite sets in one-one correspondence must have the same number of elements. This proves the "if" part. On the other hand, if X is infinite then there exist distinct elements $x_1, x_2, x_3, \ldots$ in X. The complement of $\{x_1, x_2, x_3, \ldots\}$ in X is a subset Y, so that

$$X = \{x_1, x_2, x_3, \ldots\} \cup Y \text{ and } \{x_1, x_2, x_3, \ldots\} \cap Y = \emptyset.$$

A one-one correspondence between X and its proper subset $\{x_2, x_3, \ldots\} \cup Y$ is given by the function which sends each x_n into x_{n+1} and each element of Y into itself. This proves the "only if" part, completing the proof.

The set of natural numbers can be used to give an easy definition of the notion of sequence. A *sequence of n elements in a set X*, or an *n-tuple of elements of X*, may be defined to be a function from $\{1, 2, \ldots, n\}$ into X; if the function is denoted f and we write $f(1) = x_1, f(2) = x_2, \ldots, f(n) = x_n$, then the n-tuple is often written $(x_1, x_2, \ldots, x_n)$. An *infinite sequence of elements of X* (or a *sequence of elements of X*, if there isn't any danger of confusion with finite sequences) is a function from the set of all natural numbers into X; as above this can be written $(x_1, x_2, x_3, \ldots)$, or, more conventionally, $x_1, x_2, x_3, \ldots$, or sometimes $\{x_n\}_{n=1,2,3,\ldots}$.

PROBLEMS

1. Let $\mathbf{R}$ be the set of real numbers and let the symbols $<$, $\leq$ have their conventional meanings.
   (a) Show that
   $$\{x \in \mathbf{R} : 0 \leq x \leq 3\} \cap \{x \in \mathbf{R} : -1 < x < 1\} = \{x \in \mathbf{R} : 0 \leq x < 1\}.$$
   (b) List the elements of
   $$(\{2, 3, 4\} \cup \{x \in \mathbf{R} : x^2 - 4x + 3 = 0\}) \cap \{x \in \mathbf{R} : -1 \leq x < 3\}.$$
   (c) Show that
   $$(\{x \in \mathbf{R} : -2 \leq x \leq 0\} \cup \{x \in \mathbf{R} : 2 < x < 4\}) \cap \{x \in \mathbf{R} : 0 \leq x \leq 3\} = \{x \in \mathbf{R} : 2 < x \leq 3\} \cup \{0\}.$$

2. If A is a subset of the set S, show that
   (a) $\mathcal{C}(\mathcal{C}A) = A$
   (b) $A \cup A = A \cap A = A \cup \varnothing = A$
   (c) $A \cap \varnothing = \varnothing$
   (d) $A \times \varnothing = \varnothing$.

3. Let A, B, C be subsets of a set S. Prove the following statements and illustrate them with diagrams.
   (a) $\mathcal{C}A \cup \mathcal{C}B = \mathcal{C}(A \cap B)$
   (b) $A \cap (B \cup C) = (A \cap B) \cup (A \cap C)$
   (c) $A \cup (B \cap C) = (A \cup B) \cap (A \cup C)$.

4. If A, B, C are sets, show that
   (a) $(A - B) \cap C = (A \cap C) - B$
   (b) $(A \cup B) - (A \cap B) = (A - B) \cup (B - A)$
   (c) $A - (B - C) = (A - B) \cup (A \cap B \cap C)$
   (d) $(A - B) \times C = (A \times C) - (B \times C)$.

5. Let I be a nonempty set and for each $i \in I$ let X_i be a set. Prove that
   (a) for any set B we have
   $$B \cap \bigcup_{i \in I} X_i = \bigcup_{i \in I} (B \cap X_i)$$
   (b) if each X_i is a subset of a given set S, then
   $$\mathcal{C}\left(\bigcup_{i \in I} X_i\right) = \bigcap_{i \in I} \mathcal{C}X_i.$$

6. Prove that if $f: X \to Y$, $g: Y \to Z$, and $h: Z \to W$ are functions, then
   $$h \circ (g \circ f) = (h \circ g) \circ f.$$

7. Let $f: X \to Y$ be a function, let A and B be subsets of X, and let C and D be subsets of Y. Prove that
   (a) $f(A \cup B) = f(A) \cup f(B)$
   (b) $f(A \cap B) \subset f(A) \cap f(B)$
   (c) $f^{-1}(C \cup D) = f^{-1}(C) \cup f^{-1}(D)$
   (d) $f^{-1}(C \cap D) = f^{-1}(C) \cap f^{-1}(D)$
   (e) $f^{-1}(f(A)) \supset A$
   (f) $f(f^{-1}(C)) \subset C$.

8. Under the assumptions of Problem 7, prove that f is one-one if and only if the sign $\supset$ in (e) can be replaced by $=$ for all $A \subset X$, and f is onto if and only if the sign $\subset$ in (f) can be replaced by $=$ for all $C \subset Y$.

9. How many subsets are there of the set $\{1, 2, 3, \ldots, n\}$? How many maps of this set into itself? How many maps of this set onto itself?

10. (a) How many functions are there from a nonempty set S into $\varnothing$?
(b) How many functions are there from $\varnothing$ into an arbitrary set S?
(c) Show that the notation $\{X_i\}_{i \in I}$ implicitly involves the notion of function.

CHAPTER II

The Real Number System

The real numbers are basic to analysis, so we must have a clear idea of what they are. It is possible to construct the real number system in an entirely rigorous manner, starting from careful statements of a few of the basic principles of set theory,* but we do not follow this approach here for two reasons. One is that the detailed construction of the real numbers, while not very difficult, is time-consuming and fits more properly into a course on the foundations of arithmetic, and the other reason is that we already "know" the real numbers and would like to get down to business. On the other hand we have to be sure of what we are doing. Our procedure in this book is therefore to *assume* certain basic properties (or axioms) of the real number system, all of which are in complete agreement with our intuition and all of which can be proved easily in the course of any rigorous construction of the system. We then sketch how most of the familiar properties of the real numbers are consequences of the basic properties assumed and how these properties actually completely determine the real numbers. The rest of the course will be built on this foundation.

---

* The standard procedure for constructing the real numbers is as follows: One first uses basic set theory to define the natural numbers $\{1, 2, 3, \ldots\}$ (which, to begin with, are merely a set with an order relation), then one defines the addition and multiplication of natural numbers and shows that these operations satisfy the familiar rules of algebra. Using the natural numbers, one then defines the set of integers $\{0, \pm 1, \pm 2, \ldots\}$ and extends the operations of addition and multiplication to all the integers, again verifying the rules of algebra. From the integers one next obtains the rational numbers, or fractions. Finally, from the rational numbers one constructs the real numbers, the basic idea in this last step being that a real number is something that can be approximated arbitrarily closely by rational numbers. (The manufacture of the real numbers may be witnessed in E. Landau's *Foundations of Analysis*.)

§ 1. THE FIELD PROPERTIES.

We define the *real number system* to be a set **R** together with an ordered pair of functions from $\mathbf{R} \times \mathbf{R}$ into **R** that satisfy the seven properties listed in this and the succeeding two sections of this chapter. The elements of **R** are called *real numbers,* or just *numbers.* The two functions are called *addition* and *multiplication,* and they make correspond to an element $(a, b) \in \mathbf{R} \times \mathbf{R}$ specific elements of **R** that are denoted by $a + b$ and $a \cdot b$ respectively.

We speak of *the* real number system, rather than *a* real number system, because it will be shown at the end of this chapter that the listed properties completely determine the real numbers, in the sense that if we have two systems which satisfy our properties then the two underlying sets **R** can be put into a unique one-one correspondence in such a way that the functions $+$ and $\cdot$ agree. Thus the basic assumption made in this chapter is that *a* system of real numbers exists.

The five properties listed in this section are called the *field properties* because of the mathematical convention calling a *field* any set, together with two functions $+$ and $\cdot$, satisfying these properties. They express the fact that the real numbers are a field.

PROPERTY I. (COMMUTATIVITY). *For every* $a, b \in \mathbf{R}$, *we have* $a + b = b + a$ and $a \cdot b = b \cdot a$.

PROPERTY II. (ASSOCIATIVITY). For every $a, b, c \in \mathbf{R}$, *we have* $(a + b) + c = a + (b + c)$ and $(a \cdot b) \cdot c = a \cdot (b \cdot c)$.

PROPERTY III. (DISTRIBUTIVITY). *For every* $a, b, c \in \mathbf{R}$, *we have* $a \cdot (b + c) = a \cdot b + a \cdot c$.

PROPERTY IV. (EXISTENCE OF NEUTRAL ELEMENTS). *There are distinct elements* 0 *and* 1 *of* **R** *such that for all* $a \in \mathbf{R}$ *we have* $a + 0 = a$ *and* $a \cdot 1 = a$.

PROPERTY V. (EXISTENCE OF ADDITIVE AND MULTIPLICATIVE INVERSES). *For any* $a \in \mathbf{R}$ *there is an element of* **R**, *denoted* $-a$, *such that* $a + (-a) = 0$, *and for any nonzero* $a \in \mathbf{R}$ *there is an element of* **R**, *denoted* a^{-1}, *such that* $a \cdot a^{-1} = 1$.

Most of the rules of elementary algebra can be justified by these five properties of the real number system. The main consequences of the field properties are given in paragraphs F 1 through F 10 immediately below, together with brief demonstrations. We shall employ the common notational conventions of elementary algebra when no confusion is possible.

For example, we often write ab for $a \cdot b$. One such convention is already implicit in the statement of the distributive property (Property III above), where the expression $a \cdot b + a \cdot c$ is meaningless unless we know the order in which the various operations are to be performed, that is how parentheses should be inserted; by $a \cdot b + a \cdot c$ we of course mean $(a \cdot b) + (a \cdot c)$.

F 1. In a sum or product of several real numbers parentheses can be omitted. That is, the way parentheses are inserted is immaterial. Thus if $a, b, c, d \in \mathbf{R}$, the expression $a + b + c + d$ may be defined to be the common value of $(a + (b + c)) + d = ((a + b) + c) + d = (a + b) + (c + d) = a + (b + (c + d)) = \cdots$; that these expressions with parentheses indeed possess a common value can be shown by repeated application of the associative property. The general fact (with perhaps more than four summands or factors) can be proved by starting with any meaningful expression involving elements of $\mathbf{R}$, parentheses, and several $+$'s or several $\cdot$'s, and repeatedly shoving as many parentheses as possible all the way to the left, always ending up with an expression of the type $((a + b) + c) + d$.

F 2. In a sum or product of several real numbers the order of the terms is immaterial. For example

$$a \cdot b \cdot c = b \cdot a \cdot c = c \cdot b \cdot a = \cdots.$$

This is shown by repeated application of the commutative property (together with F 1).

F 3. For any $a, b \in \mathbf{R}$ the equation $x + a = b$ has one and only one solution. For if $x \in \mathbf{R}$ is such that $x + a = b$, then $x = x + 0 = x + (a + (-a)) = (x + a) + (-a) = b + (-a)$, so $x = b + (-a)$ is the only possible solution; that this is indeed a solution is immediate. One consequence is that the element 0 of Property IV is unique; another is that for any $a \in \mathbf{R}$, the element $-a$ of Property V is unique.

For convenience, instead of $b + (-a)$ one usually writes $b - a$. (This is a definition of the symbol "$-$" between two elements of $\mathbf{R}$.) Thus $-a = 0 - a$.

We take the opportunity to reiterate here the important role of convention. $a + b + c$ has been defined (and by F 1 there is only one reasonable way to define it), but we have not yet defined $a - b - c$. Of course by the latter expression we understand $(a - b) - c$, but it is important to realize that this is merely convention, and reading aloud the words "a minus b minus c" with a sufficient pause after the first "minus" points out that our convention could equally well have defined $a - b - c$ to be $a - (b - c)$. In this connection note the absence of *any* standard convention for $a \div b \div c$. In a similar connection, note that a^{b^c} could be taken to mean $(a^b)^c$ if it were not conventionally taken to mean $a^{(b^c)}$. As stated above we use all the ordinary notational conventions when no confusion can result. For example, without further ado we shall interpret an expression like $\log a^b$ to mean $\log (a^b)$ and not $(\log a)^b$, ab^{-1} does not mean $(ab)^{-1}$, etc.

F 4. For any $a, b \in \mathbf{R}$, with $a \neq 0$, the equation $xa = b$ has one and only one solution. In fact from $xa = b$ follows $x = xaa^{-1} = ba^{-1}$, and from $x = ba^{-1}$ follows $xa = b$. Thus the element 1 of Property IV is unique and, given any $a \in \mathbf{R}$, $a \neq 0$, the element a^{-1} of Property V is unique.

For $a, b \in \mathbf{R}$, $a \neq 0$, we define b/a, in accord with convention, to be $b \cdot a^{-1}$. In particular, $a^{-1} = 1/a$.

F 5. For any $a \in \mathbf{R}$ we have $a \cdot 0 = 0$. This is true since $a \cdot 0 + a \cdot 0 = a \cdot (0 + 0) = a \cdot 0 = a \cdot 0 + 0$, so that $a \cdot 0$ and 0 are both solutions of the equation $x + a \cdot 0 = a \cdot 0$, hence equal, by F 3. From this it follows immediately that if a product of several elements of $\mathbf{R}$ is 0 then one of the factors must be 0: for if $ab = 0$ and $a \neq 0$ we can multiply both sides by a^{-1} to get $b = 0$. Hence the illegitimacy of division by zero.

F 6. $-(-a) = a$ for any $a \in \mathbf{R}$. For both $-(-a)$ and a are solutions of the equation $x + (-a) = 0$, hence equal, by F 3.

F 7. $(a^{-1})^{-1} = a$ for any nonzero $a \in \mathbf{R}$. In fact since $a \cdot a^{-1} = 1$, by F 5 we know that $a^{-1} \neq 0$, so $(a^{-1})^{-1}$ exists, and F 4 implies that $(a^{-1})^{-1}$ and a are equal, since both are solutions of the equation $x \cdot a^{-1} = 1$.

F 8. $-(a + b) = (-a) + (-b)$ for all $a, b \in \mathbf{R}$. For both are solutions of the equation $x + (a + b) = 0$.

F 9. $(ab)^{-1} = a^{-1}b^{-1}$ if a, b are nonzero elements of $\mathbf{R}$. For $ab \neq 0$ by F 5, so $(ab)^{-1}$ exists, and both $(ab)^{-1}$ and $a^{-1}b^{-1}$ are solutions of the equation $x(ab) = 1$.

The usual rules for operating with fractions follow easily from F 9:

$$\frac{ac}{bc} = (ac)(bc)^{-1} = acb^{-1}c^{-1} = ab^{-1} = \frac{a}{b},$$

$$\frac{a}{b} \cdot \frac{c}{d} = (ab^{-1})(cd^{-1}) = ac(bd)^{-1} = \frac{ac}{bd},$$

$$\begin{aligned}\frac{a}{b} + \frac{c}{d} = \frac{ad}{bd} + \frac{bc}{bd} &= (ad)(bd)^{-1} + (bc)(bd)^{-1} \\ &= (ad + bc)(bd)^{-1} = \frac{ad + bc}{bd}.\end{aligned}$$

F 10. $-a = (-1) \cdot a$ for all $a \in \mathbf{R}$. For $(-1) \cdot a + a = a \cdot ((-1) + 1) = a \cdot 0 = 0$, so that $(-1) \cdot a$ and $-a$ are both solutions of the equation $x + a = 0$, hence are equal. Two immediate consequences are $a \cdot (-b) = a \cdot (-1) \cdot b = (-1) \cdot a \cdot b = (-a) \cdot b = -ab$ and $(-a) \cdot (-b) = -(a \cdot (-b)) = -(-ab) = ab$.

Notice that all five field properties of the real numbers, and therefore all consequences of them, are satisfied by the rational numbers, or by the complex numbers. That is, the rational numbers and the complex numbers are also fields. In fact there exist fields with only a finite number of elements, the simplest one being a field with just the two elements 0 and 1. To describe the real numbers completely, more properties are needed.

§ 2. ORDER.

The order property of the real number system is the following:

PROPERTY VI. *There is a subset* $\mathbf{R}_+$ *of* $\mathbf{R}$ *such that*
(1) *if* $a, b \in \mathbf{R}_+$, *then* $a + b, a \cdot b \in \mathbf{R}_+$
(2) *for any* $a \in \mathbf{R}$, *one and only one of the following statements is true*

$$a \in \mathbf{R}_+$$
$$a = 0$$
$$-a \in \mathbf{R}_+.$$

The elements $a \in \mathbf{R}$ such that $a \in \mathbf{R}_+$ will of course be called *positive*, those such that $-a \in \mathbf{R}_+$ *negative*. From the above property of $\mathbf{R}_+$ we shall deduce all the usual rules for working with inequalities.

To be able to express the consequences of Property VI most conveniently we introduce the relations "$>$" and "$<$". For $a, b \in \mathbf{R}$, either of the expressions

$$a > b \quad \text{or} \quad b < a$$

(read respectively as "a is greater than b" and "b is less than a") will mean that $a - b \in \mathbf{R}_+$. Either of the expressions

$$a \geq b \quad \text{or} \quad b \leq a$$

will mean that $a > b$ or $a = b$.

Clearly $a \in \mathbf{R}_+$ if and only if $a > 0$. An element $a \in \mathbf{R}$ is negative if and only if $a < 0$.

The following are the consequences of the order property.

O 1. **(Trichotomy).** If $a, b \in \mathbf{R}$ then one and only one of the following statements is true:

$$a > b$$
$$a = b$$
$$a < b.$$

For if we apply part (2) of the order property to the number $a - b$ then exactly one of three possibilities holds, $a - b \in \mathbf{R}_+$, $a - b = 0$, or $b - a \in \mathbf{R}_+$, which are the three cases of the assertion O 1.

O 2. **(Transitivity).** If $a > b$ and $b > c$ then $a > c$. For we are given $a - b \in \mathbf{R}_+$ and $b - c \in \mathbf{R}_+$; it therefore follows that $a - c = (a - b) + (b - c) \in \mathbf{R}_+$, so $a > c$.

O 3. If $a > b$ and $c \geq d$ then $a + c > b + d$. In fact, the hypotheses mean $a - b \in \mathbf{R}_+$, $c - d \in \mathbf{R}_+ \cup \{0\}$, and as a consequence $(a + c) - (b + d) = (a - b) + (c - d) \in \mathbf{R}_+$, proving the assertion.

O 4. If $a > b > 0$ (meaning that $a > b$ and $b > 0$) and $c \geq d > 0$, then $ac > bd$. For $a - b \in \mathbf{R}_+$ and $c \in \mathbf{R}_+$, so $ac - bc = (a - b)c \in \mathbf{R}_+$, and similarly $c - d \in \mathbf{R}_+ \cup \{0\}$ and $b \in \mathbf{R}_+$ together imply that $bc - bd \in \mathbf{R}_+ \cup \{0\}$; it necessarily follows that $ac - bd = (ac - bc) + (bc - bd) \in \mathbf{R}_+$, that is $ac > bd$.

Note that the assumptions that b and d are positive are essential; the assertion O 4 does not hold, for example, with $a = 1$, $b = -1$, $c = 2$, $d = -3$.

O 5. The following rules of sign for adding and multiplying real numbers hold:

(positive number) + (positive number) = (positive number)
(negative number) + (negative number) = (negative number)
(positive number) · (positive number) = (positive number)
(positive number) · (negative number) = (negative number)
(negative number) · (negative number) = (positive number).

These are immediate from F 10 and Property VI.

O 6. For any $a \in \mathbf{R}$ we have $a^2 \geq 0$, with the equality holding only if $a = 0$; more generally the sum of the squares of several elements of $\mathbf{R}$ is always greater than or equal to zero, with equality only if all the elements in question are zero. For by O 5, the statement $a \neq 0$ implies $a^2 > 0$, and a sum of positive elements is positive. Note the special consequence $1 = 1^2 > 0$.

O 7. If $a > 0$, then $1/a > 0$. In fact $a \cdot (1/a) = 1 > 0$, which would contradict the rules of sign if we had $1/a \leq 0$.

O 8. If $a > b > 0$, then $1/a < 1/b$. For $ab > 0$, hence $(ab)^{-1} > 0$, so $(ab)^{-1}a > (ab)^{-1}b$, which simplifies to $1/b > 1/a$.

O 9. We now show how the computational rules of elementary arithmetic work out as consequences of our assumptions. Let us make the definitions $2 = 1 + 1$, $3 = 2 + 1$, $4 = 3 + 1$, etc., and let us define the *natural numbers* to be the set $\{1, 2, 3, \ldots\}$. Since $1 > 0$ it follows that $0 < 1 < 2 < 3 < \cdots$. The set of natural numbers is ordered exactly as we would like it to be—in particular, the natural numbers have the following properties: for any natural numbers a, b, exactly one of the statements $a < b$, $a = b$, $b < a$ holds; if a, b, c are natural numbers and $a < b$ and $b < c$ then also $a < c$; any natural number has an immediate successor (a least natural number that is greater than it); different natural numbers have different immediate successors; and there is a natural number 1 with the property that any set of natural numbers that includes 1 and with each element also its immediate successor consists of all natural numbers. For any natural number n, n is the sum of a set of 1's that is in one-one correspondence with the elements of the set $\{1, 2, 3, \ldots, n\}$. This implies that in whatever order we count off the elements of a set of n objects (that is, a set in one-one correspondence with the set $\{1, 2, 3, \ldots, n\}$) we arrive at the final count n, and if a proper subset of a set of n objects has m objects, then $m < n$. The usual rules for adding natural numbers come from such computations as

$$2 + 3 = (1 + 1) + (1 + 1 + 1) = 1 + 1 + 1 + 1 + 1 = 5,$$

while the rules for multiplication follow from the fact that sums of equal terms may be written as products; for example, for any $a \in \mathbf{R}$ we have $a + a + a = (1 + 1 + 1) \cdot a = 3a$. Thus $3 \cdot 4 = 4 + 4 + 4 = 12$, so we can verify the entire multiplication table, as high as we care to go. The *integers*, that is the subset $\{0, \pm 1, \pm 2, \pm 3, \ldots\}$ of $\mathbf{R}$, are also ordered in the correct way $\cdots < -2 < -1 < 0 < 1 < 2 < \cdots$. It is easy to check that the integers add according to the ordinary rules; that they multiply in the usual way is implied by F 10 and the corresponding fact for the natural numbers. The *rational numbers*, that is the elements of $\mathbf{R}$ which can be written a/b, with a, b integers and $b \neq 0$, are also ordered in the usual way; indeed the order relation of two rational numbers can be determined by writing the two numbers with a positive common denominator and comparing the numerators. Addition and multiplication of rational numbers are also determined by the same operations for the integers. Thus the rational numbers, a certain subset of $\mathbf{R}$, have all the arithmetic and order properties with which we are familiar.

Here is as good a place as any to introduce into our logical discussion of the real number system the notion of exponentiation with integral exponents. If $a \in \mathbf{R}$ and n is some positive integer we define a^n to be

$a \cdot a \cdot a \cdots a$ (n times), and if $a \neq 0$ we define $a^0 = 1$, $a^{-n} = 1/a^n$. From these definitions we immediately derive the usual rules of exponentiation, in particular

$$a^m \cdot a^n = a^{m+n}$$
$$(a^m)^n = a^{mn}$$
$$(ab)^n = a^n b^n$$

The definition of the absolute value of a real number is most conveniently introduced at this point: if $a \in \mathbf{R}$, the *absolute value of a*, denoted $|a|$, is given by

$$\begin{aligned} |a| &= a \quad \text{if} \quad a > 0, \\ |a| &= 0 \quad \text{if} \quad a = 0, \\ |a| &= -a \quad \text{if} \quad a < 0. \end{aligned}$$

The absolute value has the following properties:

(1) $|a| \geq 0$ for all $a \in \mathbf{R}$, and $|a| = 0$ if and only if $a = 0$
(2) $|ab| = |a| \cdot |b|$ for all $a, b \in \mathbf{R}$
(3) $|a|^2 = a^2$ for all $a \in \mathbf{R}$
(4) $|a + b| \leq |a| + |b|$ for all $a, b \in \mathbf{R}$
(5) $|a - b| \geq \big||a| - |b|\big|$ for all $a, b \in \mathbf{R}$.

The first three properties above are trivial consequences of the definition of $|a|$. To prove (4) note first that

$$\pm a \leq |a|$$

(meaning that $a \leq |a|$ and $-a \leq |a|$) and

$$\pm b \leq |b|,$$

so adding gives

$$\pm(a + b) \leq |a| + |b|,$$

or

$$|a + b| \leq |a| + |b|.$$

To prove (5), note that $|a| = |(a - b) + b| \leq |a - b| + |b|$, so that

$$|a - b| \geq |a| - |b|.$$

Interchanging a and b,

$$|a - b| \geq |b| - |a|,$$

and the last two inequalities combine into (5).

It is useful to note that repeated application of (4) gives

$$|a_1 + a_2 + \cdots + a_n| \leq |a_1| + |a_2| + \cdots + |a_n|.$$

We also note the trivial but very useful fact that if $x, a, \epsilon \in \mathbf{R}$, then

$$|x - a| < \epsilon$$

if and only if

$$a - \epsilon < x < a + \epsilon.$$

For $|x - a| < \epsilon$ is precisely equivalent to $x - a < \epsilon$ and $-(x - a) < \epsilon$, or $-\epsilon < x - a < \epsilon$, which in turn is equivalent to $a - \epsilon < x < a + \epsilon$.

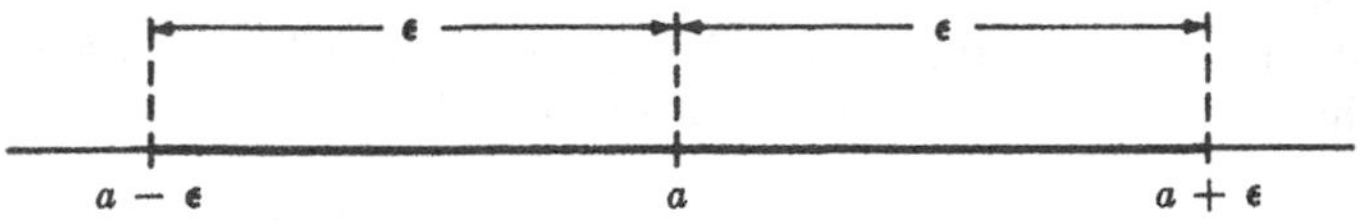

FIGURE 5. The points x such that $|x - a| < \epsilon$.

At the end of the previous section a number of other systems were given which satisfy the first five properties of the real number system. The order property excludes two of the systems given there: the field consisting of just the two elements 0 and 1 (since then $1 + 1 = 0$, contradicting $1 + 1 > 0$), and the complex numbers (since any number must have a nonnegative square). But the rational numbers satisfy all the properties given so far. Since it is known that there exist real numbers which are not rational (this will be proved shortly), still more properties are needed to describe the real numbers completely.

§ 3. THE LEAST UPPER BOUND PROPERTY.

To introduce the last fundamental property of the real number system we need the following concepts. If $S \subset \mathbf{R}$, then *an upper bound for the set S* is a number $a \in \mathbf{R}$ such that $s \leq a$ for each $s \in S$. If the set S has an upper bound, we say that S is *bounded from above*. We call a real number y a *least upper bound of the set S* if

(1) y is an upper bound for S

and

(2) if a is any upper bound for S, then $y \leq a$.

From this definition it follows that two least upper bounds of a set $S \subset \mathbf{R}$ must be less than or equal to each other, hence equal. Thus a set $S \subset \mathbf{R}$ can have at most one least upper bound and we may speak of *the* least upper bound of S (if one exists). Note also the following important fact: if y is the least upper bound of S and $x \in \mathbf{R}$, $x < y$, then there exists an element $s \in S$ such that $x < s$.

A nonempty finite subset $S \subset \mathbf{R}$ always has a least upper bound; in this case the least upper bound is simply the greatest element of S. More generally any subset $S \subset \mathbf{R}$ that has a greatest element (usually denoted max S) has max S as a least upper bound. But an infinite subset of $\mathbf{R}$ need not have a least upper bound, for example, $\mathbf{R}$ itself has no upper bound at all. Furthermore, if a subset S of $\mathbf{R}$ has a least upper bound it does not necessarily follow that this least upper bound is in S; for example, if S is the set of all negative numbers then S has no greatest element, but any $a \geq 0$ is an upper bound of S and zero (a number *not* in S) is the least upper bound of S.

The last axiom for the real number system is the following, which gives a further condition on the ordering of Property VI.

PROPERTY VII. (LEAST UPPER BOUND PROPERTY). *A nonempty set of real numbers that is bounded from above has a least upper bound.*

If we look at the real numbers geometrically, imagining them plotted on a straight line in the usual manner of analytic geometry, Property VII becomes quite plausible. For if $S \subset \mathbf{R}$ is nonempty and bounded from above then either S has a greatest element or, if we try to pick a point in S as far to the right as possible, we can find a point in S such that no point in S is more than a distance of one unit to the right of the chosen point. Then we can pick a point in S farther to the right than the first chosen point and such that no point in S is more than one-half unit to the right of this second chosen point, then a point of S still farther to the right such that no point of S is more than one-third unit to the right of the last chosen point, etc. It is intuitively clear that the sequence of chosen points in S must "gang up" toward some point of $\mathbf{R}$, and this last point will be the least upper bound of S. (See Figure 6.)

Another way to justify Property VII in our minds is to look upon the real numbers as represented by infinite decimals, i.e., symbols of the form

$$\text{(integer)} + .a_1a_2a_3\ldots,$$

where each of the symbols $a_1, a_2, a_3, \ldots$ is one of the integers $0, 1, 2, \ldots, 9$, with the symbols $<, >, +, \cdot$ being interpreted for infinite decimals in the standard way. (Note that any terminating decimal can be considered an

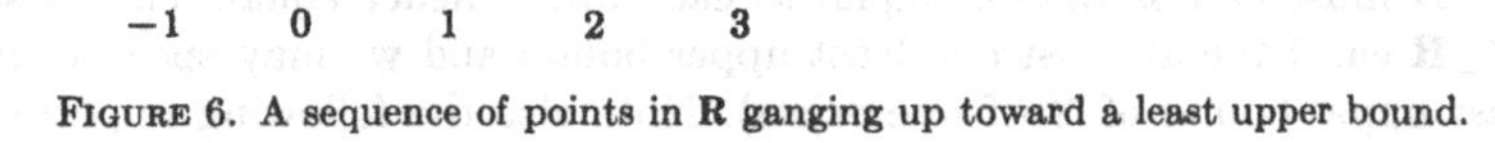

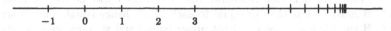

FIGURE 6. A sequence of points in $\mathbf{R}$ ganging up toward a least upper bound.

infinite decimal by adding an infinite string of zeros.) If S is a nonempty set of infinite decimals that is bounded from above, then we can find an element of S whose integral part is maximal, then an element of S having the same integral part and with a_1 maximal, then an element of S having the same integral part and same a_1 with a_2 maximal, and we can continue this process indefinitely, ending up with an infinite decimal (which may or may not be in S) which is clearly a least upper bound of S.*

The least upper bound of a subset S of $\mathbf{R}$ will be denoted l.u.b. S; another common notation is sup S (sup standing for the Latin *supremum*). Property VII says that l.u.b. S exists whenever $S \subset \mathbf{R}$ is nonempty and bounded from above. Conversely, if $S \subset \mathbf{R}$ and l.u.b. S exists, then S must be nonempty (for *any* real number is an upper bound for the empty set and there is no least real number) and bounded from above.

Analogous to the above there are the notions of lower bound and greatest lower bound: $a \in \mathbf{R}$ is a *lower bound* for the subset $S \subset \mathbf{R}$ if $a \leq s$ for each $s \in S$, and a is a *greatest lower bound of* S if a is a lower bound of S and there exists no larger one. S is called *bounded from below* if it has a lower bound. It follows from Property VII that every set S of real numbers that is nonempty and bounded from below has a greatest lower bound: as a matter of fact, a set $S \subset \mathbf{R}$ is bounded from below if and only if the set $S' = \{x : -x \in S\}$ is bounded from above, and if S is nonempty and bounded from below then $-$l.u.b. S' is the greatest lower bound of S. The greatest lower bound of a subset S of $\mathbf{R}$ is denoted g.l.b. S; another notation is inf S (inf abbreviating the Latin *infimum*). If S has a smallest element (for example, if S is finite and nonempty) then g.l.b. S is simply this smallest element, often denoted min S.

We proceed to draw some consequences of Property VII. Among other things we shall show that the real numbers are not very far from the rational numbers, in the sense that any real number may be "approximated as closely as we wish" by rational numbers. The way to view the situation is that the rational numbers are in many ways very nice, but there are certain "gaps" among them that may prevent us from doing all the things we would like to do with numbers, such as solving equations (e.g., extracting roots), or measuring geometric objects, and the introduction of the real numbers that are not rational amounts to closing the gaps.

Here are the consequences of the least upper bound property:

---

* Let us remark here that once the set of integers is known, together with their addition and multiplication, it is possible to construct the real number system by defining real numbers by means of infinite decimals. This is in fact the way real numbers are usually introduced in elementary arithmetic, and we know how easy it is to compute with decimals. But there are a few inconveniences in this method stemming from the fact that some numbers have more than one decimal representation (e.g., $.999\ldots = 1.000\ldots$). There is also the esthetic inconvenience of giving a preferred status to the number 10—almost a biological accident. In any case we shall discuss later in this section how the seven properties of real numbers imply that they can indeed be represented by infinite decimals, thus completing the circle with elementary arithmetic.

LUB 1. For any real number x, there is an integer n such that $n > x$. (In other words, there exist arbitrarily large integers.) To prove this, assume we have a real number x for which the assertion is wrong. Then $n \leq x$ for each integer n, so that the set of integers is bounded from above. Since the set of integers is nonempty it has a least upper bound, say a. But for any integer n, $n + 1$ is also an integer, so $n + 1 \leq a$ and thus $n \leq a - 1$, showing that $a - 1$ is also an upper bound for the set of integers. Since $a - 1 < a$, a is not a least upper bound. This is a contradiction.

LUB 2. For any positive real number ϵ there exists an integer n such that $1/n < \epsilon$. (In other words, there are arbitrarily small positive rational numbers.) For the proof it suffices to choose an integer $n > 1/\epsilon$, which is possible by LUB 1, then use O 8, which is permissible since by O 7 we have $1/\epsilon > 0$.

LUB 3. For any $x \in \mathbf{R}$ there is an integer n such that $n \leq x < n + 1$. To prove this, choose an integer $N > |x|$, so that $-N < x < N$. The integers from $-N$ to N form the finite set $\{-N, -N + 1, \ldots, 0, 1, \ldots, N\}$ and all we need do is take n to be the greatest of these that is less than or equal to x.

LUB 4. For any $x \in \mathbf{R}$ and positive integer N, there is an integer n such that

$$\frac{n}{N} \leq x < \frac{n+1}{N}.$$

To show this we merely have to apply LUB 3 to the number Nx, getting an integer n such that $n \leq Nx < n + 1$.

LUB 5. If $x, \epsilon \in \mathbf{R}$, $\epsilon > 0$, then there exists a rational number r such that $|x - r| < \epsilon$. (In other words, a real number may be approximated as closely as we wish by a rational number.) To prove this, use LUB 2 to find a positive integer N such that $1/N < \epsilon$, then use LUB 4 to find an integer n such that $n/N \leq x < (n + 1)/N$. Then $0 \leq x - n/N < 1/N < \epsilon$, so $|x - n/N| < \epsilon$.

We now discuss the decimal representation of real numbers. First consider finite decimals. If a_0 is any integer, n any positive integer, and $a_1, a_2, \ldots, a_n$ any integers chosen from among $0, 1, 2, \ldots, 9$, the symbol

$$a_0.a_1 a_2 \ldots a_n$$

will mean, as usual, the rational number

$$a_0 + \frac{a_1}{10} + \frac{a_2}{10^2} + \cdots + \frac{a_n}{10^n}.$$

If m is a positive integer less than n, then

$$\begin{aligned} a_0.a_1\ldots a_m \leq a_0.a_1\ldots a_n &= a_0.a_1\ldots a_m + a_{m+1}\cdot 10^{-(m+1)} + \cdots + a_n \cdot 10^{-n} \\ &\leq a_0.a_1\ldots a_m + 9\cdot 10^{-(m+1)} + \cdots + 9\cdot 10^{-n}. \end{aligned}$$

If we add 10^{-n} to this last number a lot of cancellation occurs, resulting in

$$a_0.a_1\ldots a_m \leq a_0.a_1\ldots a_n < a_0.a_1\ldots a_m + 10^{-m}.$$

This last inequality is at the base of most rounding-off procedures in approximate calculations and in addition shows that two numbers in the above decimal form are equal only if (except for the possible addition of a number of zeros to the right, which doesn't change the value of the symbol) they have the same digits in corresponding places. It also enables us to tell at a glance which of two numbers in the given form is larger. The ordinary rules for adding and multiplying numbers in this form are clearly legitimate.

By an *infinite decimal* we mean a formal expression

$$a_0.a_1a_2a_3\ldots$$

(this is just another way of writing a sequence) where a_0 is an integer and each of $a_1, a_2, a_3, \ldots$ is one of the integers $0, 1, \ldots, 9$. The set $\{a_0.a_1\ldots a_n : n = \text{positive integer}\}$ is nonempty and bounded from above (for any integer $m > 0$, $a_0.a_1\ldots a_m + 10^{-m}$ is an upper bound) hence has a least upper bound. The symbol $a_0.a_1a_2a_3\ldots$ is called a *decimal expansion* for this least upper bound and we say that the least upper bound is *represented by* the infinite decimal. Thus every infinite decimal is a decimal expansion for a definite real number and we may use the infinite decimal itself as a symbol for the number. Thus

$$a_0.a_1a_2a_3\ldots = \text{l.u.b. } \{a_0.a_1\ldots a_n : n = \text{positive integer}\},$$

and for any positive integer n we have the inequality

$$a_0.a_1\ldots a_n \leq a_0.a_1a_2a_3\ldots \leq a_0.a_1\ldots a_n + 10^{-n}.$$

This enables us to tell immediately which of two infinite decimals represents the larger real number. Note that two different infinite decimals may be decimal expansions for the same real number, for example $5.1399999\ldots = 5.1400000\ldots$, but the last inequality shows that different infinite decimals are decimal expansions for the same real number only in this case, that is when we can get one infinite decimal from the other by replacing one of the digits $0, 1, \ldots, 8$ followed by an infinite sequence of nines by the next higher digit followed by a sequence of zeros.

Any real number is represented by at least one infinite decimal. To see this, apply LUB 4 to the case $N = 10^m$, where m is any positive integer: we get a finite decimal $a_0.a_1\ldots a_m$ such that

$$a_0.a_1\ldots a_m \leq x < a_0.a_1\ldots a_m + 10^{-m}.$$

If we try doing this for $m + 1$ in place of m, then a_0 and the digits $a_1, \ldots, a_m$ will not change, and we simply get another digit a_{m+1}. Letting m get larger and larger, we get more and more digits of an infinite decimal, and this is our desired decimal expansion for x. Note that the addition or multiplication of two infinite decimals goes according to the usual rules: we round off each decimal and add or multiply the corresponding finite decimals to get a decimal approximation of the desired sum or product. We obtain as many digits as we wish of the decimal expansion of the sum or product by rounding off the given infinite decimals to a sufficiently large number of places.

Using decimal expansions of real numbers, it is very easy to exhibit real numbers which are not rational. One such number is

$$.101001000100001000001\ldots .$$

Multiply this by any positive integer and one gets a number which is not an integer, so this number cannot be rational.

§ 4. THE EXISTENCE OF SQUARE ROOTS.

It is convenient to prove here a special result, even though this can be derived as a consequence of a much more general theorem to be proved later.

A square root of a given number is a number whose square is the given number. Since the square of any nonzero number is positive, only nonnegative numbers can have square roots. The number zero has one square root, which is zero itself.

Proposition. *Every positive number has a unique positive square root.*

If $0 < x_1 < x_2$ then $x_1^2 < x_2^2$. That is, bigger positive numbers have bigger squares. Thus any given real number can have at most one positive square root. It remains to show that if $a \in \mathbf{R}$, $a > 0$, then a has at least one positive square root. For this purpose consider the set

$$S = \{x \in \mathbf{R} : x \geq 0,\ x^2 \leq a\}.$$

This set is nonempty, since $0 \in S$, and bounded from above, since if $x > \max\{a, 1\}$ we have $x^2 = x \cdot x > x \cdot 1 = x > a$. Hence $y = \text{l.u.b. } S$ exists. We proceed to show that $y^2 = a$. First, $y > 0$, for $\min\{1, a\} \in S$, since $(\min\{1, a\})^2 \leq \min\{1, a\} \cdot 1 = \min\{1, a\} \leq a$. Next, for any ϵ such that $0 < \epsilon < y$ we have $0 < y - \epsilon < y < y + \epsilon$, so

$$(y - \epsilon)^2 < y^2 < (y + \epsilon)^2,$$

since bigger positive numbers have bigger squares. By the definition of y there are numbers greater than $y - \epsilon$ in S, but $y + \epsilon \notin S$. Again using the fact that bigger positive numbers have bigger squares, we get

$$(y - \epsilon)^2 < a < (y + \epsilon)^2.$$

Hence

$$(y-\epsilon)^2 - (y+\epsilon)^2 < y^2 - a < (y+\epsilon)^2 - (y-\epsilon)^2,$$

so

$$|y^2 - a| < (y+\epsilon)^2 - (y-\epsilon)^2 = 4y\epsilon.$$

The inequality $|y^2 - a| < 4y\epsilon$ holds for any ϵ such that $0 < \epsilon < y$, and by choosing ϵ small enough we can make $4y\epsilon$ less than any preassigned positive number. Thus $|y^2 - a|$ is less than any positive number. Since $|y^2 - a| \geq 0$, we must have $|y^2 - a| = 0$, proving $y^2 = a$.

If $a > 0$, the unique positive square root of a is denoted $\sqrt{a}$; thus a has exactly two square roots, namely $\sqrt{a}$ and $-\sqrt{a}$. We also write $\sqrt{0} = 0$.

We now know that the positive real numbers are precisely the squares of the nonzero real numbers. This shows that the set of positive numbers $\mathbf{R}_+$ whose existence is affirmed by Property VI is completely determined by the multiplication function of $\mathbf{R}$. A priori, it might seem that there could be several possible subsets $\mathbf{R}_+$ of $\mathbf{R}$ for which Properties VI and VII hold and that in any discussion of the ordering of $\mathbf{R}$ the subset $\mathbf{R}_+$ would have to be specified, but we now know this to be unnecessary. The set $\mathbf{R}$, together with the functions $+$ and $\cdot$, determine the ordering of $\mathbf{R}$. It therefore follows that the decimal expansions of elements of $\mathbf{R}$ are completely determined by the triple $\{\mathbf{R}, +, \cdot\}$. Since the addition and multiplication of decimals follow the usual rules of arithmetic, *the real number system is completely determined by Properties I–VII*, in the sense that if we have another triple $\{\mathbf{R}', +', \cdot'\}$ satisfying these properties then there will exist a unique one-one correspondence between $\mathbf{R}$ and $\mathbf{R}'$ preserving sums and products. Thus we may speak of *the* real number system. In fact one often speaks of "the real numbers $\mathbf{R}$", meaning the real number *system*; this is strictly speaking erroneous, since $\mathbf{R}$ is merely a set and we also have to know what the operations $+$ and $\cdot$ on this set are, but when there is no danger of confusion this is a convenient abbreviation.

PROBLEMS

1. Show that there exists one and (essentially) only one field with three elements.
2. Prove in detail that for any $a, b, c, d \in \mathbf{R}$
   (a) $-(a - b) = b - a$
   (b) $(a - b)(c - d) = (ac + bd) - (ad + bc)$.
3. Prove that if $a, b \in \mathbf{R}$ and $a < b < 0$, then $1/a > 1/b$.
4. (a) Is 223/71 greater than 22/7?
   (b) Is 265/153 greater than 1351/780?

5. For which $x \in \mathbf{R}$ are the following inequalities true?
   (a) $3(x+2) < x+5$
   (b) $x^2 - 5x - 6 \geq 0$
   (c) $\frac{2}{x} > x - 1$
   (d) $\frac{7}{x-3} > x + 3 > 0$.

6. Show that if $a, b, x, y \in \mathbf{R}$ and $a < x < b$, $a < y < b$, then $|y - x| < b - a$.

7. Show that for any $a, b \in \mathbf{R}$,

$$\max\{a, b\} = \frac{a + b + |a - b|}{2}$$

$$\min\{a, b\} = -\max\{-a, -b\} = \frac{a + b - |a - b|}{2}.$$

8. The *complex number system* is defined to be the set $\mathbf{C} = \mathbf{R} \times \mathbf{R}$ (called the *complex numbers*) together with the two functions from $\mathbf{C} \times \mathbf{C}$ into $\mathbf{C}$, denoted by $+$ and $\cdot$, that are given by $(a, b) + (c, d) = (a + c, b + d)$ and $(a, b) \cdot (c, d) = (ac - bd, ad + bc)$ for all $a, b, c, d \in \mathbf{R}$.
   (a) Show that $\mathbf{C}$, together with the functions $+$ and $\cdot$, is a field.
   (b) Show that the map from $\mathbf{R}$ into $\mathbf{C}$ which sends each $a \in \mathbf{R}$ into $(a, 0)$ is one-one and "preserves addition and multiplication" (being careful to define the meaning of the words in quotes).
   (c) Identifying $\mathbf{R}$ with a subset of $\mathbf{C}$ by means of part (b) (so that we can consider $\mathbf{R} \subset \mathbf{C}$) and setting $i = (0, 1)$, show that $i^2 = -1$ and that each element of $\mathbf{C}$ can be written in a unique way as $a + bi$, with $a, b \in \mathbf{R}$.

9. Is the subset $\varnothing$ of $\mathbf{R}$ bounded from above or below? Does it have an l.u.b. or a g.l.b.?

10. Find the g.l.b. and l.u.b. of the following sets, giving reasons if you can.
   (a) $\left\{1, \frac{1}{2}, \frac{1}{3}, \frac{1}{4}, \ldots\right\}$
   (b) $\left\{\frac{1}{3}, \frac{4}{9}, \frac{13}{27}, \frac{40}{81}, \ldots\right\}$
   (c) $\left\{\sqrt{2}, \sqrt{2 + \sqrt{2}}, \sqrt{2 + \sqrt{2 + \sqrt{2}}}, \ldots\right\}$.

11. Prove that if $a \in \mathbf{R}$, $a > 1$, then the set $\{a, a^2, a^3, \ldots\}$ is not bounded from above. $\left(\textit{Hint:}\right.$ First find a positive integer n such that $a > 1 + \frac{1}{n}$ and prove that $a^n > \left(1 + \frac{1}{n}\right)^n \geq 2.\left.\right)$

12. Let X and Y be nonempty subsets of $\mathbf{R}$ whose union is $\mathbf{R}$ and such that each element of X is less than each element of Y. Prove that there exists $a \in \mathbf{R}$ such that X is one of the two sets

$$\{x \in \mathbf{R} : x \leq a\} \quad \text{or} \quad \{x \in \mathbf{R} : x < a\}.$$

13. If S_1, S_2 are nonempty subsets of $\mathbf{R}$ that are bounded from above, prove that

$$\text{l.u.b. } \{x + y : x \in S_1, y \in S_2\} = \text{l.u.b. } S_1 + \text{l.u.b. } S_2.$$

14. Let $a, b \in \mathbf{R}$, with $a < b$. Show that there exists a number $x \in \mathbf{R}$ such that $a < x < b$, with x rational or not rational, as we wish.

15. "A real number is rational if and only if it has a periodic decimal expansion." Define the present usage of the word *periodic* and prove the statement.

16. Decimal (10-nary) expansions of real numbers were defined by special reference to the number 10. Show that real numbers have b-nary expansions with analogous properties, where b is any integer greater than 1.

CHAPTER III

Metric Spaces

Most of elementary analysis is concerned with functions of one or more real variables, that is functions defined on a subset of the real line, or the plane, or ordinary 3-space or, more generally, n-dimensional Euclidean space. The real line, plane, etc. are special cases of the general concept of "metric space" which is introduced in this chapter. We also introduce a convenient geometric language for dealing with so-called "topological" questions, which are questions associated with the notion of "points near each other", a notion that is a priori rather vague. The ideas we develop can be applied to any metric space at all. This not only provides a great economy of thought, since it will be necessary to introduce a new idea only once instead of having to define it for each special case that may occur, but will also lead to productive new ways of looking at familiar objects. Thus one single proof will cover all cases simultaneously and a certain result, applied in a manner slightly out of the ordinary, may have rather unexpected and far-reaching consequences, as will be seen later when other examples of metric spaces arise.

§ 1. DEFINITION OF METRIC SPACE. EXAMPLES.

A metric space is a set together with a rule which associates with each pair of elements of the set a real number such that certain axioms are satisfied. The axioms are chosen in such a manner that it is reasonable to think of the set as a "space" (a word we don't bother to define in isolation), the elements of the set as "points", and the real number associated with two elements of the set as the "distance between two points". Here is the precise definition:

Definition. A *metric space* is a set E, together with a rule which associates with each pair $p, q \in E$ a real number $d(p, q)$ such that

(1) $d(p, q) \geq 0$ for all $p, q \in E$
(2) $d(p, q) = 0$ if and only if $p = q$
(3) $d(p, q) = d(q, p)$ for all $p, q \in E$
(4) $d(p, r) \leq d(p, q) + d(q, r)$ for all $p, q, r \in E$ (triangle inequality).

Thus a metric space is an ordered pair (E, d), where E is a set and d a function $d: E \times E \to \mathbf{R}$ satisfying properties (1)–(4). In dealing with a metric space (E, d) it is often understood from the context what d is, or that a certain specific d is to be borne in mind, and then one often speaks simply of "the metric space E"; this is logically incorrect but very convenient. The elements $p, q, r, \ldots$ of a metric space E (to be absolutely correct we should say "the elements of the underlying set E of the metric space (E, d)", but let us not be too pedantic) are called the *points* of E, and if $p, q \in E$ we call $d(p, q)$ the *distance between p and q*; d itself is called the *distance function*, or *metric*.

Here are some examples of metric spaces:

(1) $E = \mathbf{R}$ (the set of real numbers), $d(p, q) = |p - q|$.
The first three metric space axioms obviously hold. The fourth follows from the computation

$$d(p, r) = |p - r| = |(p - q) + (q - r)| \leq |p - q| + |q - r| = d(p, q) + d(q, r).$$

(2) For any positive integer n we define a metric space E^n, called *n-dimensional Euclidean space*, by taking the underlying set of E^n to be all n-tuples of real numbers $\{(a_1, \ldots, a_n) : a_1, \ldots, a_n \in \mathbf{R}\}$, and defining, for $p = (x_1, \ldots, x_n)$, $q = (y_1, \ldots, y_n)$,

$$d(p, q) = \sqrt{(x_1 - y_1)^2 + (x_2 - y_2)^2 + \cdots + (x_n - y_n)^2}.$$

We must prove that E^n is actually a metric space. The first three metric space axioms are trivial to verify, so it remains to prove the triangle inequality. We need two preliminary results.

***Proposition* (*Schwarz inequality*).** *For any real numbers* $a_1, a_2, \ldots, a_n, b_1, b_2, \ldots, b_n$, *we have*

$$|a_1b_1 + a_2b_2 + \cdots + a_nb_n| \\ \leq \sqrt{a_1^2 + a_2^2 + \cdots + a_n^2}\ \sqrt{b_1^2 + b_2^2 + \cdots + b_n^2}.$$

The proof starts with the remark that for any $\alpha, \beta \in \mathbf{R}$ we have

$$0 \leq (\alpha a_1 - \beta b_1)^2 + (\alpha a_2 - \beta b_2)^2 + \cdots + (\alpha a_n - \beta b_n)^2 \\ = \alpha^2(a_1^2 + a_2^2 + \cdots + a_n^2) - 2\alpha\beta(a_1b_1 + a_2b_2 + \cdots + a_nb_n) \\ + \beta^2(b_1^2 + b_2^2 + \cdots + b_n^2).$$

If we set $\alpha = \sqrt{b_1^2 + b_2^2 + \cdots + b_n^2}$ and $\beta = \pm\sqrt{a_1^2 + a_2^2 + \cdots + a_n^2}$, the last inequality becomes

$$0 \leq 2(a_1^2 + \cdots + a_n^2)(b_1^2 + \cdots + b_n^2) \\ \mp 2\sqrt{a_1^2 + \cdots + a_n^2}\ \sqrt{b_1^2 + \cdots + b_n^2}\ (a_1b_1 + \cdots + a_nb_n),$$

or

$$\pm\sqrt{a_1^2 + \cdots + a_n^2}\ \sqrt{b_1^2 + \cdots + b_n^2}\ (a_1b_1 + \cdots + a_nb_n) \\ \leq (a_1^2 + \cdots + a_n^2)(b_1^2 + \cdots + b_n^2),$$

or

$$\sqrt{a_1^2 + \cdots + a_n^2}\ \sqrt{b_1^2 + \cdots + b_n^2}\ |a_1b_1 + \cdots + a_nb_n| \\ \leq (a_1^2 + \cdots + a_n^2)(b_1^2 + \cdots + b_n^2).$$

If $\sqrt{a_1^2 + \cdots + a_n^2}$ and $\sqrt{b_1^2 + \cdots + b_n^2}$ are both nonzero we can divide by their product to get the desired inequality. If, on the other hand, either $\sqrt{a_1^2 + \cdots + a_n^2} = 0$ or $\sqrt{b_1^2 + \cdots + b_n^2} = 0$, then either $a_1 = \cdots = a_n = 0$ or $b_1 = \cdots = b_n = 0$, and the desired inequality reduces to $0 \leq 0$.

Corollary. *For any real numbers* $a_1, a_2, \ldots, a_n, b_1, b_2, \ldots, b_n$ we have

$$\sqrt{(a_1 + b_1)^2 + (a_2 + b_2)^2 + \cdots + (a_n + b_n)^2} \\ \leq \sqrt{a_1^2 + a_2^2 + \cdots + a_n^2} + \sqrt{b_1^2 + b_2^2 + \cdots + b_n^2}.$$

To prove this write

$$(a_1 + b_1)^2 + (a_2 + b_2)^2 + \cdots + (a_n + b_n)^2 \\ = (a_1^2 + a_2^2 + \cdots + a_n^2) + 2(a_1b_1 + a_2b_2 + \cdots + a_nb_n) \\ + (b_1^2 + b_2^2 + \cdots + b_n^2).$$

By the Schwarz inequality the last expression is less than or equal to

$$(a_1^2 + a_2^2 + \cdots + a_n^2) + 2\sqrt{a_1^2 + a_2^2 + \cdots + a_n^2}\sqrt{b_1^2 + b_2^2 + \cdots + b_n^2} + (b_1^2 + b_2^2 + \cdots + b_n^2)$$

which equals

$$(\sqrt{a_1^2 + a_2^2 + \cdots + a_n^2} + \sqrt{b_1^2 + b_2^2 + \cdots + b_n^2})^2.$$

Thus

$$(a_1 + b_1)^2 + (a_2 + b_2)^2 + \cdots + (a_n + b_n)^2 \le (\sqrt{a_1^2 + a_2^2 + \cdots + a_n^2} + \sqrt{b_1^2 + b_2^2 + \cdots + b_n^2})^2.$$

The desired result comes from this last inequality and the comment that if $0 \le x \le y$ then $\sqrt{x} \le \sqrt{y}$.

We can now verify the triangle inequality for E^n. Let $p = (x_1, \ldots, x_n)$, $q = (y_1, \ldots, y_n)$, $r = (z_1, \ldots, z_n)$. Then

$$\begin{aligned} d(p, r) &= \sqrt{(x_1 - z_1)^2 + \cdots + (x_n - z_n)^2} \\ &= \sqrt{((x_1 - y_1) + (y_1 - z_1))^2 + \cdots + ((x_n - y_n) + (y_n - z_n))^2} \\ &\le \sqrt{(x_1 - y_1)^2 + \cdots + (x_n - y_n)^2} + \sqrt{(y_1 - z_1)^2 + \cdots + (y_n - z_n)^2} \end{aligned}$$

by the Corollary, so that

$$d(p, r) \le d(p, q) + d(q, r).$$

Thus E^n is a metric space.

We note that E^n is a generalization of the first example, since E^1 is simply **R** with $d(p, q) = |p - q|$.

(3) If E is a metric space and E_1 is a subset of E then E_1 can be made a metric space in an obvious way: the distance between two points of E_1 is the same as the distance between them when they are considered points of E. That E_1, together with its metric, satisfies our four axioms is immediate. E_1, with its metric, is called a *subspace* of E. Note that by taking subsets of Euclidean space we get an infinite number of metric spaces, in a tremendous variety of sizes and shapes.

(4) Let E be an arbitrary set and, for $p, q \in E$, define $d(p, q) = 0$ if $p = q$, $d(p, q) = 1$ if $p \neq q$. This is clearly a metric space. It is a very special kind of metric space, quite unlike the previous examples, but illustrates nicely the generality of the concept with which we are dealing.

In this text we shall for the most part be interested in E^n and its subspaces, but other important metric spaces will also appear and indeed will sometimes be introduced to prove things about E^n itself.

Proposition. *If $p_1, p_2, \ldots, p_n$ are points of the metric space E, then*

$$d(p_1, p_n) \leq d(p_1, p_2) + d(p_2, p_3) + \cdots + d(p_{n-1}, p_n).$$

This comes from repeated application of the triangle inequality:

$$d(p_1, p_n) \leq d(p_1, p_2) + d(p_2, p_n) \leq d(p_1, p_2) + d(p_2, p_3) + d(p_3, p_n) \leq \cdots$$
$$\leq d(p_1, p_2) + d(p_2, p_3) + \cdots + d(p_{n-1}, p_n).$$

Proposition. *If p, q, r are points of the metric space E, then*

$$|d(p, r) - d(q, r)| \leq d(p, q).$$

This is essentially the well-known fact of elementary geometry that the difference of two sides of a triangle is less than the third side. To prove it note that

$$d(p, r) \leq d(p, q) + d(q, r) \quad \text{and} \quad d(q, r) \leq d(q, p) + d(p, r),$$

which can be rewritten

$$d(p, r) - d(q, r) \leq d(p, q) \quad \text{and} \quad d(q, r) - d(p, r) \leq d(q, p),$$

which combine to

$$|d(p, r) - d(q, r)| \leq d(p, q).$$

§ 2. OPEN AND CLOSED SETS.

Definitions. Let E be a metric space, $p_0 \in E$, and $r > 0$ a real number. Then the *open ball in E of center p_0 and radius r* is the subset of E given by

$$\{p \in E : d(p_0, p) < r\}.$$

The *closed ball in E of center p_0 and radius r* is

$$\{p \in E : d(p_0, p) \leq r\}.$$

If there can be no misunderstanding about what the metric space E is, one often speaks of the "open (or closed) ball of center p_0 and radius r". When one speaks of an "open (or closed) ball" one means an open (or closed) ball of some center p_0 in the metric space and some radius $r > 0$. By a "ball" is meant an open ball or a closed ball. If our metric space E is ordinary 3-space E^3 then the preceding terminology is in accord with everyday language: an open ball in E^3 is the inside of some sphere while a closed ball is the inside of some sphere together with the points on the sphere, the centers and radii of the balls being the centers and radii respectively of the spheres. In the plane E^2 an open ball is the inside of some circle while a closed ball is the inside of a circle together with the points on the circle. In other metric spaces balls may look even less like an ordinary "ball" (cf. Fig. 7).

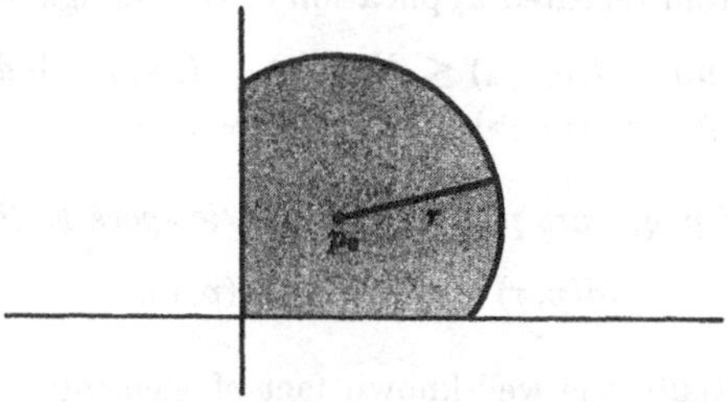

FIGURE 7. A closed ball in the subspace E of the plane E^2 given by $E = \{(x, y) \in E^2 : x > 0, y > 0\}$.

We recall that if $a, b \in \mathbf{R}$, $a < b$, then the *open interval with extremities a and b* is the set

$$(a, b) = \{x \in \mathbf{R} : a < x < b\}^*$$

while the *closed interval with extremities a and b* is the set

$$[a, b] = \{x \in \mathbf{R} : a \leq x \leq b\}.$$

The inequalities

$$a < x < b$$

are equivalent to the inequalities

$$a - \frac{a+b}{2} < x - \frac{a+b}{2} < b - \frac{a+b}{2}$$

or

$$\frac{a-b}{2} < x - \frac{a+b}{2} < \frac{b-a}{2}$$

or

$$\left| x - \frac{a+b}{2} \right| < \frac{b-a}{2},$$

* Note the possible notational confusion involved here. The symbol (a, b) can stand for an open interval in $\mathbf{R}$ or an ordered pair of elements of $\mathbf{R}$ (or, which is the same thing, a point of E^2). The meaning to be given to (a, b) should always be clear from the context.

so that (a, b) is the open ball in the metric space $\mathbf{R}$ with center $(a + b)/2$ and radius $(b - a)/2$. Similarly $[a, b]$ is the closed ball with the same center and radius. If $p_0 \in \mathbf{R}$ and $r \in \mathbf{R}$, $r > 0$, then the open (or closed) ball in $\mathbf{R}$ with center p_0 and radius r is $(p_0 - r, p_0 + r)$ (or $[p_0 - r, p_0 + r]$), so that the open (closed) balls in $\mathbf{R}$ are just the open (closed) intervals.

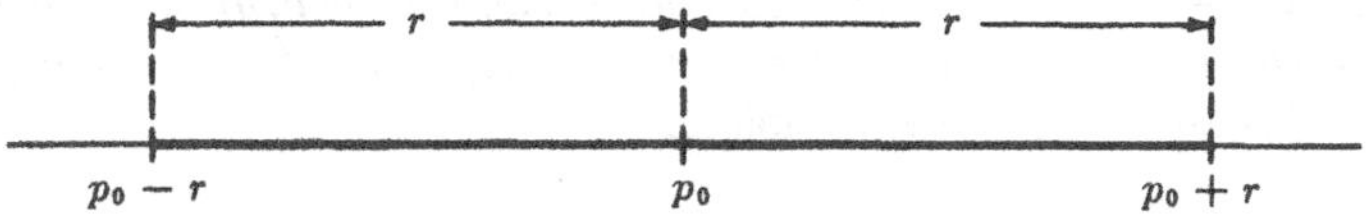

FIGURE 8. Ball in $\mathbf{R}$ with center p_0 and radius r.

Definition. A subset S of a metric space E is *open* if, for each $p \in S$, S contains some open ball of center p.

Intuitively, a subset S of the metric space E is open if S contains all points of E that are sufficiently near any given point of S, but this property can be stated precisely only by repeating the definition just given. Note that it makes no sense to say that a set S is open or not unless S is a subset of some specific metric space E.

Proposition. *For any metric space E,*
(1) the subset $\varnothing$ is open
(2) the subset E is open
(3) the union of any collection of open subsets of E is open
(4) the intersection of a finite number of open subsets of E is open.

The proof of the first item is trivial, though perhaps tricky for beginners: we have to show that "for any $p \in \varnothing$ there is an open ball such that . . .", a statement that is automatically true since there is *no* p such that $p \in \varnothing$. The second item is equally trivial; indeed any ball in E is contained in E. Item (3) is also clear. To prove (4), let $S_1, \ldots, S_n$ be open subsets of E and $p \in S_1 \cap \cdots \cap S_n$. For $i = 1, \ldots, n$, each S_i is open, so there exists a real number $r_i > 0$ such that the open ball of center p and radius r_i is entirely contained in S_i. Then the open ball of center p and radius $\min\{r_1, \ldots, r_n\}$ is contained in each S_i and is contained therefore in $S_1 \cap \cdots \cap S_n$. Thus $S_1 \cap \cdots \cap S_n$ is open.

The word "open" so far occurs in two contexts: we have open balls and open sets. That no errors can arise is a consequence of the following result.

Proposition. *In any metric space, an open ball is an open set.*

For let E be a metric space and consider the open ball S of center p_0 and radius r. We have to show that if $p \in S$, then some open ball of center p is entirely contained in S. Figure 9 gives the idea of the proof: if $p \in S$ then the open ball of center p and radius $r - d(p_0, p)$ is contained in S. To fill in the details of the proof, note first that $d(p_0, p) < r$ since $p \in S$, so that $r - d(p_0, p) > 0$ and there actually exists an open ball of center p and radius $r - d(p_0, p)$. If q is in this latter ball then $d(p, q) < r - d(p_0, p)$ and therefore $d(p_0, q) \leq d(p_0, p) + d(p, q) < d(p_0, p) + (r - d(p_0, p)) = r$. Thus $q \in S$ and the proof is complete.

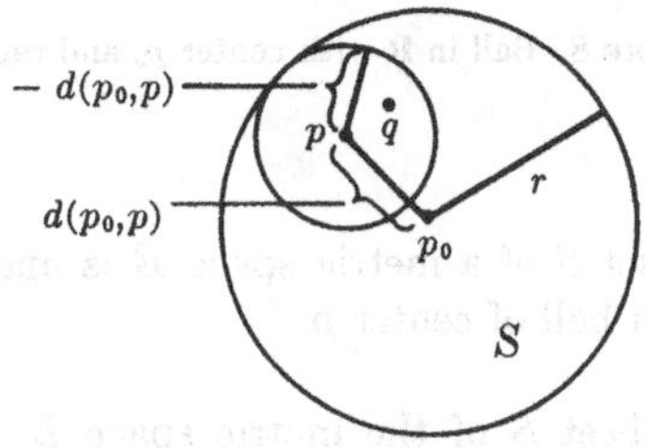

FIGURE 9. An open ball is an open set.

As a consequence of the last two propositions we can assert that the open subsets of E are precisely the unions of open balls of E, that is, any open subset is such a union (in fact it is the union of all the open balls it contains) and any such union is an open set.

The difference between (3) and (4) in the first proposition is to be noted seriously. Item (4) is no longer true if the word "finite" is dropped. For example, consider in E^n the open balls with center the origin $(0, 0, \ldots, 0)$ and radii $1, \frac{1}{2}, \frac{1}{3}, \frac{1}{4}, \ldots$; the intersection of these open balls is just the origin itself, a set which is clearly not open.

Definition. A subset S of a metric space E is *closed* if its complement $\complement S$ (that is, all points in E which are not in S) is open.

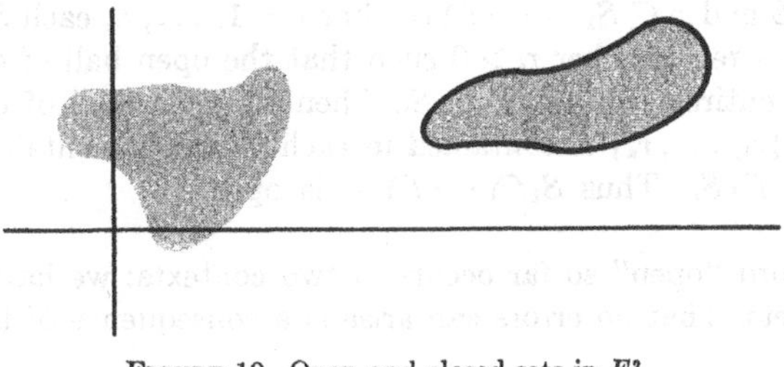

FIGURE 10. Open and closed sets in E^2.

As before, to avoid trouble we should show that a closed ball is a closed set. Here are the statement and proof, quite analogous to what was done above. Figure 11 illustrates the argument.

Proposition. *In any metric space, a closed ball is a closed set.*

Let S be the closed ball of center $p_0 \in E$ and radius r and let $p \in \mathfrak{C}S$. Then $d(p_0, p) > r$, so $d(p_0, p) - r > 0$ and we can consider the open ball of center p and radius $d(p_0, p) - r$. For any point q in the latter open ball we have $d(p, q) < d(p_0, p) - r$, so that

$$d(p_0, q) = d(p_0, q) + d(q, p) - d(p, q) \geq d(p_0, p) - d(p, q) > r.$$

Thus the open ball of center p and radius $d(p_0, p) - r$ is entirely contained in $\mathfrak{C}S$, so that $\mathfrak{C}S$ is open. Thus S is closed.

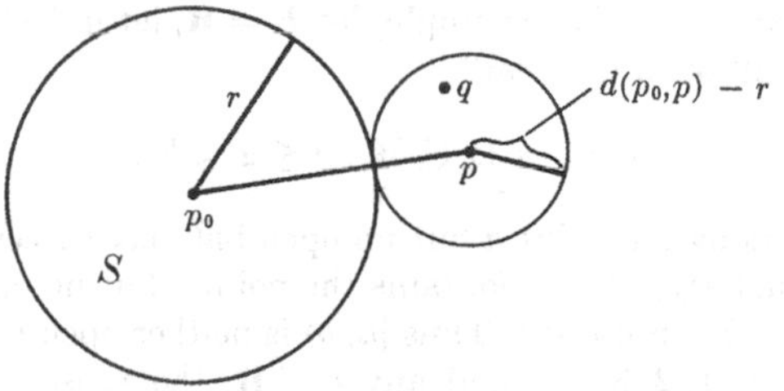

FIGURE 11. A closed ball is a closed set.

Analogous to the first proposition on open sets we have the following result.

Proposition. *For any metric space E,*
(1) the subset E is closed
(2) the subset $\varnothing$ is closed
(3) the intersection of any collection of closed subsets of E is closed
(4) the union of a finite number of closed subsets of E is closed.

This result follows immediately from the analogous proposition for open subsets. For E and $\varnothing$ are the complements of $\varnothing$ and E respectively, which are open, so that E and $\varnothing$ are closed. Parts (3) and (4) result from the fact that the complement of an intersection is a union, and vice versa (bottom EXERCISE, page 6). For (3), we note that the complement of the intersection of any collection of closed subsets of E is the union of the complements of these closed subsets, that is the union of certain open sets, which is open. For (4), note that the complement of the union of a finite number of closed subsets is the intersection of the complements of this finite number of closed sets, that is the intersection of a finite number of open sets, which is open.

If p, q are distinct points of the metric space E then there are balls (either open or closed) containing one point but not the other, for example any ball (either open or closed) centered at p or q and of radius less than $d(p, q)$. Thus any point is the intersection of all the closed balls containing it, proving (by (3) above) that a point is a closed set. By (4), *any finite subset of a metric space is closed.*

The complement of an open ball is a closed set, as is the closed ball of the same center and radius, so again by (3) the intersection of these two sets is closed. Thus for any $p_0 \in E$ and any $r > 0$, the "sphere of center p_0 and radius r", that is, the set $\{x \in E : d(p_0, p) = r\}$, is closed.

It is easy to give examples of subsets of a metric space which are neither open nor closed. For example, let $E = \mathbf{R}$, let $a, b \in \mathbf{R}$, $a < b$, and consider the "half-open interval"

$$[a, b) = \{x \in \mathbf{R} : a \leq x < b\}.$$

Then $[a, b)$ contains the point a but no open ball having center a, so $[a, b)$ is not open. Similarly, $\mathcal{C}[a, b)$ contains the point b but no open ball having center b, so $\mathcal{C}[a, b)$ is not open. Thus $[a, b)$ is neither open nor closed.

For any $n = 1, 2, 3, \ldots$ and any $a \in \mathbf{R}$, the subset of E^n given by

$$\{(x_1, \ldots, x_n) \in E^n : x_1 > a\}$$

is open. For let $p = (x_1, x_2, \ldots, x_n)$ be in this subset and consider the open ball of center p and radius $x_1 - a$. If $q = (y_1, y_2, \ldots, y_n)$ is in the latter ball, then

$$|x_1 - y_1| \leq \sqrt{(x_1 - y_1)^2 + \cdots + (x_n - y_n)^2} = d(p, q) < x_1 - a,$$

so $y_1 = x_1 - (x_1 - y_1) \geq x_1 - |x_1 - y_1| > x_1 - (x_1 - a) = a$, so that the point q is in the set $\{(x_1, \ldots, x_n) \in E^n : x_1 > a\}$. Hence this latter set is open. Similarly the set $\{(x_1, \ldots, x_n) \in E^n : x_1 < a\}$ is open, and for any $i = 1, 2, \ldots, n$, the sets

$$\{(x_1, \ldots, x_n) \in E^n : x_i > a\} \quad \text{and} \quad \{(x_1, \ldots, x_n) \in E^n : x_i < a\}$$

are open. Consequently their complements

$$\{(x_1, \ldots, x_n) \in E^n : x_i \leq a\} \quad \text{and} \quad \{(x_1, \ldots, x_n) \in E^n : x_i \geq a\}$$

are closed.

If $a_1, a_2, \ldots, a_n, b_1, b_2, \ldots, b_n \in \mathbf{R}$ and $a_1 < b_1$, $a_2 < b_2$, $\ldots$, $a_n < b_n$, then the set of points

$$\{(x_1, \ldots, x_n) \in E^n : a_i < x_i < b_i \text{ for each } i = 1, \ldots, n\}$$

is called an *open interval in* E^n and the set of points

$$\{(x_1, \ldots, x_n) \in E^n : a_i \leq x_i \leq b_i \text{ for each } i = 1, \ldots, n\}$$

is called a *closed interval in* E^n. (This generalizes the usual notion of open and closed interval in $\mathbf{R} = E^1$.) *An open (or closed) interval in* E^n *is an open (or closed, respectively) subset.* For such a set is the intersection of the $2n$ subsets given by the separate conditions $a_1 < x_1, \ldots, a_n < x_n, b_1 > x_1, \ldots, b_n > x_n$ (or $a_1 \leq x_1, \ldots, a_n \leq x_n, b_1 \geq x_1, \ldots, b_n \geq x_n$) and these latter subsets are all open (or closed), hence so is their intersection.

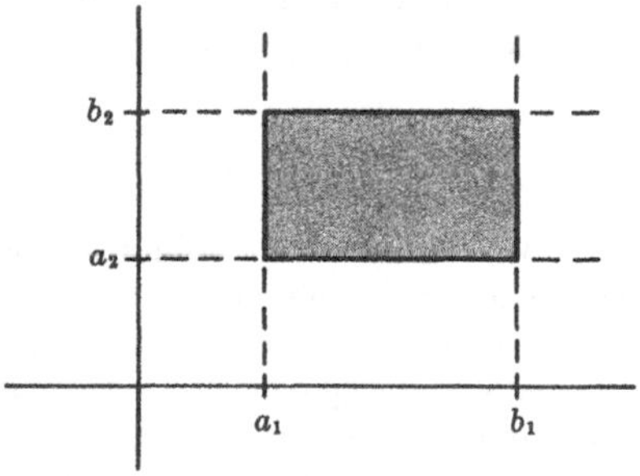

FIGURE 12. A closed interval in E^2.

Definition. A subset S of a metric space E is *bounded* if it is contained in some ball.

In this definition the ball in question may be either open or closed, for any open ball is contained in a closed ball (for example the closed ball of the same center and radius) and any closed ball is contained in an open ball (for example the open ball of the same center and any larger radius). As an example of a bounded set, consider the open and closed intervals in E^n discussed above; these are all bounded since the set

$$\{(x_1, \ldots, x_n) \in E^n : a_i \leq x_i \leq b_i \text{ for each } i = 1, \ldots, n\}$$

is contained in the closed ball of center $(a_1, \ldots, a_n)$ and radius of magnitude $\sqrt{(b_1 - a_1)^2 + \cdots + (b_n - a_n)^2}$.

If S is a bounded subset of the metric space E then S is contained in some ball (either open or closed) with center p_0, where p_0 is *any* point of E. For since S is contained in some ball, say the closed ball of center p_1 and radius r, then S is also contained in the closed ball of center p_0 and radius $r + d(p_0, p_1)$. This makes obvious the fact that the union of a finite number of bounded subsets of E is bounded.

It is clear from the definition that a subset of $\mathbf{R}$ is bounded if and only if it is both bounded from above and bounded from below.

Proposition. *A nonempty closed subset of* $\mathbf{R}$, *if it is bounded from above, has a greatest element and if it is bounded from below has a least element.*

Let S be a nonempty subset of $\mathbf{R}$ and suppose, for definiteness, that S is bounded from above (the proof being almost the same if S is bounded from below). Let $a = \text{l.u.b. } S$. If $a \notin S$, then $a \in \mathcal{C}S$ and since $\mathcal{C}S$ is open there exists a number $\epsilon > 0$ such that the open ball in $\mathbf{R}$ of center a and radius ϵ is contained in $\mathcal{C}S$. This means that no element of S is greater than $a - \epsilon$. Therefore, $a - \epsilon$ is an upper bound for S, contrary to the assumption $a = \text{l.u.b. } S$. We conclude that $a \in S$, as was to be shown.

§ 3. CONVERGENT SEQUENCES.

Let $p_1, p_2, p_3, \ldots$ be a sequence of points in the metric space E. It may happen that as we go out in the sequence the points of the sequence "get arbitrarily close to" some point p of E. This is illustrated in Figure 13, where the various terms of the sequence at first oscillate irregularly, then proceed to get closer and closer to p, in fact "gang up on" p, or "get arbitrarily close to" p. The purpose of the following definition is to give some precise sense to the intuitive words "get arbitrarily close to".

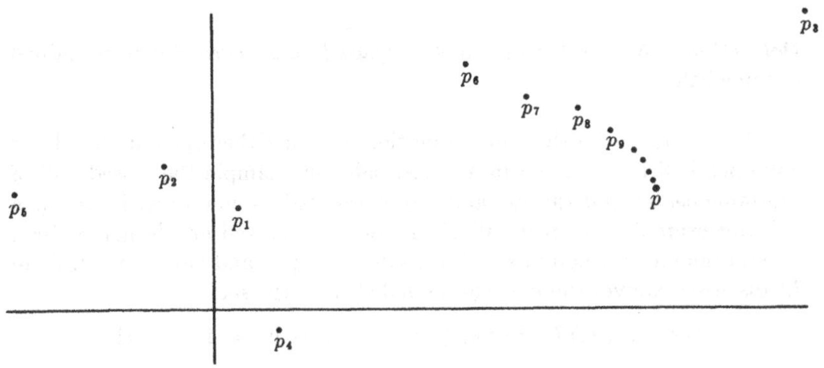

FIGURE 13. A convergent sequence of points in E^2.

Definition. Let $p_1, p_2, p_3, \ldots$ be a sequence of points in the metric space E. A point $p \in E$ is called a *limit* of the sequence $p_1, p_2, p_3, \ldots$ if, given any real number $\epsilon > 0$, there is a positive integer N such that $d(p, p_n) < \epsilon$ whenever $n > N$. If the sequence $p_1, p_2, p_3, \ldots$ has a limit, we call the sequence *convergent*, and if p is a limit of the sequence we say that *the sequence converges to* p.

Let us make a few observations concerning this definition. The main one is that given any $\epsilon > 0$ there exists a positive integer N having a certain property involving ϵ. Thus N usually depends on ϵ, and it would have been more precise to write $N(\epsilon)$ instead of N. However the extra precision obtained by writing $N(\epsilon)$ instead of N results in unnecessary notational confusion, so from now on we just tacitly understand that N depends on ϵ and stick to our shorter notation. Note that we do not even care what specific N goes with each $\epsilon > 0$, the important thing being that for each $\epsilon > 0$ there exists *some* N with the desired property; if for some $\epsilon > 0$ we have a certain N with the desired property, then any larger N would do equally well for our given ϵ. Thus in deciding whether or not a sequence of points is convergent, only the terms far out count; that is, if we obtain a new sequence by lopping off the first few terms of our original sequence, the two sequences we have are either both convergent (with the same limits) or both not convergent.

Another observation on the definition of limit and convergence is that these concepts are always relative to some specific metric space E. Thus it might happen that for the given sequence of points in E, $p_1, p_2, p_3, \ldots$, the condition in the definition of limit holds for a certain point p of E', where E' is some metric space of which E is a subspace; the sequence would then be convergent in E', but we could not call it convergent in E unless we knew that $p \in E$. Thus in using the notion of convergence a specific metric space must be borne in mind. As an easy example, the sequence 3, 3.1, 3.14, 3.141, 3.1415, 3.14159, ... converges in $\mathbf{R}$ (to the limit π), but not in the subspace $\mathcal{C}\{\pi\}$ of $\mathbf{R}$.

In speaking informally of a convergent sequence of points $p_1, p_2, p_3, \ldots$ in a metric space E, one often says that "the points of the sequence get closer and closer to the limit", but this need not be literally true. For example, as we move along in the sequence the points of the sequence may at first get nearer and nearer to the limit, then move somewhat away, then get nearer and nearer again, then move somewhat away, etc. Thus while the terms of the sequence of points in $\mathbf{R}$

$$1, 1/2, 1/3, 1/4, 1/5, \ldots$$

do get closer and closer at each stage to the limit zero, this is not literally true of the convergent sequence

$$1, 1/3, 1/2, 1/5, 1/4, 1/7, 1/6, \ldots.$$

The points of a convergent sequence may also happen to be all *equally* near the limit, since for any $p \in E$, the sequence

$$p, p, p, \ldots$$

converges to p.

Proposition. *A sequence $p_1, p_2, p_3, \ldots$ of points in a metric space E has at most one limit.*

For suppose that $p, q \in E$ are both limits of the sequence. For any $\epsilon > 0$ there are positive integers N, N' such that $d(p, p_n) < \epsilon$ if $n > N$ and $d(q, p_n) < \epsilon$ if $n > N'$. If we choose an integer $n > \max\{N, N'\}$ we must have $d(p, p_n) < \epsilon$, $d(q, p_n) < \epsilon$, so that

$$d(p, q) \leq d(p, p_n) + d(p_n, q) < \epsilon + \epsilon = 2\epsilon.$$

If $d(p, q) \neq 0$ we get a contradiction by choosing $\epsilon \leq d(p, q)/2$. Therefore we must have $d(p, q) = 0$, so that $p = q$.

Thus a convergent sequence has a unique limit, and it makes sense to speak of *the* limit of a convergent sequence. A sequence which is not convergent of course has *no* limit.

The statement that the sequence of points $p_1, p_2, p_3, \ldots$ (in a metric space E) converges to the point p (also in E) is written concisely as

$$\lim_{n\to\infty} p_n = p.$$

For a nonconvergent sequence $p_1, p_2, p_3, \ldots$ of points in a metric space E the expression $\lim_{n\to\infty} p_n$ is meaningless. In this case we also say that $\lim_{n\to\infty} p_n$ "does not exist".

If $a_1, a_2, a_3, \ldots$ is a sequence (of any kind of objects) and if $n_1, n_2, n_3, \ldots$ is a strictly increasing sequence of positive integers (that is, $n_1, n_2, n_3, \ldots$ are positive integers and $n_1 < n_2 < n_3 < \cdots$) then the sequence $a_{n_1}, a_{n_2}, a_{n_3}, \ldots$ is called a *subsequence* of the sequence $a_1, a_2, a_3, \ldots$.

Proposition. *Any subsequence of a convergent sequence of points in a metric space converges to the same limit.*

For let $\lim_{n\to\infty} p_n = p$ and let $n_1, n_2, n_3, \ldots$ be a strictly increasing sequence of positive integers. Given any $\epsilon > 0$ there is a positive integer N such that $d(p, p_n) < \epsilon$ whenever $n > N$. Since $n_m \geq m$ for all positive integers m, we have $d(p, p_{n_m}) < \epsilon$ whenever $m > N$. This means that $\lim_{m\to\infty} p_{n_m} = p$, which is what we wanted to show. [Note that m in the last formula is just a "dummy variable", as was n in the expression $\lim_{n\to\infty} p_n$.]

The preceding result generalizes the fact that if a finite number of terms are lopped off the beginning of a convergent sequence, the resulting sequence converges to the same limit.

Call a sequence of points $p_1, p_2, p_3, \ldots$ in a metric space *bounded* if the set of points $\{p_1, p_2, p_3, \ldots\}$ is bounded. It is easy to show that any convergent sequence of points in a metric space is bounded: If the sequence $p_1, p_2, p_3, \ldots$ converges to the point p, pick any $\epsilon > 0$ and then find a positive integer N such that $d(p, p_n) < \epsilon$ whenever $n > N$. Then $\{p_1, p_2, p_3, \ldots\}$ is contained in the closed ball of center p and radius

$$\max\{\epsilon, d(p, p_1), d(p, p_2), \ldots, d(p, p_N)\}.$$

We recall that a closed subset of a metric space was defined as the complement of an open subset. Thus the knowledge of all the closed subsets of a metric space is equivalent to the knowledge of all the open subsets. It is also true that the knowledge of all the open subsets of a metric space determines which sequences of points are convergent, and to which limits. For it is immediate from the definitions that the sequence $p_1, p_2, p_3, \ldots$ converges to the limit p if and only if, for any open set U that contains the point p, there exists a positive integer N such that for any integer $n > N$ we have $p_n \in U$. The next result will tell us that knowledge of all the convergent sequences of points in the metric space, together with their limits, determines the closed subsets of the metric space. Thus any statement concerning the open subsets of a metric space can be translated into a statement concerning the closed subsets, which itself can be translated into another statement concerning convergent sequences of points and their limits. Thus there are three languages capable of making essentially the same statements: the language of open sets, that of closed sets, and that of convergent sequences. As one would expect, however, simple statements in one language may translate into complicated statements in another. We shall therefore use all three languages simultaneously, or rather a common language including all three, always striving for simplicity (as opposed to purity!) of expression.

Theorem. *Let S be a subset of the metric space E. Then S is closed if and only if, whenever $p_1, p_2, p_3, \ldots$ is a sequence of points of S that is convergent in E, we have*

$$\lim_{n\to\infty} p_n \in S.$$

There are two parts to the proof. First suppose that S is closed and that $p_1, p_2, p_3, \ldots$ is a sequence of points of S that converges to a point p of E. We must show that $p \in S$. If this is not so, we have $p \in \mathcal{C}S$. Since $\mathcal{C}S$ is open, there is some $\epsilon > 0$ such that $\mathcal{C}S$ contains the entire open ball of center p and radius ϵ. Thus if N is a positive integer such that $d(p, p_n) < \epsilon$ whenever $n > N$, we have $p_n \in \mathcal{C}S$ whenever $n > N$, a contradiction. This shows that $p \in S$ and proves the "only if" part of the theorem.

To prove the "if" part, suppose $S \subset E$ is not closed. Then $\mathcal{C}S$ is not open, and there exists a point $p \in \mathcal{C}S$ such that any open ball of center p contains points of S. Hence for each positive integer n we can choose $p_n \in S$ such that $d(p, p_n) < 1/n$. Then $\lim_{n\to\infty} p_n = p$, with each $p_n \in S$ and $p \notin S$. This shows that if the hypothesis on convergent sequences holds, then S must be closed, completing the proof of the "if" part, and hence of the whole theorem.

The metric space **R** has special properties that are not shared by all metric spaces: its elements can be added and multiplied, and they are ordered. For want of a better place, we insert here the relevant properties of sequences of real numbers.

Proposition. *If $a_1, a_2, a_3, \ldots$ and $b_1, b_2, b_3, \ldots$ are convergent sequences of real numbers, with limits a and b respectively, then*

$$\lim_{n\to\infty} (a_n + b_n) = a + b$$

$$\lim_{n\to\infty} (a_n - b_n) = a - b$$

$$\lim_{n\to\infty} a_n b_n = ab$$

and, in case b and each b_n are nonzero,

$$\lim_{n\to\infty} \frac{a_n}{b_n} = \frac{a}{b}.$$

We prove each part separately. Recall that for $x, y \in \mathbf{R}$, $d(x, y) = |x - y|$. For the first part, given $\epsilon > 0$ we also have $\epsilon/2 > 0$, so that we can find a positive integer N_1 such that $|a - a_n| < \epsilon/2$ whenever $n > N_1$ and we can also find a positive integer N_2 such that $|b - b_n| < \epsilon/2$ whenever $n > N_2$. If we set $N = \max\{N_1, N_2\}$, then whenever $n > N$ we have

$$\begin{aligned} |(a + b) - (a_n + b_n)| &= |(a - a_n) + (b - b_n)| \\ &\leq |a - a_n| + |b - b_n| < \frac{\epsilon}{2} + \frac{\epsilon}{2} = \epsilon. \end{aligned}$$

This proves the first part.

The second part, about subtraction, can be proved in a similar manner. As a matter of fact, a few changes of sign in the above proof will prove the result for subtraction. Here is an alternate proof: By the third part (which is proved in the next paragraph), observing that the sequence $-1, -1, -1, \ldots$ converges to -1, we have $\lim_{n\to\infty} (-b_n) = -b$, so by the first part

$$\begin{aligned} \lim_{n\to\infty} (a_n - b_n) &= \lim_{n\to\infty} (a_n + (-b_n)) = \lim_{n\to\infty} a_n + \lim_{n\to\infty} (-b_n) \\ &= a + (-b) = a - b. \end{aligned}$$

To prove the third part we start with the fact that convergent sequences are bounded to get a number $M \in \mathbf{R}$ such that $|a_n| < M$ and $|b_n| < M$ for all positive integers n. Clearly $M > 0$. Since a closed ball is a closed set, the preceding theorem implies $|a|, |b| \leq M$. Given any $\epsilon > 0$ we also have $\epsilon/2M > 0$, so there exists a positive integer N such that $|a - a_n| < \epsilon/2M$ and $|b - b_n| < \epsilon/2M$ whenever $n > N$. Hence for any integer $n > N$ we have

$$\begin{aligned} |ab - a_nb_n| &= |ab - ab_n + ab_n - a_nb_n| = |a(b - b_n) + b_n(a - a_n)| \\ &\leq |a| \cdot |b - b_n| + |b_n| \cdot |a - a_n| < M \cdot \frac{\epsilon}{2M} + M \cdot \frac{\epsilon}{2M} = \epsilon. \end{aligned}$$

This proves the third part.

To prove the last part, about division, first consider the special case where $a_n = 1$ for all n. We want to show that $1/b_n$ converges to $1/b$, that is that

$$\left| \frac{1}{b} - \frac{1}{b_n} \right| = \frac{|b_n - b|}{|b| \cdot |b_n|}$$

is small if n is large. The numerator is clearly small if n is large, but we also need to have the $|b_n|$ in the denominator bounded away from zero, or the total fraction may not be small. The latter objective is easily accomplished by taking n so large that $|b - b_n| < |b|/2$. The formal details of the proof that $\lim_{n\to\infty} 1/b_n = 1/b$ we give as follows:

Given $\epsilon > 0$, choose a positive integer N such that if $n > N$ then

$$|b - b_n| < \min\left\{\frac{|b|}{2}, \frac{|b|^2\epsilon}{2}\right\}.$$

Then if $n > N$ we have

$$|b_n| = |b - (b - b_n)| \geq |b| - |b - b_n| > |b| - \frac{|b|}{2} = \frac{|b|}{2},$$

so that

$$\left| \frac{1}{b} - \frac{1}{b_n} \right| = \frac{|b_n - b|}{|b| \cdot |b_n|} < \frac{|b|^2 \frac{\epsilon}{2}}{|b| \cdot \left(\frac{|b|}{2}\right)} = \epsilon.$$

This proves that $\lim_{n\to\infty} 1/b_n = 1/b$. To complete the proof of the last part of the proposition we use the third part together with this last result as follows:

$$\lim_{n\to\infty} \frac{a_n}{b_n} = \lim_{n\to\infty} \left(a_n \cdot \frac{1}{b_n}\right) = \lim_{n\to\infty} a_n \cdot \lim_{n\to\infty} \frac{1}{b_n} = a \cdot \frac{1}{b} = \frac{a}{b}.$$

Note the special cases of the last proposition when one or the other of the sequences $a_1, a_2, a_3, \ldots$ or $b_1, b_2, b_3, \ldots$ is constant:

$$\lim_{n\to\infty} (a_n + b) = \lim_{n\to\infty} a_n + b, \quad \lim_{n\to\infty} ab_n = a \lim_{n\to\infty} b_n, \text{ etc.}$$

The next very easy result expresses the compatibility of convergence of sequences of real numbers with order.

Proposition. *If $a_1, a_2, a_3, \ldots$ and $b_1, b_2, b_3, \ldots$ are convergent sequences of real numbers, with limits a and b respectively, and if $a_n \leq b_n$ for all n, then $a \leq b$.*

For $b - a = \lim_{n\to\infty} b_n - \lim_{n\to\infty} a_n = \lim_{n\to\infty} (b_n - a_n)$, and since each $b_n - a_n$ is nonnegative the theorem implies that the limit also is nonnegative, the set $\{x \in \mathbf{R} : x \geq 0\}$ being closed.

Definition. A sequence of real numbers $a_1, a_2, a_3, \ldots$ is *increasing* if $a_1 \leq a_2 \leq a_3 \leq \cdots$, *decreasing* if $a_1 \geq a_2 \geq a_3 \geq \cdots$, and *monotonic* if it is either increasing or decreasing.

Proposition. *A bounded monotonic sequence of real numbers is convergent.*

Suppose first that $a_1, a_2, a_3, \ldots$ is a bounded increasing sequence of real numbers. We shall prove that the sequence converges to the limit $a = \text{l.u.b.}\ \{a_1, a_2, a_3, \ldots\}$. We have $a \geq a_n$ for all n, by the definition of upper bound. For any $\epsilon > 0$ we have $a - \epsilon < a$, so by the definition of least upper bound there is a positive integer N such that $a_N > a - \epsilon$. Since the sequence is increasing, it follows that if $n > N$ then $a_n > a - \epsilon$, so that

$$a - \epsilon < a_n \leq a < a + \epsilon, \quad \text{or} \quad |a - a_n| < \epsilon.$$

Thus the sequence converges to a. Essentially the same proof shows that a bounded decreasing sequence of real numbers converges; we have only to replace l.u.b. by g.l.b. and change the sense of some of the inequalities. Or we can apply the part already proved to get the convergence of the increasing sequence $-a_1, -a_2, -a_3, \ldots$, which implies the convergence of $a_1, a_2, a_3, \ldots$.

EXAMPLE. The last result gives an easy proof that if $a \in \mathbf{R}$, $|a| < 1$, then

$$\lim_{n\to\infty} a^n = 0.$$

First note that $|a^n| = |a|^n$, so that we may suppose that $a \geq 0$. Then the sequence $a, a^2, a^3, a^4, \ldots$ is decreasing, since $a^n - a^{n+1} = a^n(1 - a) \geq 0$.

The terms of the sequence are all nonnegative, so the sequence is bounded, hence convergent. Let $\lim_{n\to\infty} a^n = x$. Then

$$ax = a \lim_{n\to\infty} a^n = \lim_{n\to\infty} (a \cdot a^n) = \lim_{n\to\infty} a^{n+1} = \lim_{n\to\infty} a^n = x.$$

Since $ax = x$, we have $(a - 1)x = 0$, so $x = 0$.

As a consequence of what we have just shown, if $a \in \mathbf{R}$, $|a| > 1$, then the sequence $a, a^2, a^3, a^4, \ldots$ is unbounded. (Another proof of this fact is indicated in Problem 11, Chapter II.)

§ 4. COMPLETENESS.

The definition of convergence of a sequence of points in a metric space enables us to verify the convergence of a given sequence only if the limit is known. It is desirable to be able to state that a given sequence is convergent without actually having to find the limit. We can already do this in certain cases, for the last proposition of the previous section states that a bounded monotonic sequence of real numbers is always convergent. However we need a more general "inner" criterion for the convergence of a sequence of points in a metric space. For this reason we introduce below the concept of "Cauchy sequence", a sequence of points in a metric space that satisfies a certain property depending only on the terms of the sequence. It will turn out that all convergent sequences are Cauchy sequences and that, at least for certain important metric spaces, any Cauchy sequence is convergent. Thus we shall often be able to state that a sequence is convergent without having to determine the limit. We very often are not at all interested in computing the limit of a sequence but rather in verifying that the limit possesses certain properties, and once it is known that the limit exists such properties can often be inferred directly from the sequence.

Definition. A sequence of points $p_1, p_2, p_3, \ldots$ in a metric space is a *Cauchy sequence* if, given any real number $\epsilon > 0$, there is a positive integer N such that $d(p_n, p_m) < \epsilon$ whenever $n, m > N$.

The number N in the definition above of course depends on ϵ. The important point is that given *any* $\epsilon > 0$ there exists *some* N with the desired property.

Proposition. *A convergent sequence of points in a metric space is a Cauchy sequence.*

For if $p_1, p_2, p_3, \ldots$ converges to p then for any $\epsilon > 0$ there is an integer N such that $d(p, p_n) < \epsilon/2$ whenever $n > N$. Hence if $n, m > N$ we have

$$d(p_n, p_m) \leq d(p_n, p) + d(p, p_m) < \frac{\epsilon}{2} + \frac{\epsilon}{2} = \epsilon.$$

However, not every Cauchy sequence is convergent. For example the sequence 1, ½, ⅓, ¼, ⅕, ... is a Cauchy sequence in the metric space $E = \mathbf{R} - \{0\}$ (the complement of $\{0\}$ in $\mathbf{R}$), but it is not convergent in E. More generally, if we take any sequence of points in a metric space which converges to a limit which is not one of the terms of the sequence and then delete the limit from the metric space, we get a Cauchy sequence which is not convergent.

The following two easy propositions give known properties of convergent sequences that generalize to Cauchy sequences. The first proposition is trivial so the proof is omitted.

Proposition. *Any subsequence of a Cauchy sequence is a Cauchy sequence.*

Proposition. *A Cauchy sequence of points in a metric space is bounded.*

For if the sequence is $p_1, p_2, p_3, \ldots$ and ϵ is any positive number and N an integer such that $d(p_n, p_m) < \epsilon$ if $n, m > N$, then for any fixed $m > N$ the entire sequence is contained in the closed ball of center p_m and radius

$$\max \{d(p_m, p_1), d(p_m, p_2), \ldots, d(p_m, p_N), \epsilon\}.$$

Proposition. *A Cauchy sequence that has a convergent subsequence is itself convergent.*

Let $p_1, p_2, p_3, \ldots$ be the Cauchy sequence, p the limit of a convergent subsequence. For $\epsilon > 0$, let N be such that $d(p_n, p_m) < \epsilon/2$ if $n, m > N$. Fix an integer $m > N$ so that p_m is in the convergent subsequence and so far out that $d(p, p_m) < \epsilon/2$. Then for $n > N$ we have

$$d(p, p_n) \leq d(p, p_m) + d(p_m, p_n) < \frac{\epsilon}{2} + \frac{\epsilon}{2} = \epsilon.$$

Definition. A metric space E is *complete* if every Cauchy sequence of points of E converges to a point of E.

It will be proved shortly that E^n is complete. Other useful examples of complete metric spaces will appear later. The following proposition gives us infinitely more examples.

Proposition. *A closed subset of a complete metric space is a complete metric space.*

The subset is of course considered to be a subspace, so that a Cauchy sequence in the subset is a fortiori one in the original metric space. The result is an immediate consequence of the theorem of the last section.

Theorem. **R** *is complete.*

Let $a_1, a_2, a_3, \ldots$ be a Cauchy sequence of real numbers. We must show that this sequence converges to a real number. Consider the set

$$S = \{x \in \mathbf{R} : x \leq a_n \text{ for an infinite number of positive integers } n\}.$$

Since the Cauchy sequence is bounded, S is nonempty and bounded from above. Therefore $a = \text{l.u.b. } S$ exists. We proceed to prove that $a_1, a_2, a_3, \ldots$ converges to a. Given any $\epsilon > 0$ choose a positive integer N such that $|a_n - a_m| < \epsilon/2$ if $n, m > N$. Since $a = \text{l.u.b. } S$ we have $a + \epsilon/2 \notin S$, but $a - \epsilon/2 \in S$. This means that for only a finite number of positive integers n is it true that $a + \epsilon/2 \leq a_n$, but we have $a - \epsilon/2 \leq a_n$ for an infinite number of n. Hence we can find a specific integer $m > N$ such that $a + \epsilon/2 > a_m$ and $a - \epsilon/2 \leq a_m$, so that $|a - a_m| \leq \epsilon/2$. Therefore if $n > N$ we have

$$\begin{aligned} |a - a_n| &= |(a - a_m) + (a_m - a_n)| \\ &\leq |a - a_m| + |a_m - a_n| < \frac{\epsilon}{2} + \frac{\epsilon}{2} = \epsilon. \end{aligned}$$

This proves that $a_1, a_2, a_3, \ldots$ converges to a.

Corollary. *For any positive integer n, E^n is complete.*

For simplicity of notation let $n = 3$; the proof for any n will be essentially the same. We have to show that a Cauchy sequence $p_1, p_2, p_3, \ldots$ of points of E^3 has a limit. Let $p_1 = (x_1, y_1, z_1)$, $p_2 = (x_2, y_2, z_2)$, $p_3 = (x_3, y_3, z_3)$, etc. Given $\epsilon > 0$ there exists a positive integer N such that $d(p_n, p_m) < \epsilon$ if $n, m > N$. Since

$$\begin{aligned} d(p_n, p_m) &= \sqrt{(x_n - x_m)^2 + (y_n - y_m)^2 + (z_n - z_m)^2} \\ &\geq |x_n - x_m|, |y_n - y_m|, |z_n - z_m| \end{aligned}$$

we have $|x_n - x_m|, |y_n - y_m|, |z_n - z_m| < \epsilon$ whenever $n, m > N$. This means that each of the sequences $x_1, x_2, x_3, \ldots, y_1, y_2, y_3, \ldots,$ and $z_1, z_2, z_3, \ldots$ is a Cauchy sequence in **R**. By the theorem these sequences are convergent, and we denote their limits by x, y, z respectively. We shall prove that $p_1, p_2, p_3, \ldots$ converges to the point $p = (x, y, z)$. To do this, given $\epsilon > 0$ choose N such that for any integer $n > N$ we have $|x - x_n|$, $|y - y_n|, |z - z_n| < \epsilon/\sqrt{3}$. Then for $n > N$ we have

$$\begin{aligned} d(p, p_n) &= \sqrt{(x - x_n)^2 + (y - y_n)^2 + (z - z_n)^2} \\ &< \sqrt{\frac{\epsilon^2}{3} + \frac{\epsilon^2}{3} + \frac{\epsilon^2}{3}} = \epsilon. \end{aligned}$$

This proves that $\lim_{n \to \infty} p_n = p$.

§ 5. COMPACTNESS.

Definition. A subset S of a metric space E is *compact* if, whenever S is contained in the union of a collection of open subsets of E, then S is contained in the union of a finite number of these open subsets.

Clearly any finite subset of a metric space is compact. Unfortunately we have to wait before we can give a less trivial example of a compact subset of a metric space. An example of a noncompact subset of a metric space is, however, easy to give: the open interval $(0, 1)$ is not a compact subset of the metric space $\mathbf{R}$, since $(0, 1)$ is contained in the union of all open subsets of $\mathbf{R}$ of the form $(1/n, 1)$, where n is a positive integer, but is not contained in the union of any finite number of these open subsets. It will be seen as we go along that compactness is a property related to completeness, but much stronger.

The definition just given of a compact subset of a metric space can be applied to E itself: the metric space E is called *compact* if it is a compact subset of itself. This means that whenever E is the union of a collection of open subsets, it is the union of some finite subcollection of these open subsets.

Let S be an arbitrary subset of a metric space E. When we consider S as a subspace of E, an open ball in S is simply the set of points in S of an open ball in E whose center is in S, that is the intersection with S of an open ball in E whose center is in S. Thus the open subsets of the metric space S are precisely the intersections with S of the open subsets of E. Hence S is a compact subset of E if and only if the metric space S is compact.

One example of a compact metric space will give us many more, by means of the following result.

Proposition. *Any closed subset of a compact metric space is compact.*

For let the closed subset S of the compact metric space E be contained in the union of a collection of open subsets of E, say $S \subset \bigcup_{i \in I} U_i$, where each U_i is an open subset of E, i ranging over an indexing family I. Then $E \subset (\bigcup_{i \in I} U_i) \cup \complement S$. Since S is closed, $\complement S$ is open, so by the compactness of E we can find a finite subset $J \subset I$ such that $E \subset (\bigcup_{i \in J} U_i) \cup \complement S$. Hence $S \subset \bigcup_{i \in J} U_i$. This shows that S is compact.

Here are some of the basic properties of compact sets.

Proposition. *A compact subset of a metric space is bounded. In particular, a compact metric space is bounded.*

For any metric space is the union of its open balls, and the union of any finite number of balls is a bounded set.

***Proposition* (*Nested set property*).** *Let $S_1, S_2, S_3, \ldots$ be a sequence of nonempty closed subsets of a compact metric space, with the property that $S_1 \supset S_2 \supset S_3 \supset \ldots$. Then there is at least one point that belongs to each of the sets $S_1, S_2, S_3, \ldots$.*

If not, we must have $\bigcap_{n=1,2,3,\ldots} S_n = \emptyset$, implying that $\bigcup_{n=1,2,3,\ldots} \mathcal{C}S_n$ is the entire metric space E. Since E is compact it is the union of a finite number of the open subsets $\mathcal{C}S_1, \mathcal{C}S_2, \mathcal{C}S_3, \ldots$. Since $\mathcal{C}S_1 \subset \mathcal{C}S_2 \subset \mathcal{C}S_3 \subset \ldots$, we must have $E = \mathcal{C}S_n$ for some n, which produces the contradiction $S_n = \emptyset$.

The above proposition does not hold if the word "compact" is replaced by "complete": for example let $E = \mathbf{R}$, taking $S_n = \{x \in \mathbf{R} : x \geq n\}$, $n = 1, 2, 3, \ldots$.

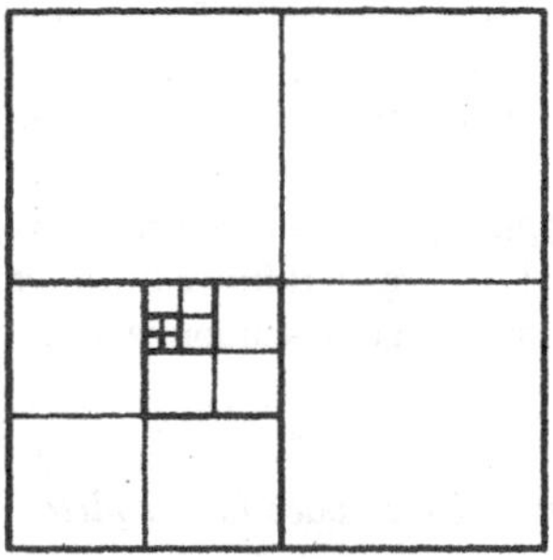

FIGURE 14. Nested set property. There is a point common to all the squares, each of which has half the dimensions of its predecessor. (A closed square is compact, by a theorem to be proved shortly.)

For a better insight into the meaning of compactness, we introduce another definition.

Definition. If E is a metric space, S a subset of E, and p a point of E, then p is a *cluster point of S* if any open ball with center p contains an infinite number of points of S.

Theorem. *An infinite subset of a compact metric space has at least one cluster point.*

If this were false, then for any given point of a certain compact metric space E we could find an open ball having the given point as center and containing only a finite number of points of the infinite subset S of E. E is the union of all such open balls. Since E is compact, it is the union of a finite number of such open balls. This implies that S is finite, a contradiction.

Corollary 1. *Any sequence of points in a compact metric space has a convergent subsequence.*

Let $p_1, p_2, p_3, \ldots$ be the sequence, E the metric space. We must separate cases, according as the set $\{p_1, p_2, p_3, \ldots\}$ is infinite or not.

Case 1. The set $\{p_1, p_2, p_3, \ldots\}$ is infinite. In this case the set $S = \{p_1, p_2, p_3, \ldots\}$ has at least one cluster point, say $p \in E$. Pick a positive integer n_1 such that p_{n_1} is in the open ball of center p and radius 1, then pick an integer $n_2 > n_1$ such that p_{n_2} is in the open ball of center p and radius $1/2$, then pick an integer $n_3 > n_2$ such that p_{n_3} is in the open ball of center p and radius $1/3$, etc.; this process can be continued indefinitely since each open ball of center p contains an infinite number of points of S. We end up with a subsequence of $p_1, p_2, p_3, \ldots$ whose n^{th} term has distance less than $1/n$ from p, for all $n = 1, 2, 3, \ldots$. This subsequence converges to p.

Case 2. The set $\{p_1, p_2, p_3, \ldots\}$ is finite. In this case at least one point $p \in \{p_1, p_2, p_3, \ldots\}$ occurs an infinite number of times in the sequence $p_1, p_2, p_3, \ldots$. Thus the convergent sequence $p, p, p, \ldots$ is a subsequence of the given sequence.

Corollary 2. *A compact metric space is complete.*

For any Cauchy sequence has a convergent subsequence and therefore is itself convergent.

Corollary 3. *A compact subset of a metric space is closed.*

For any convergent sequence of points in the compact subset must have its limit in the compact subset, by Corollary 2. Thus the theorem of § 3 implies that the subset is closed.

We now know that any compact subset of a metric space is both closed and bounded. We proceed to prove the fundamental fact that any closed bounded subset of E^n is compact.

Lemma. *Let S be a bounded subset of E^n. Then for any $\epsilon > 0$, S is contained in the union of a finite number of closed balls of radius ϵ.*

We begin the proof by showing that E^n itself is the union of a set of evenly spaced closed balls of radius ϵ, as illustrated in Figure 15.

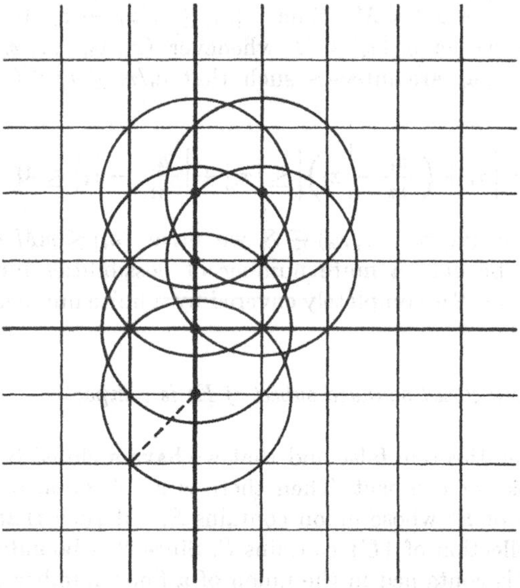

FIGURE 15. E^2 is the union of evenly spaced small closed balls.

To be concrete, consider the points in E^n of the form $(a_1/m, a_2/m, \ldots, a_n/m)$, where $a_1, a_2, \ldots, a_n$ are integers and m is a fixed positive integer, to be determined shortly. For any $(x_1, x_2, \ldots, x_n) \in E^n$ there are integers $a_1, a_2, \ldots, a_n$ such that

$$\frac{a_i}{m} \leq x_i < \frac{a_i + 1}{m}, \qquad i = 1, 2, \ldots, n$$

(by LUB 4 of Chapter II) and we then have

$$d\left((x_1, \ldots, x_n), \left(\frac{a_1}{m}, \ldots, \frac{a_n}{m}\right)\right) = \sqrt{\left(x_1 - \frac{a_1}{m}\right)^2 + \cdots + \left(x_n - \frac{a_n}{m}\right)^2}$$
$$< \sqrt{\frac{1}{m^2} + \cdots + \frac{1}{m^2}} = \frac{\sqrt{n}}{m}.$$

Now suppose that the fixed integer m had been chosen greater than $\sqrt{n}/\epsilon$, so that $\sqrt{n}/m < \epsilon$. Then $(x_1, \ldots, x_n)$ is contained in the closed ball of radius ϵ and center $(a_1/m, \ldots, a_n/m)$. In particular E^n is the union of all these balls, for varying integral $a_1, a_2, \ldots, a_n$. It remains to show that if $S \subset E^n$ is bounded then a finite number of values of $a_1, a_2, \ldots, a_n$ suffice to give us balls of radius ϵ whose union contains S. To do this note that since S is bounded it is contained in some ball with center at the origin. If this ball has radius M, then for each point $(x_1, x_2, \ldots, x_n) \in S$ we have $\sqrt{x_1^2 + x_2^2 + \cdots + x_n^2} \leq M$. Since $|x_i| \leq \sqrt{x_1^2 + x_2^2 + \cdots + x_n^2}$ for $i = 1, 2, \ldots, n$, we have $|x_i| \leq M$ whenever $(x_1, x_2, \ldots, x_n) \in S$. If, as above, $a_1, a_2, \ldots, a_n$ are integers such that $a_i/m \leq x_i < (a_i + 1)/m$, we have

$$\left|\frac{a_i}{m}\right| = \left|x_i + \left(\frac{a_i}{m} - x_i\right)\right| \leq |x_i| + \left|\frac{a_i}{m} - x_i\right| \leq M + \frac{1}{m}.$$

Hence whenever $(x_1, x_2, \ldots, x_n) \in S$ we have $|a_i| \leq mM + 1$, showing that there will be only a finite number of possibilities for the integers $a_1, a_2, \ldots, a_n$. Thus S is completely covered by a finite number of our closed balls of radius ϵ.

Theorem. *Any closed bounded subset of E^n is compact.*

Suppose the theorem false and that we have a closed bounded subset S of E^n that is not compact. Then there is a collection of open subsets $\{U\} = \{U_i\}_{i \in I}$ of E^n whose union contains S, but such that the union of no finite subcollection of $\{U\}$ contains S. Since S is bounded, the lemma tells us that S is contained in the union of a finite number of closed balls of radius 1/2, say $B_1, B_2, \ldots, B_r$. Thus

$$S = (S \cap B_1) \cup (S \cap B_2) \cup \cdots \cup (S \cap B_r).$$

At least one of the sets $S \cap B_1, S \cap B_2, \ldots, S \cap B_r$, say $S \cap B_j$, is not contained in the union of any finite subcollection of $\{U\}$. $S \cap B_j$ is closed (being the intersection of closed sets) and if $p, q \in S \cap B_j$ then $d(p, q) \leq 1$ (since each of the points p, q has distance at most 1/2 from the center of the ball B_j). Set $S_1 = S \cap B_j$. Then S_1 is closed and bounded, not contained in the union of any finite subcollection of $\{U\}$, and if $p, q \in S_1$ then $d(p, q) \leq 1$. Now apply the lemma to S_1 and $\epsilon = 1/4$ to get S_1 contained in the union of a finite number of closed balls of radius 1/4 and repeat the above argument to get a closed subset S_2 of S_1, not contained in the union of any finite subcollection of $\{U\}$, such that if $p, q \in S_2$ then $d(p, q) \leq 1/2$. Repeating the argument with $\epsilon = 1/6, 1/8, 1/10, \ldots$ we obtain a sequence of closed sets $S \supset S_1 \supset S_2 \supset S_3 \supset \cdots$, none of which is contained in the union of any finite subcollection of $\{U\}$, and such that if $p, q \in S_N$ then $d(p, q) \leq 1/N$, for each $N = 1, 2, 3, \ldots$. No S_N is empty, so let $p_N \in S_N$,

$N = 1, 2, 3, \ldots$. The sequence $p_1, p_2, p_3, \ldots$ is a Cauchy sequence, since if α, β, N are positive integers and $\alpha, \beta > N$ then $p_\alpha, p_\beta \in S_N$, so that $d(p_\alpha, p_\beta) \leq 1/N$. Since E^n is complete, the sequence $p_1, p_2, p_3, \ldots$ converges to a point p_0 of E^n. Since each point $p_1, p_2, p_3, \ldots$ is in the closed set S, we have also $p_0 \in S$. Hence p_0 is in one of the sets of the collection $\{U\}$, say $p_0 \in U_0 \in \{U\}$. Since U_0 is open there is an open ball of center p_0 and some radius $\epsilon > 0$ that is entirely contained in U_0. Pick some N so large that $1/N < \epsilon/2$ and $d(p_0, p_N) < \epsilon/2$. Then for any point $p \in S_N$ we have

$$d(p_0, p) \leq d(p_0, p_N) + d(p_N, p) < \frac{\epsilon}{2} + \frac{1}{N} < \frac{\epsilon}{2} + \frac{\epsilon}{2} = \epsilon.$$

Thus S_N is contained in the open ball of center p_0 and radius ϵ, giving $S_N \subset U_0$. This contradicts the statement that S_N is not contained in the union of any finite subcollection of $\{U\}$, completing the proof.

§ 6. CONNECTEDNESS.

The intuitive idea of connectedness is simple enough. It is illustrated in Figure 16. But of course we need a precise definition.

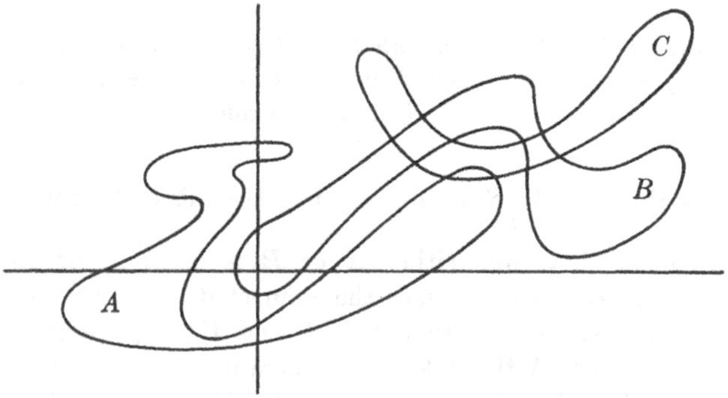

FIGURE 16. $A, B, C, A \cup C, B \cup C, A \cup B \cup C$ are connected. $A \cup B$ is not.

Definition. A metric space E is *connected* if the only subsets of E which are both open and closed are E and $\varnothing$. A subset S of a metric space is a *connected subset* if the subspace S is connected.

Thus a metric space E that is not connected has a subset $A \neq E, \emptyset$ that is both open and closed. Then also $\complement A \neq E, \emptyset$ is both open and closed. Setting $B = \complement A$, we have $E = A \cup B$, where A and B are disjoint nonempty open subsets of E. Conversely if a metric space E can be written $E = A \cup B$, where A and B are disjoint nonempty open subsets of E, then E is not connected: for $A \neq \emptyset$ (given), $A \neq E$ (since A and B are disjoint and $B \neq \emptyset$) and A is both open (given) and closed (since its complement B is open).

If $a, b, c \in \mathbf{R}$, we say that c *is between* a *and* b if either $a < c < b$ or $b < c < a$.

Proposition. *Any subset of* $\mathbf{R}$ *which contains two distinct points* a *and* b *and does not contain all points between* a *and* b *is not connected.*

For suppose that $a < c < b$ and that S is a subset of $\mathbf{R}$ with $a, b \in S$, $c \notin S$. Then

$$S = (S \cap \{x \in \mathbf{R} : x < c\}) \cup (S \cap \{x \in \mathbf{R} : x > c\})$$

expresses S as the union of two disjoint nonempty open subsets.

It will be shown shortly that any closed interval in $\mathbf{R}$ is connected. In the next chapter it will be seen how to deduce from this fact that certain other metric spaces, for example balls in E^n, or the subsets A, B, C of Figure 16, are connected. The following is one of the principal arguments one uses to prove these and other connectedness results for subsets of E^n.

Proposition. *Let* $\{S_i\}_{i \in I}$ *be a collection of connected subsets of a metric space* E. *Suppose there exists* $i_0 \in I$ *such that for each* $i \in I$ *the sets* S_i *and* S_{i_0} *have a nonempty intersection. Then* $\bigcup_{i \in I} S_i$ *is connected.*

Suppose that $S = \bigcup_{i \in I} S_i$ is the union of two disjoint open subsets A and B. We must show that either A or B is empty. For any $i \in I$, $S_i = (A \cap S_i) \cup (B \cap S_i)$ expresses the connected set S_i as the union of two disjoint open subsets, so that $A \cap S_i$ and $B \cap S_i$ are just the sets $S_i, \emptyset$ in some order. Without loss of generality we may assume that $A \cap S_{i_0} = S_{i_0}$. Then $A \supset S_{i_0}$, so that for each $i \in I$ we have $A \cap S_i \neq \emptyset$. Thus $A \cap S_i = S_i$ and $B \cap S_i = \emptyset$. Since $B \cap S_i = \emptyset$ for all $i \in I$ we have $B = \emptyset$, proving the proposition.

Theorem. $\mathbf{R}$, *or any open or closed interval in* $\mathbf{R}$, *is connected.*

A somewhat more general statement is just as easy to prove, namely that any subset S of $\mathbf{R}$ which contains all points between any two of its points is connected. To prove this, suppose that such a subset S is not connected, so that we may write $S = A \cup B$, where A and B are disjoint nonempty open subsets of S. Choose $a \in A$, $b \in B$ and assume, as we may, that $a < b$. By assumption $[a, b] \subset S$. Set $A_1 = A \cap [a, b]$, $B_1 = B \cap [a, b]$. Then A_1, B_1 are disjoint open subsets of $[a, b]$, $a \in A_1$, $b \in B_1$, and $[a, b] = A_1 \cup B_1$. From these facts we derive a contradiction, as follows. Since B_1 is an open subset of $[a, b]$, its complement A_1 is a closed subset of $[a, b]$, hence a closed subset of $\mathbf{R}$. Since A_1 is also nonempty and bounded from above, the last result of § 2 tells us that there is a greatest element c in A_1. Since $b \in B_1$ we must have $c < b$. But since A_1 is an open subset of $[a, b]$ it must contain the intersection of $[a, b]$ with some open ball in $\mathbf{R}$ of center c, hence it must contain points greater than c. This is a contradiction, and this proves the theorem.

PROBLEMS

1. Verify that the following are metric spaces:
   (a) all n-tuples of real numbers, with

$$d((x_1, \ldots, x_n), (y_1, \ldots, y_n)) = \sum_{i=1}^{n} |x_i - y_i|$$

   (b) all bounded infinite sequences $x = (x_1, x_2, x_3, \ldots)$ of elements of $\mathbf{R}$, with $d(x, y) = \text{l.u.b.}\,\{|x_1 - y_1|, |x_2 - y_2|, |x_3 - y_3|, \ldots\}$
   (c) $(E_1 \times E_2, d)$, where (E_1, d_1), (E_2, d_2) are metric spaces and d is given by $d((x_1, x_2), (y_1, y_2)) = \max\,\{d_1(x_1, y_1), d_2(x_2, y_2)\}$.
2. Show that $(\mathbf{R} \times \mathbf{R}, d)$ is a metric space, where

$$d((x, y), (x', y')) = \begin{cases} |y| + |y'| + |x - x'| & \text{if } x \neq x' \\ |y - y'| & \text{if } x = x'. \end{cases}$$

   Illustrate by diagrams in the plane E^2 what the open balls of this metric space are.
3. What are the open and closed balls in the metric space of example (4), § 1? Show that two balls of different centers and radii may be equal. What are the open sets in this metric space?
4. Show that the subset of E^2 given by $\{(x_1, x_2) \in E^2 : x_1 > x_2\}$ is open.
5. Prove that any bounded open subset of $\mathbf{R}$ is the union of disjoint open intervals.
6. Show that the subset of E^2 given by $\{(x_1, x_2) \in E^2 : x_1x_2 = 1, x_1 > 0\}$ is closed.
7. Give the details of the proof of the last proposition of § 2 for sets bounded from below.
8. Prove that if the points of a convergent sequence of points in a metric space are reordered, then the new sequence converges to the same limit.

9. Prove that $\lim_{n\to\infty} p_n = p$ in a given metric space if and only if the sequence $p_1, p, p_2, p, p_3, p, \ldots$ is convergent.

10. Prove that if $\lim_{n\to\infty} p_n = p$ in a given metric space then the set of points $\{p, p_1, p_2, p_3, \ldots\}$ is closed.

11. Show that if $a_1, a_2, a_3, \ldots$ is a sequence of real numbers that converges to a, then

$$\lim_{n\to\infty}\left(\frac{\sum_{i=1}^{n} a_i}{n}\right) = a.$$

12. Prove that the sequence $x_1, x_2, x_3, \ldots$ of real numbers given by $x_1 = 1$ and $x_{n+1} = x_n + \dfrac{1}{x_n^2}$ for each $n = 1, 2, 3, \ldots$ is unbounded.

13. Consider the sequence of real numbers

$$\frac{1}{2}, \quad \frac{1}{2+\frac{1}{2}}, \quad \frac{1}{2+\cfrac{1}{2+\frac{1}{2}}}, \quad \ldots.$$

Show that this sequence is convergent and find its limit by first showing that the two sequences of alternate terms are monotonic and finding their limits.

14. Prove that any sequence in **R** has a monotonic subsequence. (*Hint:* This is easy if there exists a subsequence with no least term, hence we may suppose that each subsequence has a least term.) (Note that this result and the theorem on the convergence of bounded monotonic sequences gives another proof that **R** is complete.)

15. Let S be a subset of the metric space E. A point $p \in S$ is called an *interior point of* S if there is an open ball in E of center p which is contained in S. Prove that the set of interior points of S is an open subset of E (called the *interior* of S) that contains all other open subsets of E that are contained in S.

16. Let S be a subset of the metric space E. Define the *closure* of S, denoted $\overline{S}$, to be the intersection of all closed subsets of E that contain S. Show that
    (a) $\overline{S} \supset S$, and S is closed if and only if $\overline{S} = S$
    (b) $\overline{S}$ is the set of all limits of sequences of points of S that converge in E
    (c) a point $p \in E$ is in $\overline{S}$ if and only if any ball in E of center p contains points of S, which is true if and only if p is not an interior point of $\mathcal{C}S$ (cf. Prob. 15).

17. Let S be a subset of the metric space E. The *boundary* of S is defined to be $\overline{S} \cap \overline{\mathcal{C}S}$ (cf. Prob. 16). Show that
    (a) E is the disjoint union of the interior of S, the interior of $\mathcal{C}S$, and the boundary of S
    (b) S is closed if and only if S contains its boundary
    (c) S is open if and only if S and its boundary are disjoint.

18. If $a_1, a_2, a_3, \ldots$ is a bounded sequence of real numbers, define $\limsup_{n\to\infty} a_n$ (also denoted $\overline{\lim_{n\to\infty}}\, a_n$) to be

$$\text{l.u.b. } \{x \in \mathbf{R} : a_n > x \text{ for an infinite number of integers } n\}$$

and define $\liminf_{n\to\infty} a_n$ (also denoted $\underline{\lim}_{n\to\infty}\, a_n$) to be

$$\text{g.l.b. } \{x \in \mathbf{R} : a_n < x \text{ for an infinite number of integers } n\}.$$

Prove that $\liminf_{n\to\infty} a_n \leq \limsup_{n\to\infty} a_n$, with the equality holding if and only if the sequence converges.

19. Let $a_1, a_2, a_3, \ldots$ and $b_1, b_2, b_3, \ldots$ be bounded sequences of real numbers. Show that

$$\limsup_{n\to\infty} (a_n + b_n) \leq \limsup_{n\to\infty} a_n + \limsup_{n\to\infty} b_n,$$

with the equality holding if one of the original sequences converges (cf. Prob. 18).

20. The complex numbers $\mathbf{C} = \mathbf{R} \times \mathbf{R}$ (cf. Prob. 8, Chap. II) are the underlying set of the metric space E^2. The metric in E^2 therefore makes $\mathbf{C}$ itself a metric space. If $z \in \mathbf{C}$, the *absolute value* of z, denoted by $|z|$, is defined by $|z| = d(z, 0)$. Show that
    (a) $|x + iy| = \sqrt{x^2 + y^2}$ if $x, y \in \mathbf{R}$ (and therefore $|z|$ agrees with the previously defined $|z|$ if $z \in \mathbf{R}$)
    (b) $|z_1 + z_2| \leq |z_1| + |z_2|$ for all $z_1, z_2 \in \mathbf{C}$
    (c) $|z_1 z_2| = |z_1| \cdot |z_2|$ for all $z_1, z_2 \in \mathbf{C}$.

21. Show that the proposition of page 48 remains true if the word "real" is replaced by "complex" (cf. Prob. 20).

22. A *normed vector space* is a vector space V over $\mathbf{R}$, together with a real-valued function on V, called the *norm* and indicated by $\| \ \|$, the value of $\| \ \|$ at any element $x \in V$ being indicated $\|x\|$, having the following properties:
    (i) $\|x\| \geq 0$ for all $x \in V$
    (ii) $\|x\| = 0$ if and only if $x = 0$
    (iii) $\|cx\| = |c| \cdot \|x\|$ for any $x \in V$ and any $c \in \mathbf{R}$
    (iiii) $\|x + y\| \leq \|x\| + \|y\|$ for any $x, y \in V$.

    Show that a normed vector space V becomes a metric space if for any $x, y \in V$ we take $d(x, y) = \|x - y\|$. Recall that the set of all n-tuples of real numbers is a vector space $\mathbf{R}^n$ over $\mathbf{R}$ if addition and scalar multiplication are defined by

$$(x_1, \ldots, x_n) + (y_1, \ldots, y_n) = (x_1 + y_1, \ldots, x_n + y_n),$$
$$c(x_1, \ldots, x_n) = (cx_1, \ldots, cx_n).$$

Show that $\mathbf{R}^n$ becomes a normed vector space if $\|(x_1, \ldots, x_n)\|$ is taken to be $\sqrt{x_1^2 + \cdots + x_n^2}$ and that in this case the resulting metric space is just E^n; show that if $n = 2$ and $\mathbf{C}$ is identified with E^2 as in Prob. 20 then $\mathbf{C}$ becomes a normed vector space with $\|z\| = |z|$. Show that $\mathbf{R}^n$ becomes another normed vector space if $\|(x_1, \ldots, x_n)\|$ is defined to be $|x_1| + \cdots + |x_n|$, and yet another normed vector space if $\|(x_1, \ldots, x_n)\|$ is defined to be $\max\{|x_1|, \ldots, |x_n|\}$. (Note: What we have for brevity called a normed vector space is more properly called a *real normed vector space;* there is also the notion of a *complex normed vector space*, got by altering the above definition by twice replacing the symbol **R** by **C**.)

23. Prove that if V is a normed vector space (cf. Prob. 22) and $a_1, a_2, a_3, \ldots$ and $b_1, b_2, b_3, \ldots$ are convergent sequences of elements of V with limits a and b respectively, then
$$\lim_{n\to\infty} (a_n + b_n) = a + b \quad \text{and} \quad \lim_{n\to\infty} (a_n - b_n) = a - b,$$
and if furthermore $c_1, c_2, c_3, \ldots$ is a sequence of real numbers converging to c, then
$$\lim_{n\to\infty} c_n a_n = ca.$$

24. Show that a complete subspace of a metric space is a closed subset.

25. Write down in all detail the proof that E^m is complete. (*Hint:* A convenient notation is $p_n = (x_1^{(n)}, \ldots, x_m^{(n)})$.)

26. Find all cluster points of the subset of **R** given by
$$\left\{\frac{1}{n} + \frac{1}{m} : n, m = \text{positive integers}\right\}.$$

27. Let S be a nonempty subset of **R** that is bounded from above but has no greatest element. Prove that l.u.b. S is a cluster point of S.

28. Prove that a subset of a metric space is closed if and only if it contains all its cluster points.

29. Let S be a subset of a metric space E and let $p \in E$. Show that p is a cluster point of S if and only if p is the limit of a Cauchy sequence of points in $S \cap \mathcal{C}\{p\}$.

30. Give an example of each of the following:
    (a) an infinite subset of **R** with no cluster point
    (b) a complete metric space that is bounded but not compact
    (c) a metric space none of whose closed balls is complete.

31. Let $a, b \in \mathbf{R}$, $a < b$. The following outlines a proof that $[a, b]$ is compact. Rewrite this proof, filling in all details: Let $\{U_i\}_{i\in I}$ be a collection of open subsets of **R** whose union contains $[a, b]$. Let $S = \{x \in [a, b] : x > a$ and $[a, x]$ is contained in the union of a finite number of the sets $\{U_i\}_{i\in I}\}$. Then l.u.b. $S \in U_i$ for some $i \in I$. Since U_i is open we must have l.u.b. $S = b \in S$.

32. Show that the union of a finite number of compact subsets of a metric space is compact.

33. Let E be a compact metric space, $\{U_i\}_{i\in I}$ a collection of open subsets of E whose union is E. Show that there exists a real number $\epsilon > 0$ such that any closed ball in E of radius ϵ is entirely contained in at least one set U_i. (*Hint:* If not, take bad balls of radii 1, ½, ⅓, ... and a cluster point of their centers.)

34. If $(x_1, \ldots, x_n) \in E^n$ and $(y_1, \ldots, y_m) \in E^m$, then $(x_1, \ldots, x_n, y_1, \ldots, y_m) \in E^{n+m}$. Therefore if S and T are subsets of E^n and E^m respectively, we may identify $S \times T$ with a subset of E^{n+m}. Prove that if S and T are nonempty, then $S \times T$ is bounded, or open, or closed, or compact, if and only if both S and T are bounded, or open, or closed, or compact, respectively.

35. Call a metric space *sequentially compact* if every sequence has a convergent subsequence. Prove that a metric space is sequentially compact if and only if every infinite subset has a cluster point.

36. Call a metric space *totally bounded* if, for every $\epsilon > 0$, the metric space is the union of a finite number of closed balls of radius ϵ. Prove that a metric space is totally bounded if and only if every sequence has a Cauchy subsequence.

37. Prove that the following three conditions on a metric space E (cf. Probs. 35, 36) are equivalent:
   (i) E is compact
   (ii) E is sequentially compact
   (iii) E is totally bounded and complete.
   (*Hint:* That (i) implies (ii) occurs in the text. That (ii) implies (iii) is easy. That (iii) implies (i) follows from the argument of the last proof of § 5.)

38. Prove that an open (closed) subset of a metric space E is connected if and only if it is not the disjoint union of two nonempty open (closed) subsets of E.

CHAPTER IV

Continuous Functions

Elementary analysis is largely concerned with real-valued functions of a real variable. Most first courses in calculus quote and use, but refrain from proving, such theorems about continuous real-valued functions on a closed interval in **R** as the attaining of a maximum and the intermediate value theorem. Among other things the present chapter proves a number of such fundamental facts on real-valued functions of a real variable. But it would not be reasonable to restrict ourselves to such functions alone, for elementary calculus also involves real-valued functions of more than one real variable, that is real-valued functions on subsets of some Euclidean space, and it also involves finite sets of real-valued functions of one or several variables, as for example when a curve or surface in some Euclidean space is given parametrically or when complex-valued functions are considered. Thus the natural class of functions to consider would appear to be functions on one metric space with values in another metric space. It is for such functions that we define the notion of continuity, deriving from this generality the usual advantages of clarity of concept and obligation to do a thing only once. Our basic theorems will be general results on continuous functions from one metric space to another, from which the basic results needed for elementary calculus can be read off by taking both metric spaces to be subsets of **R**. At the same time we go forward, developing a number of useful concepts that usually do not appear in elementary courses, such as uniform continuity. A final section on sequences of functions will illustrate the previous concepts and will provide us with further examples of metric spaces, useful in the sequel. This last section could more logically be placed at the beginning of Chapter VII than here, but its flavor is more like that of the present chapter.

§ 1. DEFINITION OF CONTINUITY. EXAMPLES.

By a function on a metric space E we of course mean a function on the set of points of E, and by a function with values in a metric space E' we mean a function with values in the set of points of E'. Thus if f is a function from the metric space E into the metric space E', written

$$f: E \rightarrow E',$$

then to each point $p \in E$ is associated a point $f(p) \in E'$. The function $f: E \rightarrow E'$ will be called continuous at a point $p_0 \in E$ if, roughly speaking, points of E that are near p_0 are mapped by f into points of E' that are near $f(p_0)$. Here is the precise definition:

Definition. Let E and E' be metric spaces, with distances denoted d and d' respectively, let $f: E \rightarrow E'$ be a function, and let $p_0 \in E$. Then f is said to be *continuous at* p_0 if, given any real number $\epsilon > 0$, there exists a real number $\delta > 0$ such that if $p \in E$ and $d(p, p_0) < \delta$, then $d'(f(p), f(p_0)) < \epsilon$.

The number δ of course depends on ϵ, so we could more accurately have written $\delta(\epsilon)$ instead of δ. We stick to the notation δ rather than $\delta(\epsilon)$ for notational simplicity, always bearing in mind that each ϵ must have its own δ.

The definition may be reformulated by saying that f is continuous at p_0 if, given any open ball in E' of center $f(p_0)$, there exists an open ball in E of center p_0 whose image under f is contained in the former ball. Another reformulation is that f is continuous at p_0 if, given any open subset of E' that contains $f(p_0)$, there exists an open subset of E that contains p_0 whose image under f is contained in the former open subset.

If E and E' are both subsets of $\mathbf{R}$ (so that we have a real-valued function of a real variable) the original definition says that f is continuous at p_0 if, given any $\epsilon > 0$, there exists a $\delta > 0$ such that $|f(p) - f(p_0)| < \epsilon$ whenever $p \in E$ and $|p - p_0| < \delta$. This is illustrated in Figure 17.

Definition. If E, E' are metric spaces and $f: E \rightarrow E'$ is a function, then f is said to be *continuous on* E or, more briefly, *continuous*, if f is continuous at all points of E.

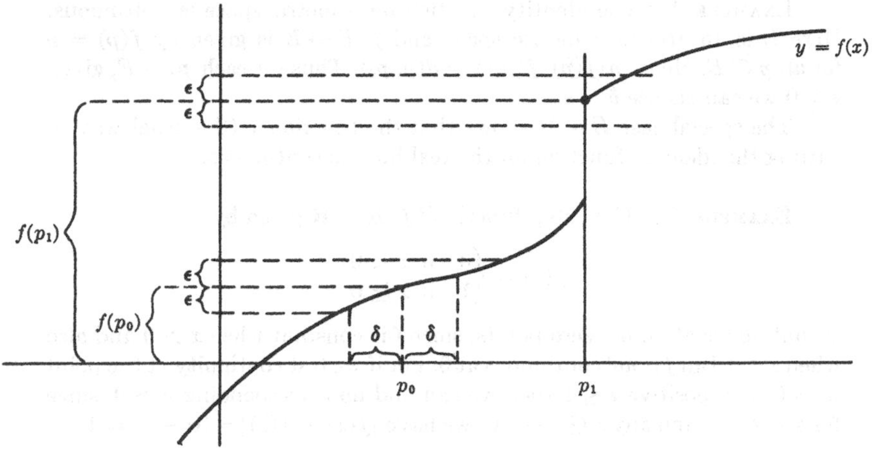

FIGURE 17. f is continuous at p_0, not at p_1.

EXAMPLE 1. The function $f\colon \mathbf{R} \to \mathbf{R}$ given by $f(x) = x^2$ for each $x \in \mathbf{R}$ is continuous. To prove this we have to show that f is continuous at each $x_0 \in \mathbf{R}$. We have to show that for any $\epsilon > 0$ we can find a $\delta > 0$ such that $|x^2 - x_0^2| < \epsilon$ whenever $|x - x_0| < \delta$. But since

$$\begin{aligned}|x^2 - x_0^2| &= |(x + x_0)(x - x_0)|\\ &= |(x - x_0 + 2x_0)(x - x_0)|\\ &\le (|x - x_0| + 2|x_0|)\,|x - x_0|,\end{aligned}$$

we will have $|x^2 - x_0^2| < \epsilon$ if $|x - x_0| < 1$ and $|x - x_0| < \epsilon/(1 + 2|x_0|)$. Hence we may take $\delta = \min\{1, \epsilon/(1 + 2|x_0|)\}$.

EXAMPLE 2. Let E be any metric space, p_0 a fixed point of E. Then the function $f\colon E \to \mathbf{R}$ given by $f(p) = d(p, p_0)$ for all $p \in E$ is continuous. To prove this we have to show that f is continuous at any given point p_1 of E. But

$$|f(p) - f(p_1)| = |d(p, p_0) - d(p_1, p_0)| \le d(p, p_1),$$

so if we have $d(p, p_1) < \epsilon$ then also $|f(p) - f(p_1)| < \epsilon$. Thus for any point p_1, corresponding to any $\epsilon > 0$ we can choose $\delta = \epsilon$.

The special case $E = \mathbf{R}$, $p_0 = 0$ shows that the function $|x|$ is continuous.

EXAMPLE 3. Any constant function is continuous. In this case E and E' are arbitrary and the point $f(p) \in E'$ is the same for all $p \in E$. Thus we always have $d'(f(p), f(p_0)) = 0$, so given any $p_0 \in E$ and any $\epsilon > 0$ we can take δ to be *any* positive number.

EXAMPLE 4. The identity function on a metric space is continuous. Here E is an arbitrary metric space and $f: E \to E$ is given by $f(p) = p$ for all $p \in E$. Hence $d(f(p), f(p_0)) = d(p, p_0)$. Thus for each $p_0 \in E$, given $\epsilon > 0$ we can choose $\delta = \epsilon$.

The special case $E = \mathbf{R}$ shows that the function x (the usual way of writing the identity function on the real line) is continuous.

EXAMPLE 5. The "step function" $f: \mathbf{R} \to \mathbf{R}$ given by

$$f(x) = \begin{cases} 0 & \text{if } x < 0 \\ 1 & \text{if } x \geq 0 \end{cases}$$

is continuous at all nonzero points, since f is constant when $x > 0$ and also when $x < 0$. But f is *not* continuous at 0. For if we test continuity at the point $p_0 = 0$ with positive $\epsilon \leq 1$ then we can find no corresponding $\delta > 0$, since for any $\delta > 0$ and any $x \in (-\delta, 0)$ we have $|f(x) - f(0)| = |0 - 1| = 1$.

EXAMPLE 6. The function $f: \mathbf{R} \to \mathbf{R}$ given by

$$f(x) = \begin{cases} 1 & \text{if } x \text{ is rational} \\ 0 & \text{if } x \text{ is not rational} \end{cases}$$

is continuous at no point. For any open ball in $\mathbf{R}$ contains both numbers that are rational and numbers that are not (we already know we can find a rational number a in the ball, and if b is any fixed irrational number and N a sufficiently large integer then the irrational number $a + b/N$ will also be in the ball). Thus for any $p_0 \in \mathbf{R}$ and any positive $\epsilon \leq 1$, a corresponding $\delta > 0$ cannot be found.

EXAMPLE 7. If $f: E \to E'$ is continuous and S is a subspace of E, then the restriction of f to S is continuous on S. This is clear from the definitions.

The following criterion for the continuity of a function from one metric space into another is often useful.

Proposition. *Let E, E' be metric spaces and $f: E \to E'$ a function. Then f is continuous if and only if, for every open subset U of E', the inverse image*

$$f^{-1}(U) = \{p \in E : f(p) \in U\}$$

is an open subset of E.

To prove this, first suppose that f is continuous. We have to show that if $U \subset E'$ is open, then also $f^{-1}(U)$ is open. Let $p_0 \in f^{-1}(U)$. Then $f(p_0) \in U$. Since U is open, it contains the open ball in E' of center $f(p_0)$ and some radius $\epsilon > 0$. Since f is continuous at p_0 there is a $\delta > 0$ such that if $p \in E$ and $d(p, p_0) < \delta$ then $d'(f(p), f(p_0)) < \epsilon$. This means that if p is contained in the open ball in E of center p_0 and radius δ then $f(p)$ is contained in the open ball in E' of center $f(p_0)$ and radius ϵ, so that $f(p) \in U$. That is, $f^{-1}(U)$ contains the open ball in E of center p_0 and radius δ. Since p_0 was any point of $f^{-1}(U)$, the set $f^{-1}(U)$ is open. Conversely, suppose that for every open $U \subset E'$, the set $f^{-1}(U)$ is an open subset of E. We must show that f is continuous at any point $p_0 \in E$. For any $\epsilon > 0$ the set

$$f^{-1}(\{\text{open ball in } E' \text{ of center } f(p_0) \text{ and radius } \epsilon\})$$

is an open subset of E that contains p_0, hence contains the open ball in E of center p_0 and some radius $\delta > 0$. Thus if $p \in E$ and $d(p, p_0) < \delta$, then $d'(f(p), f(p_0)) < \epsilon$. This means that f is continuous at p_0, and this completes the proof.

Corollary. *If f is a continuous real-valued function on the metric space E then for any $a \in \mathbf{R}$*

$$\{p \in E : f(p) > a\} \quad \textit{and} \quad \{p \in E : f(p) < a\}$$

are open subsets of E.

For the sets $\{x \in \mathbf{R} : x > a\}$ and $\{x \in \mathbf{R} : x < a\}$ are open subsets of $\mathbf{R}$.

The following result is usually paraphrased "a continuous function of a continuous function is a continuous function".

Proposition. *Let E, E', E'' be metric spaces, $f: E \to E'$, $g: E' \to E''$ functions. Then if f and g are continuous, so is the function $g \circ f: E \to E''$. More precisely, if $p_0 \in E$ and f is continuous at p_0 and g is continuous at $f(p_0) \in E'$, then $g \circ f$ is continuous at p_0.*

We need only prove the latter, more precise, part. Let d, d', d'' denote the three metrics. Suppose $\epsilon > 0$ is given. Then, since g is continuous at $f(p_0)$, there exists $\delta > 0$ such that if $q \in E'$ and $d'(q, f(p_0)) < \delta$ then $d''(g(q), g(f(p_0))) < \epsilon$. But since f is continuous at p_0, corresponding to this δ there exists a number $\eta > 0$ such that if $p \in E$ and $d(p, p_0) < \eta$ then $d'(f(p), f(p_0)) < \delta$. Therefore if $p \in E$ and $d(p, p_0) < \eta$ we have $d''(g(f(p)), g(f(p_0))) < \epsilon$, proving $g \circ f$ continuous at p_0.

The above argument is illustrated in Figure 18 on the next page.

The weak version of the last proposition, where $f: E \to E'$ and $g: E' = E''$ are assumed to be continuous everywhere, can be proved very simply by means of the previous criterion for continuity, as follows. If U is any open subset of E'', the continuity of g implies that $g^{-1}(U)$ is an open subset of E', so the continuity of f implies that $f^{-1}(g^{-1}(U))$ is an open subset of U. But for any $U \subset E''$, if $p \in E$ then $p \in f^{-1}(g^{-1}(U))$ if and only if

$f(p) \in g^{-1}(U)$, which is true if and only if $g(f(p)) \in U$. Thus $f^{-1}(g^{-1}(U)) = (g \circ f)^{-1}(U)$. We have therefore shown that if U is an open subset of E'', then $(g \circ f)^{-1}(U)$ is an open subset of E, which implies that $g \circ f$ is continuous.

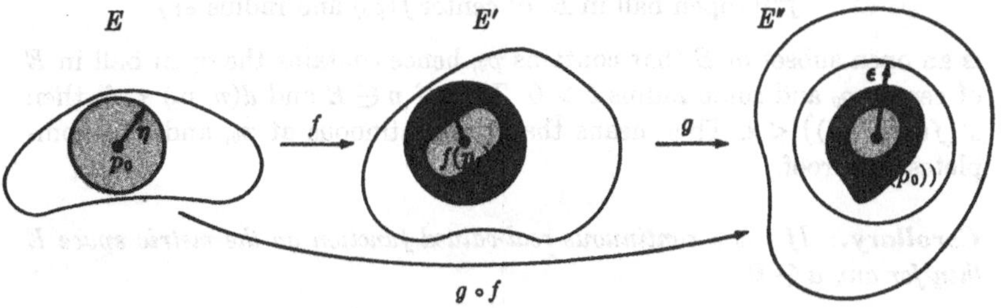

FIGURE 18. A continuous function of a continuous function is continuous.

§ 2. CONTINUITY AND LIMITS.

Consider the following question: Let E, E' be metric spaces, let $p_0 \in E$, let $\mathcal{C}\{p_0\}$ be the complement of $\{p_0\}$ in E, and let $f\colon \mathcal{C}\{p_0\} \to E'$ be a function. Is it possible to extend the definition of f to all of E in such a way as to obtain a function from E into E' that is continuous at p_0?

In one case the answer to this question is trivial, and that is the case where p_0 is not a cluster point of E. For if p_0 is not a cluster point of E then *any* function from E into E' is continuous at p_0. This is so since there exists a ball in E of center p_0 that contains only a finite number of points of E, hence a ball of center p_0 and smaller radius that contains only the point p_0. If δ is the radius of the latter ball then the statements that $p \in E$ and $d(p, p_0) < \delta$ imply that $p = p_0$. Thus for any $f\colon E \to E'$ and any $p \in E$ such that $d(p, p_0) < \delta$ we have $d'(f(p), f(p_0)) = 0$.

In the case where p_0 is a cluster point of E we are led to the following definition.

Definition. Let E, E' be metric spaces, let p_0 be a cluster point of E, and let $f\colon \mathcal{C}\{p_0\} \to E'$ be a function. A point $q \in E'$ is called a *limit of f at p_0* if the function from E into E' which is the same as f on $\mathcal{C}\{p_0\}$ and which takes on the value q at p_0 is continuous at p_0.

It is useful to reword this, without directly using the notion of continuity.

Definition'. Let E, E' be metric spaces, let p_0 be a cluster point of E, and let $f: \mathcal{C}\{p_0\} \to E'$ be a function. A point $q \in E'$ is called a *limit of f at p_0* if, given any $\epsilon > 0$, there exists a $\delta > 0$ such that if $p \in E$, $p \neq p_0$, and $d(p, p_0) < \delta$, then $d'(f(p), q) < \epsilon$.

Given E, E', a cluster point p_0 of E and $f: \mathcal{C}\{p_0\} \to E'$, there can exist at most one limit of f at p_0. The argument is the same as that for the uniqueness of the limit of a convergent sequence: If $q, q' \in E'$ are two limits of f at p_0, then given any $\epsilon > 0$ there exists a $\delta > 0$ such that if $p \in E, p \neq p_0$ and $d(p, p_0) < \delta$ then $d'(f(p), q) < \epsilon$ and $d'(f(p), q') < \epsilon$. Since p_0 is a cluster point of E, there actually exist points $p \in E$ such that $p \neq p_0$ and $d(p, p_0) < \delta$, so that we can deduce that $d'(q, q') < 2\epsilon$. Since this is true for any $\epsilon > 0$ we have $d'(q, q') = 0$, so that $q = q'$.

If, under the above conditions, a limit of f at p_0 exists, then since the limit is unique we may speak of *the* limit of f at p_0 and we denote this

$$\lim_{p \to p_0} f(p).$$

The statement that $\lim_{p \to p_0} f(p)$ exists implies that we have metric spaces E and E' in mind, that p_0 is a cluster point of E, and that we have a function $f: \mathcal{C}\{p_0) \to E'$ such that for some point $q \in E'$ q is a limit of f at p_0. In discussing $\lim_{p \to p_0} f(p)$ we may be explicitly given a function f that happens to be defined at p_0, but this is immaterial: the limit of f at p_0 does not depend on whether or not f is defined at p_0 nor, if it is, on what its value at p_0 is, but rather on the values of $f(p)$ for p near, but distinct from, p_0.

In the above work we started with a function f which was defined at all points of a metric space but one, but it is possible to discuss limits of functions which are defined on relatively small subsets of a metric space. For example suppose E, E' are metric spaces and that we have an arbitrary subset $S \subset E$ that has at least one cluster point in E, together with a function $f: S \to E'$. If $p_0 \in E$ is a cluster point of S we can consider $\lim_{p \to p_0} f(p)$ *relative to the subspace* $S \cup \{p_0\}$ *of* E and one then speaks of *the limit of* $f(p)$ *as* p *approaches* p_0 *on* S. Thus a specific metric space E must be borne in mind in considering $\lim_{p \to p_0} f(p)$, and in the last case the space to be borne in mind is actually not E, but the subspace $S \cup \{p_0\}$. The most frequently arising case is that in which f is defined on a part of E that includes all points of an open ball in E of center p_0, with the possible exception of p_0. Here we maintain the same notation

$$\lim_{p \to p_0} f(p)$$

without any reference to the fact that f may not be defined far away from p_0; it is enough that f be defined near p_0, except possibly at p_0 itself.

We gave our first definition of the limit of a function in terms of continuity, but one can equally well define continuity in terms of limits of functions. If $f: E \to E'$ is a function from one metric space into another and $p_0 \in E$, then f is continuous at p_0 if and only if, if p_0 is a cluster point of E, then

$$\lim_{p \to p_0} f(p) = f(p_0).$$

This fact is clear from the definitions.

In the preceding chapter we discussed the notion of a limit of a sequence of points in a metric space. This too can easily be defined in terms of the notion of the limit of a function, as follows: If $p_1, p_2, p_3, \ldots$ is a sequence of points in a metric space E and $q \in E$ then

$$\lim_{n \to \infty} p_n = q$$

if and only if, for the subspace $\{0, 1, \frac{1}{2}, \frac{1}{3}, \frac{1}{4}, \ldots\}$ of $\mathbf{R}$ and the function $f: \{1, \frac{1}{2}, \frac{1}{3}, \ldots\} \to E$ given by $f(1/n) = p_n$ for all $n = 1, 2, 3, \ldots$, we have

$$\lim_{p \to 0} f(p) = q.$$

This statement is easy to verify. As a matter of fact the notion of continuity (or the equivalent notion of limit of a function) should be regarded as fundamental, with convergent sequences as useful technical devices, in spite of their earlier introduction in this text. It would have been more "natural", although a bit longer, to first prove theorems about continuity (or limits of functions) and then the analogous results on limits of sequences. Of course all of our ideas are logically interdependent, and the following proposition, which gives a useful criterion of continuity, could also have been used to give an alternate definition of continuous function (and hence of the limit of a function) in terms of convergent sequences.

Proposition. *Let E, E' be metric spaces. Then a function $f: E \to E'$ is continuous at $p_0 \in E$ if and only if, for every sequence of points $p_1, p_2, p_3, \ldots$ in E such that*

$$\lim_{n \to \infty} p_n = p_0$$

we have

$$\lim_{n \to \infty} f(p_n) = f(p_0).$$

Suppose first that f is continuous at p_0 and that $p_1, p_2, p_3, \ldots$ is a sequence of points in E that converges to p_0. We have to show that the sequence $f(p_1), f(p_2), f(p_3), \ldots$ converges to $f(p_0)$. Given $\epsilon > 0$, the continuity of f at p_0 implies that there is a $\delta > 0$ such that $d'(f(p), f(p_0)) < \epsilon$

whenever $p \in E$ and $d(p, p_0) < \delta$. Since $p_1, p_2, p_3, \ldots$ converges to p_0 there is a positive integer N such that $d(p_n, p_0) < \delta$ for all $n > N$. Hence if $n > N$ then $d'(f(p_n), f(p_0)) < \epsilon$, which shows that $f(p_1), f(p_2), f(p_3), \ldots$ converges to $f(p_0)$.

We now prove that, conversely, if it is true that whenever a sequence $p_1, p_2, p_3, \ldots$ of points of E converges to p_0 then the sequence of points $f(p_1), f(p_2), f(p_3), \ldots$ of E' converges to $f(p_0)$, then f is continuous at p_0. We do this by supposing the contrary, that is that f is not continuous at p_0, and deriving from this assumption the existence of a sequence of points $p_1, p_2, p_3, \ldots$ converging to p_0 such that $f(p_1), f(p_2), f(p_3), \ldots$ does not converge to $f(p_0)$. So suppose that f is not continuous at p_0. Then there exists some $\epsilon > 0$ such that for no number $\delta > 0$ is it true that whenever $p \in E$ and $d(p, p_0) < \delta$ then necessarily $d'(f(p), f(p_0)) < \epsilon$. Hence for any $n = 1, 2, 3, \ldots$ we can find a point $p_n \in E$ such that $d(p_n, p_0) < 1/n$ and $d'(f(p_n), f(p_0)) \geq \epsilon$. Since $d(p_n, p_0) < 1/n$ for all $n = 1, 2, 3, \ldots$, the sequence $p_1, p_2, p_3, \ldots$ converges to p_0. However $f(p_1), f(p_2), f(p_3), \ldots$ does not converge to $f(p_0)$, since $d'(f(p_n), f(p_0)) \geq \epsilon$ for all n. This completes the proof.

§ 3. THE CONTINUITY OF RATIONAL OPERATIONS. FUNCTIONS WITH VALUES IN E^n.

Real-valued functions on a metric space E (or indeed on any set) can be combined in the usual way by the rational operations of addition, subtraction, multiplication, and division. Thus if f, g are real-valued functions on E, we have the real-valued functions $f + g$, $f - g$, fg and f/g, given by

$$\begin{aligned}(f+g)(p) &= f(p) + g(p)\\ (f-g)(p) &= f(p) - g(p)\\ (fg)(p) &= f(p)g(p)\\ \left(\frac{f}{g}\right)(p) &= \frac{f(p)}{g(p)}\end{aligned}$$

for any $p \in E$; in the last case f/g is of course not defined at any point p such that $g(p) = 0$, so that f/g is a function on $\{p \in E : g(p) \neq 0\}$.

Proposition. *Let f and g be real-valued functions on a metric space E. If f and g are continuous at a point $p_0 \in E$, then so are the functions $f + g$, $f - g$, fg and f/g, the last under the proviso that $g(p_0) \neq 0$ (in which case $g(p) \neq 0$ for all points p in some open ball of center p_0).*

This can be proved directly, but it is easier to deduce it from work already done. By the last proposition, to prove that $f + g$ is continuous at p_0 it suffices to show that if $p_1, p_2, p_3, \ldots$ is a sequence of points of E that converges to p_0, then the sequence $f(p_1) + g(p_1), f(p_2) + g(p_2), f(p_3) +$

$g(p_3), \ldots$ converges to $f(p_0) + g(p_0)$. But the last proposition implies that the sequences $f(p_1), f(p_2), f(p_3), \ldots$ and $g(p_1), g(p_2), g(p_3), \ldots$ converge to $f(p_0)$ and $g(p_0)$ respectively, so the convergence of $f(p_1) + g(p_1), f(p_2) + g(p_2), f(p_3) + g(p_3), \ldots$ to $f(p_0) + g(p_0)$ is a consequence of the proposition on page 48. Exactly the same argument proves the continuity at p_0 of the function $f - g$, the function fg, and also (once a certain minor detail has been verified) the function f/g. The minor detail to be verified is the parenthetical remark in the proposition to the effect that if g is continuous at p_0 and $g(p_0) \neq 0$, then $g(p) \neq 0$ for all points p in some open ball of center p_0. But since g is continuous at p_0 and $g(p_0) \neq 0$ then for all points p in some open ball of center p_0 we have

$$|g(p) - g(p_0)| < |g(p_0)|$$

which clearly implies $g(p) \neq 0$.

Corollary. *Let p_0 be a cluster point of the metric space E and let f and g be real-valued functions on $\mathcal{C}\{p_0\}$ such that*

$$\lim_{p \to p_0} f(p) \quad \text{and} \quad \lim_{p \to p_0} g(p)$$

exist. Then

$$\begin{aligned} \lim_{p \to p_0} (f(p) + g(p)) &= \lim_{p \to p_0} f(p) + \lim_{p \to p_0} g(p) \\ \lim_{p \to p_0} (f(p) - g(p)) &= \lim_{p \to p_0} f(p) - \lim_{p \to p_0} g(p) \\ \lim_{p \to p_0} f(p)g(p) &= \lim_{p \to p_0} f(p) \cdot \lim_{p \to p_0} g(p) \end{aligned}$$

and, if $\lim_{p \to p_0} g(p) \neq 0$,

$$\lim_{p \to p_0} \frac{f(p)}{g(p)} = \frac{\lim_{p \to p_0} f(p)}{\lim_{p \to p_0} g(p)}.$$

The corollary means, first, that the left-hand limits exist and, second, that they are given by the appropriate formulas. It is an immediate consequence of the proposition and the first definition of limit of a function.

Lemma. *For each $i = 1, 2, \ldots, n$, the function $x_i \colon E^n \to \mathbf{R}$ defined by $x_i((a_1, a_2, \ldots, a_n)) = a_i$ is continuous.*

We have to prove that x_i is continuous at any given point $p_0 \in E^n$. For any $p \in E^n$ we have $p = (x_1(p), x_2(p), \ldots, x_n(p))$. The inequality $|x_i(p) - x_i(p_0)| \leq \sqrt{(x_1(p) - x_1(p_0))^2 + \cdots + (x_n(p) - x_n(p_0))^2} = d(p, p_0)$ shows that if $d(p, p_0) < \epsilon$ then also $|x_i(p) - x_i(p_0)| < \epsilon$. Thus x_i is continuous at p_0.

The previous proposition can be combined with the lemma to give many examples of continuous real-valued functions on E^n. If we recall that any constant function is continuous and apply the proposition repeatedly, we see that any polynomial in $x_1, x_2, \ldots, x_n$ with coefficients in **R** is continuous. (For example, $x_1^3 - x_5 + \sqrt{2}\,x_2 - \pi x_3 x_4$ is a continuous function on E^5.) Furthermore any rational function (a rational function is the quotient of two polynomial functions) is continuous wherever the denominator is not zero. (For example $x_1x_2/(x_1^2 + x_2^2)$ is a continuous function on $E^2 - \{(0, 0)\}$.)

If E is a metric space and $f\colon E \to E^n$ a function, then the image in E^n of any point $p \in E$ is the point

$$f(p) = (x_1(f(p)), x_2(f(p)), \ldots, x_n(f(p))).$$

Thus f is determined by the n *component functions* $x_1 \circ f, x_2 \circ f, \ldots, x_n \circ f$. Conversely, an n-tuple of real-valued functions on a metric space defines a function from the metric space into E^n. The following proposition handles all relevant continuity questions.

Proposition. *Let E be a metric space and $f\colon E \to E^n$ a function. Then f is continuous at a point $p_0 \in E$ if and only if each component function $x_1 \circ f, \ldots, x_n \circ f$ is continuous at p_0.*

The "only if" part of the proposition comes trivially from the lemma, for if f is continuous at p_0 then so is the continuous function of a continuous function $x_i \circ f$. To prove the "if" part, suppose $x_1 \circ f, \ldots, x_n \circ f$ continuous at p_0 and let $\epsilon > 0$. Noting that $(x_1 \circ f)(p) = x_1(f(p))$, $(x_2 \circ f)(p) = x_2(f(p))$, etc., for each $i = 1, \ldots, n$ we can find a $\delta_i > 0$ such that

$$|x_i(f(p)) - x_i(f(p_0))| < \frac{\epsilon}{\sqrt{n}}$$

whenever $p \in E$ and $d(p, p_0) < \delta_i$, d denoting the distance in E. If we set $\delta = \min\{\delta_1, \ldots, \delta_n\}$, then if $p \in E$ and $d(p, p_0) < \delta$ we have

$$|x_i(f(p)) - x_i(f(p_0))| < \frac{\epsilon}{\sqrt{n}}$$

for $i = 1, \ldots, n$, so that

$$\sqrt{(x_1(f(p)) - x_1(f(p_0)))^2 + \cdots + (x_n(f(p)) - x_n(f(p_0)))^2} < \sqrt{\left(\frac{\epsilon}{\sqrt{n}}\right)^2 + \cdots + \left(\frac{\epsilon}{\sqrt{n}}\right)^2} = \epsilon.$$

This proves f continuous at p_0.

Note that the lemma becomes a special case of the "only if" part of the proposition when we apply the latter to the identity function on E^n, which is known to be continuous.

§ 4. CONTINUOUS FUNCTIONS ON A COMPACT METRIC SPACE.

Theorem. *Let E, E' be metric spaces, $f\colon E \to E'$ a continuous function. Then if E is compact, so is its image $f(E)$.*

We must show that if $f(E) = \{f(p) : p \in E\}$ is contained in the union of a collection of open subsets of E' then it is contained in the union of a finite number of these open subsets. So suppose that $\{U_i\}_{i \in I}$ is a collection of open subsets of E' whose union contains $f(E)$. Since f is continuous, each inverse image $f^{-1}(U_i)$ is an open subset of E, by the first proposition of § 1. Also, for any $p \in E$ we have $f(p) \in U_i$ for some $i \in I$, in which case $p \in f^{-1}(U_i)$, so that

$$E = \bigcup_{i \in I} f^{-1}(U_i).$$

Since E is compact there is a finite subset $J \subset I$ such that

$$E = \bigcup_{i \in J} f^{-1}(U_i).$$

Therefore $f(E) = \bigcup_{i \in J} f(f^{-1}(U_i)) \subset \bigcup_{i \in J} U_i$. Thus $f(E)$ is compact.

This theorem has two extremely important immediate consequences. To state the first, let us say that a function $f\colon E \to E'$ from one metric space into another is *bounded* if the image $f(E)$ is bounded. In the special case that f is a real-valued function on E this means simply that there is a number $M \in \mathbf{R}$ such that $|f(p)| \leq M$ for all $p \in E$.

Corollary 1. *Let E, E' be metric spaces, $f\colon E \to E'$ a continuous function. Then if E is compact, f is bounded.*

Reason: any compact subset of a metric space is bounded. In particular the compact subset $f(E)$ of E' is bounded.

The last result is false if compactness is omitted; for example, consider the function $f(x) = 1/x$ on the open interval $(0, 1)$.

If f is a real-valued function on a metric space E and $p_0 \in E$ we say that f *attains a maximum at* p_0 if $f(p_0) \geq f(p)$ for all $p \in E$, and we say that f *attains a minimum at* p_0 if $f(p_0) \leq f(p)$ for all $p \in E$.

Corollary 2. *A continuous real-valued function on a nonempty compact metric space attains a maximum at some point, and also attains a minimum at some point.*

For let E be a compact metric space, $f\colon E \to \mathbf{R}$ a continuous function. Then $f(E)$ is a compact subset of $\mathbf{R}$, hence closed and bounded. If E is nonempty then so is $f(E)$, and the last proposition of § 2 of the preceding chapter tells us that $f(E)$ has a greatest element, and also a least element. If $p_0 \in E$ is chosen so that $f(p_0)$ is the greatest (least) element of $f(E)$, then f attains a maximum (minimum) at p_0.

Corollary 2 is false if the compactness condition is omitted, even for bounded functions; for example, consider the function $f(x) = x$ on the open interval $(0, 1)$.

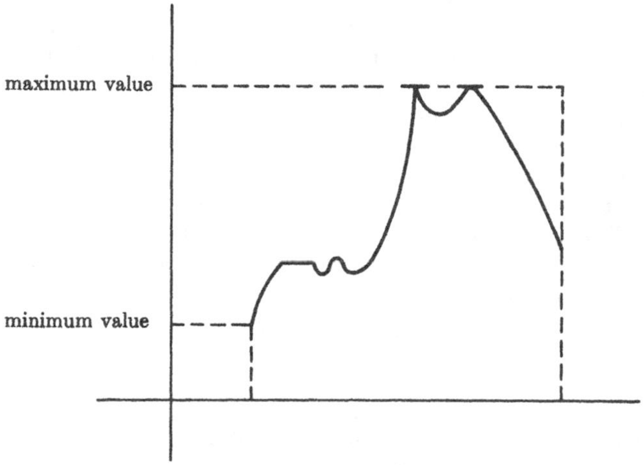

FIGURE 19. A continuous real-valued function on a closed interval in **R** attains a maximum and a minimum.

If E, E' are metric spaces and $f: E \to E'$ is a continuous function then given any $p_0 \in E$ and any $\epsilon > 0$ there exists a real number $\delta > 0$ such that if $p \in E$ and $d(p, p_0) < \delta$ then $d'(f(p), f(p_0)) < \epsilon$: this is just a literal restatement of the definition of continuity. If E, E' and f are fixed it is clear that δ will depend on both p_0 and ϵ. If p_0 is held fixed and ϵ varies, then the smaller we take ϵ the smaller δ will usually be. If on the other hand we take some fixed $\epsilon > 0$, then δ will depend on the point p_0. As p_0 varies so in general will δ, and it may or may not be true that we can find a δ that works simultaneously for *all* p_0. If it happens that we can find such a δ, and can do this for each $\epsilon > 0$, the function f will have some especially nice properties. This leads to the following definition.

Definition. Let E and E' be metric spaces, with distances denoted by d and d' respectively, and let $f: E \to E'$ be a function. Then f is said to be *uniformly continuous* if, given any real number $\epsilon > 0$, there exists a real number $\delta > 0$ such that if $p, q \in E$ and $d(p, q) < \delta$ then $d'(f(p), f(q)) < \epsilon$.

If it happens that a function $f: E \to E'$ is such that for a certain subset S of E the restriction of f to S is uniformly continuous, we say that f is *uniformly continuous on* S. Thus uniform continuity on E is the same thing as uniform continuity.

It is clear that a uniformly continuous function $f: E \to E'$ is continuous: to check continuity at a point $p_0 \in E$ just set $q = p_0$ in the definition. The next theorem will state that conversely if f is continuous then f is actually uniformly continuous, provided E is compact. If E is not compact then continuity does not in general imply uniform continuity. Here are two examples of continuous real-valued functions on the open interval $(0, 1)$ that are not uniformly continuous:

(1) The function f given by $f(x) = 1/x$ for all $x \in (0, 1)$ is continuous but not uniformly continuous. Continuity is known. Uniform continuity is disproved by showing that for any $\epsilon > 0$ and any $\delta > 0$ we can find $p, q \in (0, 1)$ such that $|p - q| < \delta$ and $|1/p - 1/q| > \epsilon$. Specific such p, q can be found, for example, by taking $q = p/2$ so that the conditions become $p/2 < \delta$, $1/p > \epsilon$, the pair of which will be satisfied if $0 < p < \min \{2\delta, 1/\epsilon, 1\}$.

(2) The function f given by $f(x) = \sin (1/x)$ for all $x \in (0, 1)$ is continuous but not uniformly continuous. To check this example we assume that the easier properties of the sine function are known (these will be rederived anyway in Chapter VII). Then f is continuous, and moreover since $|\sin (1/x)| \leq 1$ for all $x \in (0, 1)$ any δ at all will work if $\epsilon > 2$. But if $\epsilon < 1$, no δ will work. For suppose that $0 < \epsilon < 1$ and that $0 < \delta$. If we then take n a sufficiently large positive integer and set $p = 1/(2\pi n)$, $q = 1/(2\pi n + \pi/2)$, we get both $|p - q| < \delta$ and $|f(p) - f(q)| = 1 > \epsilon$.

Theorem. *Let E and E' be metric spaces and $f: E \to E'$ a continuous function. If E is compact, then f is uniformly continuous.*

It will be instructive to give two proofs of this theorem. In each proof we start with a real number $\epsilon > 0$ and try to find a number $\delta > 0$ such that if $p, q \in E$ are any points such that $d(p, q) < \delta$ then $d'(f(p), f(q)) < \epsilon$.

For the first proof we find, for each $p \in E$, a number $\delta(p) > 0$ such that if $q \in E$ and $d(p, q) < \delta(p)$ then $d'(f(p), f(q)) < \epsilon/2$; this is possible since f is continuous at p. Let $B(p)$ be the open ball in E of center p and radius $\delta(p)/2$. E is the union of the open sets $B(p)$, with p ranging over all the points of E. Since E is compact, it is the union of a finite number of these open sets. Thus there exist a finite number of points of E, say p_1, p_2, $\ldots$, p_n, such that $E = B(p_1) \cup B(p_2) \cup \cdots \cup B(p_n)$. Now define $\delta = \min\{\delta(p_1)/2, \delta(p_2)/2, \ldots, \delta(p_n)/2\}$. We claim that this δ satisfies our demands. For suppose that $p, q \in E$, with $d(p, q) < \delta$. For some $i = 1, 2, \ldots, n$ we have $p \in B(p_i)$, so that $d(p_i, p) < \delta(p_i)/2$. Also $d(p_i, q) \leq d(p_i, p) + d(p, q) < \delta(p_i)/2 + \delta \leq \delta(p_i)$. Thus $d(p_i, p)$, $d(p_i, q) < \delta(p_i)$. By the way that $\delta(p_i)$ was chosen we have $d'(f(p_i), f(p)) < \epsilon/2$ and $d'(f(p_i), f(q)) < \epsilon/2$. Therefore

$$d'(f(p), f(q)) \leq d'(f(p), f(p_i)) + d'(f(p_i), f(q)) < \frac{\epsilon}{2} + \frac{\epsilon}{2} = \epsilon$$

and the first proof is complete.

For the second proof we use the indirect method, assuming that for our given number $\epsilon > 0$ there is no $\delta > 0$ such that if $p, q \in E$ and $d(p, q) < \delta$ then $d'(f(p), f(q)) < \epsilon$, and we derive a contradiction. By assumption, for each $n = 1, 2, 3, \ldots$ the number $1/n$ is not a suitable candidate for δ, so that there exists a pair of points p_n, $q_n \in E$ such that $d(p_n, q_n) < 1/n$ and $d'(f(p_n), f(q_n)) \geq \epsilon$. Thus we have a sequence of ordered pairs of points (p_1, q_1), (p_2, q_2), (p_3, q_3), $\ldots$ with the properties that $\lim_{n\to\infty} d(p_n, q_n) = 0$ and $d'(f(p_n), f(q_n)) \geq \epsilon$ for all n. Since E is compact, the sequence $p_1, p_2, p_3, \ldots$ has a convergent subsequence. Hence we may replace the sequence (p_1, q_1), (p_2, q_2), (p_3, q_3), $\ldots$ by a subsequence in such a manner that we may assume that the sequence $p_1, p_2, p_3, \ldots$ converges to a point $p_0 \in E$, still maintaining the conditions

$$\lim_{n\to\infty} d(p_n, q_n) = 0, \quad d'(f(p_n), f(q_n)) \geq \epsilon.$$

From the inequalities

$$0 \leq d(q_n, p_0) \leq d(q_n, p_n) + d(p_n, p_0)$$

and the equations

$$\lim_{n\to\infty} (d(q_n, p_n) + d(p_n, p_0)) = \lim_{n\to\infty} d(q_n, p_n) + \lim_{n\to\infty} d(p_n, p_0) = 0$$

it follows that $\lim_{n\to\infty} d(q_n, p_0) = 0$, so that the sequence $q_1, q_2, q_3, \ldots$ also converges to p_0. Thus the continuity of f at p_0 implies that

$$\lim_{n\to\infty} f(p_n) = \lim_{n\to\infty} f(q_n) = f(p_0).$$

For n sufficiently large we therefore have

$$d'(f(p_n), f(p_0)) < \frac{\epsilon}{2}, \quad d'(f(q_n), f(p_0)) < \frac{\epsilon}{2},$$

implying that

$$d'(f(p_n), f(q_n)) \leq d'(f(p_n), f(p_0)) + d'(f(p_0), f(q_n)) < \frac{\epsilon}{2} + \frac{\epsilon}{2} = \epsilon,$$

contradicting $d'(f(p_n), f(q_n)) \geq \epsilon$. This ends the second proof.

§ 5. CONTINUOUS FUNCTIONS ON A CONNECTED METRIC SPACE.

Theorem. *Let E, E' be metric spaces, $f: E \to E'$ a continuous function. Then if E is connected, so is its image $f(E)$.*

To prove this we may without loss of generality assume that $E' = f(E)$. We shall assume that $E' = f(E)$ is not connected and derive a contradiction. Since $f(E)$ is not connected we can write $f(E) = A \cup B$, where A and B are disjoint nonempty open subsets of $f(E)$. By the first proposition of § 1, each of the sets $f^{-1}(A), f^{-1}(B)$ is an open subset of E. We therefore have the expression of E as

$$E = f^{-1}(A) \cup f^{-1}(B),$$

the union of two disjoint nonempty open subsets. This contradicts the fact that E is connected, proving the theorem.

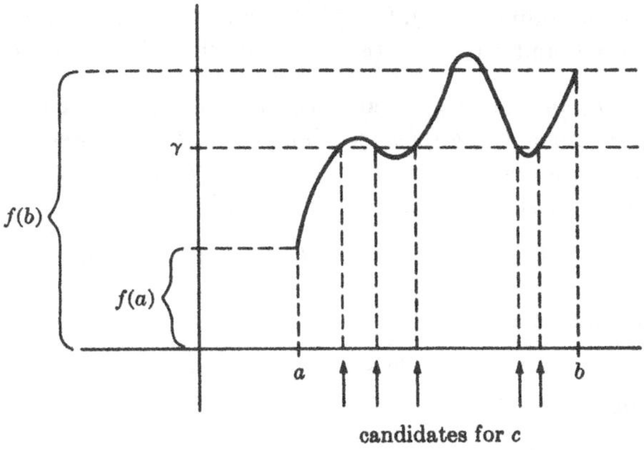

FIGURE 20. Intermediate value theorem.

Corollary (Intermediate value theorem). *If $a, b \in \mathbf{R}$, $a < b$, and f is a continuous real-valued function on the closed interval $[a, b]$, then for any real number γ between $f(a)$ and $f(b)$ there exists at least one point $c \in (a, b)$ such that $f(c) = \gamma$.*

For since $[a, b]$ is connected, so is $f([a, b])$. The (almost trivial) first proposition of the last section of the preceding chapter states that any connected subset of $\mathbf{R}$ contains all points between any two of its points. Since γ is between the points $f(a)$, $f(b)$ of $f([a, b])$, we therefore have $\gamma \in f([a, b])$. This proves the corollary.

Sometimes the name "intermediate value theorem" is applied to the slightly more general statement that if f is a continuous real-valued function on a connected metric space E, then any real number that lies between two points of $f(E)$ is itself in $f(E)$. The proof is of course the same as above. It is worth remarking that the validity of the intermediate value theorem for all continuous real-valued functions on a fixed metric space E is equivalent to E being connected; for if E is not connected we can write $E = A \cup B$, where A and B are disjoint nonempty open subsets of E, and the theorem fails for the continuous function on E which is 0 on A and 1 on B.

The previous theorem enables us to give many new examples of connected subsets of metric spaces. For example, any continuous image of an open or closed interval of $\mathbf{R}$ in another metric space (a "curve") is connected. We now apply this idea to show that all (open or closed) balls in E^n, and E^n itself, are connected: If $p = (a_1, \ldots, a_n)$ and $q = (b_1, \ldots, b_n)$ are points of E^n, define the *line segment between p and q* to be the set of points

$$\{(a_1 + (b_1 - a_1)t, \ldots, a_n + (b_n - a_n)t) : t \in [0, 1]\} \subset E^n.$$

Since the component functions $a_i + (b_i - a_i)t$ are continuous, the line segment between p and q is a continuous image of the interval $[0, 1]$, hence is connected. The distance between the point p and the point $(a_1 + (b_1 - a_1)t, \ldots, a_n + (b_n - a_n)t)$ is t times the distance between p and q, for any $t \in [0, 1]$, hence at most the distance between p and q. Thus the entire line segment between the center of any ball and any point of the ball lies entirely within the ball. Any ball in E^n is therefore the union of all line segments between its center and its various points, that is the union of connected sets that all contain the center of the ball. By the second proposition of the last section of the preceding chapter, any ball is connected. Since E^n is the union of all line segments between the origin and its various points, the same reasoning shows that E^n is connected.

§ 6. SEQUENCES OF FUNCTIONS.

Definition. Let E, E' be metric spaces and for $n = 1, 2, 3, \ldots$ let $f_n: E \to E'$ be a function. If $p \in E$, we say that the sequence $f_1, f_2, f_3, \ldots$ *converges at p* if the sequence of points $f_1(p), f_2(p), f_3(p), \ldots$ of E' converges. We say that the sequence of functions $f_1, f_2, f_3, \ldots$ *converges on E*, or *converges*, or *is convergent*, if the sequence converges at each $p \in E$. If $f_1, f_2, f_3, \ldots$ converges and $f: E \to E'$ is the function defined by

$$f(p) = \lim_{n\to\infty} f_n(p)$$

for all $p \in E$, we say that $f_1, f_2, f_3, \ldots$ *converges to f*, f is called the *limit function* of the sequence, and we write

$$f = \lim_{n\to\infty} f_n.$$

For example, for each $n = 1, 2, 3, \ldots$ let $f_n: [0, 1] \to \mathbf{R}$ be given by $f_n(x) = x - x/n$. For each $x \in [0, 1]$ we have $\lim_{n\to\infty} f_n(x) = x$. Here the limit function f is the identity function $f(x) = x$. This is illustrated in Figure 21.

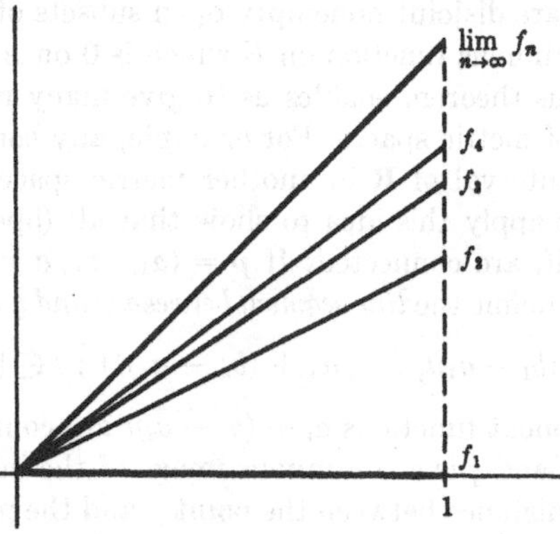

FIGURE 21. The sequence of functions $f_n(x) = x - x/n$ on $[0, 1]$.

For a second example, let $f_n: [0, 1] \to \mathbf{R}$ be given by $f_n(x) = x^n$. Since $\lim_{n\to\infty} a^n = 0$ if $|a| < 1$ (cf. end of § 3 of Chapter III), the sequence $f_1, f_2, f_3, \ldots$ converges to the limit function f given by

$$f(x) = \begin{cases} 0 & \text{if } 0 \leq x < 1 \\ 1 & \text{if } x = 1. \end{cases}$$

Notice that each f_n is continuous, but the limit function is not. This is illustrated in Figure 22.

If E, E' are metric spaces and the sequence of functions $f_1, f_2, f_3, \ldots$ from E into E' converges to f, then for any $\epsilon > 0$ and any $p \in E$ there is a positive integer N such that $d'(f(p), f_n(p)) < \epsilon$ whenever $n > N$; this is a slight amplification of the previous definition. In general the integer N depends on both ϵ and p, and for a fixed ϵ we must take N larger and larger for different points p if we want the inequality $d'(f(p), f_n(p)) < \epsilon$ to hold for all $n > N$. If it happens that for any $\epsilon > 0$ we can find an integer N that works simultaneously for all points $p \in E$ then, as we shall see, the convergence of $f_1, f_2, f_3, \ldots$ to f is especially nice in the sense that if each of the functions $f_1, f_2, f_3, \ldots$ possesses a certain kind of property (for example, continuity), then so does the limit function f. This motivates the definition on the next page.

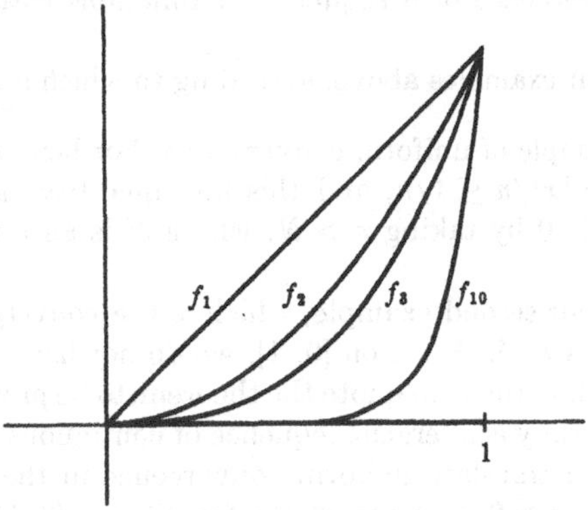

FIGURE 22. The sequence of functions $f_n(x) = x^n$ on $[0, 1]$.

Definition. Let E, E' be metric spaces, for $n = 1, 2, 3, \ldots$ let $f_n: E \to E'$ be a function, and let $f: E \to E'$ be another function. Then the sequence $f_1, f_2, f_3, \ldots$ is said to *converge uniformly to* f if, given any $\epsilon > 0$, there is a positive integer N such that $d'(f(p), f_n(p)) < \epsilon$ whenever $n > N$, for all $p \in E$.

If the sequence $f_1, f_2, f_3, \ldots$ converges uniformly to f we sometimes say, for emphasis, that $f_1, f_2, f_3, \ldots$ converges uniformly to f on E. If the restrictions of $f_1, f_2, f_3, \ldots$ to a certain subset S of E converge uniformly to some function on S, we say that $f_1, f_2, f_3, \ldots$ *converges uniformly on* S.

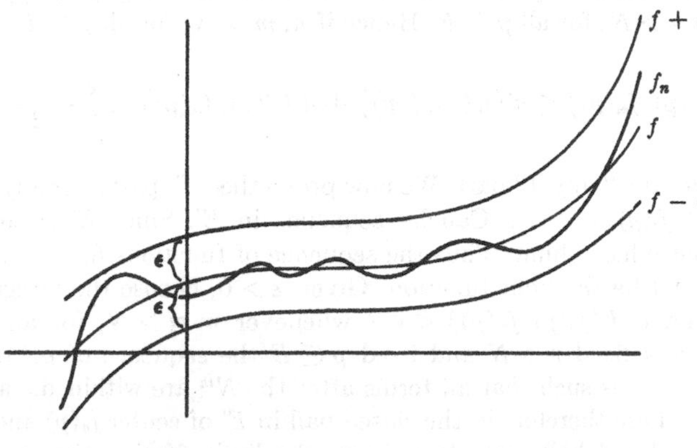

FIGURE 23. Uniform convergence of a sequence of real-valued functions of a real variable.

Uniform convergence of a sequence of functions clearly implies convergence.

The first of our examples above, according to which $\lim_{n\to\infty} (x - x/n) = x$ on $[0, 1]$, is an example of uniform convergence. For here $d'(f(x), f_n(x)) = |x - (x - x/n)| = |x|/n \leq 1/n$, and this last quantity can be made less than any given $\epsilon > 0$ by taking $n > N$, where N is an integer at least as large as $1/\epsilon$.

However, in our second example, which is the convergence of the sequence of functions $x, x^2, x^3, \ldots$ on $[0, 1]$, we do not have uniform convergence. One way to show this is to quote the theorem, to be proved shortly, that the limit of a uniformly convergent sequence of continuous functions is continuous. Or we can translate uniform convergence in the present case to mean that for any $\epsilon > 0$ we have $x^n < \epsilon$ for all $x \in (0, 1)$, provided only that n is sufficiently large, and if it happens that $\epsilon < 1$ this contradicts the continuity of the function x^n at the point 1.

As before with sequences of points, so also with sequences of functions it is important to have a criterion for convergence that does not involve foreknowledge of the limit. Here is the relevant Cauchy criterion.

Proposition. *Let E, E' be metric spaces, with E' complete, and let $f_n: E \to E'$, $n = 1, 2, 3, \ldots$. Then the sequence of functions $f_1, f_2, f_3, \ldots$ is uniformly convergent if and only if, for any $\epsilon > 0$, there is a positive integer N such that if n and m are integers greater than N then $d'(f_n(p), f_m(p)) < \epsilon$ for all $p \in E$.*

If the sequence $f_1, f_2, f_3, \ldots$ converges uniformly to f, then for any $\epsilon > 0$ there exists a positive integer N such that $d'(f(p), f_n(p)) < \epsilon/2$ whenever $n > N$, for all $p \in E$. Hence if $n, m > N$, for all $p \in E$ we have

$$d'(f_n(p), f_m(p)) \leq d'(f_n(p), f(p)) + d'(f(p), f_m(p)) < \frac{\epsilon}{2} + \frac{\epsilon}{2} = \epsilon.$$

This proves the "only if" part. We now prove the "if" part: For any $p \in E$, $f_1(p), f_2(p), f_3(p), \ldots$ is a Cauchy sequence in E'. Since E' is complete, this sequence has a limit. Thus the sequence of functions $f_1, f_2, f_3, \ldots$ converges. Let f be the limit function. Given $\epsilon > 0$, choose the integer N so that we have $d'(f_n(p), f_m(p)) < \epsilon/2$ whenever $n, m > N$, for all $p \in E$. Then for any fixed $n > N$ and fixed $p \in E$ the sequence of points $f_1(p)$, $f_2(p), f_3(p), \ldots$ is such that all terms after the Nth are within distance $\epsilon/2$ of $f_n(p)$, and are therefore in the closed ball in E' of center $f_n(p)$ and radius $\epsilon/2$. Since closed balls are closed sets, the limit $f(p)$ of the convergent sequence $f_1(p), f_2(p), f_3(p), \ldots$ is also in this closed ball, so that

$d'(f_n(p), f(p)) \leq \epsilon/2$. Hence if $n > N$ we have $d'(f_n(p), f(p)) < \epsilon$ for all $p \in E$, proving uniform convergence.

Theorem. *Let E, E' be metric spaces and let $f_1, f_2, f_3, \ldots$ be a uniformly convergent sequence of continuous functions from E into E'. Then $\lim_{n\to\infty} f_n$ is continuous.*

We must show that $f = \lim_{n\to\infty} f_n$ is continuous at each point $p_0 \in E$. So let $p_0 \in E$ be fixed. Let $\epsilon > 0$ be given. Fix a positive integer n such that $d'(f(p), f_n(p)) < \epsilon/3$ for all $p \in E$, which is possible by the uniform convergence. Since f_n is continuous at p_0, there is a number $\delta > 0$ such that if $p \in E$ and $d(p, p_0) < \delta$ then $d'(f_n(p), f_n(p_0)) < \epsilon/3$. Hence if $p \in E$ and $d(p, p_0) < \delta$ we have

$$d'(f(p), f(p_0)) \leq d'(f(p), f_n(p)) + d'(f_n(p), f_n(p_0)) + d'(f_n(p_0), f(p_0))$$
$$< \frac{\epsilon}{3} + \frac{\epsilon}{3} + \frac{\epsilon}{3} = \epsilon.$$

Thus f is continuous at p_0.

The above proof really shows something more general than is stated, namely that if we have a sequence of functions $f_1, f_2, f_3, \ldots$ from E into E' that converges uniformly on some open ball of E of center p_0 and if each f_n is continuous at p_0 then the limit function is also continuous at p_0.

If f and g are functions from a metric space E into a metric space E', it is natural to try to find some measure of the extent to which f and g differ, that is to find some sort of "distance" between f and g. For any specific $p \in E$ we may say that f and g differ at p by the distance between their values at p, that is by $d'(f(p), g(p))$, but we would really like to measure how much f and g differ over *all* points of E, not just at p. There are various ways of doing this, depending on the circumstances and purposes in mind, but the most simple-minded method turns out to be one of the most useful. It is to take the distance between f and g to be

$$\max \{d'(f(p), g(p)) : p \in E\}$$

if this maximum happens to exist. In order to develop this idea we need to digress for a simple lemma.

Lemma. *Let E and E' be metric spaces, and let f and g be continuous functions from E into E'. Then the real-valued function on E whose value at any point $p \in E$ is $d'(f(p), g(p))$ is continuous.*

We must show that this function is continuous at any given point $p_0 \in E$, that is that $|d'(f(p), g(p)) - d'(f(p_0), g(p_0))|$ is less than any given $\epsilon > 0$ for all p in some open ball in E of center p_0. But

$$\begin{aligned} &|d'(f(p), g(p)) - d'(f(p_0), g(p_0))| \\ &\quad \leq |d'(f(p), g(p)) - d'(f(p), g(p_0))| + |d'(f(p), g(p_0)) - d'(f(p_0), g(p_0))| \\ &\quad\quad \leq d'(g(p), g(p_0)) + d'(f(p), f(p_0)), \end{aligned}$$

the last step being a double application of the fact that the difference between two sides of a triangle is at most the third side. Since f and g are continuous at p_0, each of the quantities $d'(g(p), g(p_0))$, $d'(f(p), f(p_0))$ is less than $\epsilon/2$ for all p in some open ball in E of center p_0, proving that for all p in that open ball

$$|d'(f(p), g(p)) - d'(f(p_0), g(p_0))| < \epsilon,$$

as desired.

In the case of greatest interest, that where $E' = \mathbf{R}$, there was actually no need for a detour to prove the lemma, for here $d'(f(p), g(p)) = |f(p) - g(p)|$ and the continuity of this function follows from that of f and g in two easy steps: the difference of two continuous real-valued functions is continuous, and the absolute value function on $\mathbf{R}$ is continuous.

Now consider the set $\mathfrak{F}$ of all continuous functions from E into E'. We assume that E is compact. Then it is true that for any $f, g \in \mathfrak{F}$

$$\max \{d'(f(p), g(p)) : p \in E\}$$

exists, since any continuous real-valued function on a compact metric space attains a maximum. Hence we may define, for any $f, g \in \mathfrak{F}$, the distance between f and g to be

$$D(f, g) = \max \{d'(f(p), g(p)) : p \in E\}.$$

We proceed to show that $\mathfrak{F}$, together with this D, is a metric space.

For all $f, g \in \mathfrak{F}$, it is clear that $D(f, g) \geq 0$ and that $D(f, g) = 0$ if and only if $f = g$. It is also clear that $D(f, g) = D(g, f)$. It remains to prove the triangle inequality, which states that if $f, g, h \in \mathfrak{F}$ then

$$D(f, h) \leq D(f, g) + D(g, h).$$

To prove this, pick $p_0 \in E$ such that $D(f, h) = d'(f(p_0), h(p_0))$. Then

$$\begin{aligned} D(f, h) &= d'(f(p_0), h(p_0)) \leq d'(f(p_0), g(p_0)) + d'(g(p_0), h(p_0)) \\ &\leq \max \{d'(f(p), g(p)) : p \in E\} + \max \{d'(g(p), h(p)) : p \in E\} \\ &= D(f, g) + D(g, h). \end{aligned}$$

Thus $\mathfrak{F}$ is indeed a metric space. It is "abstract" in the sense that its "points" are functions on another metric space.

A sequence of points in the metric space $\mathfrak{F}$ is a sequence of functions $f_1, f_2, f_3, \ldots$ from E into E'. This sequence will converge to a point $f \in \mathfrak{F}$ if and only if

$$\lim_{n\to\infty} D(f, f_n) = 0,$$

in other words if and only if for each $\epsilon > 0$ there is a positive integer N such that for any integer $n > N$ we have

$$\max \{d'(f(p), f_n(p)) : p \in E\} < \epsilon,$$

that is

$$d'(f(p), f_n(p)) < \epsilon$$

for all $p \in E$. Thus the sequence of points $f_1, f_2, f_3, \ldots$ of $\mathfrak{F}$ converges to the point $f \in \mathfrak{F}$ if and only if the sequence of functions $f_1, f_2, f_3, \ldots$ on E converges uniformly to the function f.

Suppose that a sequence of points $f_1, f_2, f_3, \ldots$ of $\mathfrak{F}$ is a Cauchy sequence in the metric space $\mathfrak{F}$. Then for any $\epsilon > 0$ there is a positive integer N such that whenever $n, m > N$ we have

$$D(f_n, f_m) < \epsilon,$$

that is

$$\max \{d'(f_n(p), f_m(p)) : p \in E\} < \epsilon,$$

or

$$d'(f_n(p), f_m(p)) < \epsilon$$

for all $p \in E$. Assume that E' is complete. Then the proposition of this section is applicable, and it tells us that the sequence of functions

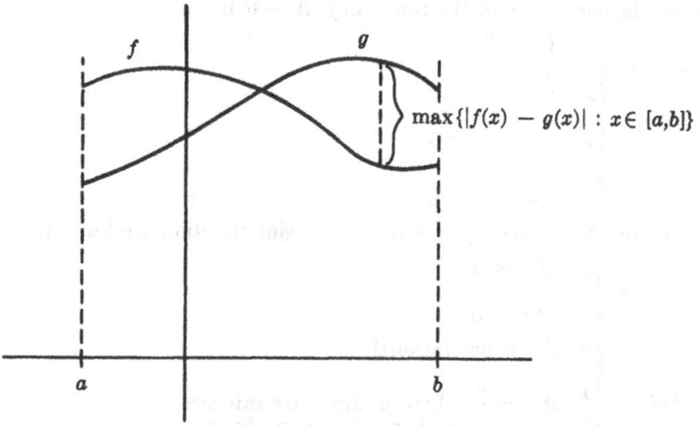

FIGURE 24. The distance between two real-valued functions on $[a, b]$.

$f_1, f_2, f_3, \ldots$ converges uniformly on E to some function $f: E \to E'$. The previous theorem tells us that f is continuous. Thus $f \in \mathfrak{F}$ and

$$\lim_{n \to \infty} f_n = f$$

in the sense of points of the metric space $\mathfrak{F}$. Thus the metric space $\mathfrak{F}$ is complete.

We have proved the following result.

Theorem. *Let E and E' be metric spaces, with E compact and E' complete. Then the set of all continuous functions from E to E', with the distance between two such functions f and g taken to be*

$$\max \{d'(f(p), g(p)) : p \in E\},$$

is a complete metric space. A sequence of points of this metric space converges if and only if it is a uniformly convergent sequence of functions on E.

If $E' = \mathbf{R}$ we have the metric space of all continuous real-valued functions on a compact metric space E, the distance between two such functions f and g being

$$\max \{ |f(p) - g(p)| : p \in E\}.$$

This metric space is important enough to be denoted by a standard symbol $C(E)$.

PROBLEMS

1. Discuss the continuity of the function $f: \mathbf{R} \to \mathbf{R}$ if

(a) $f(x) = \begin{cases} 0 & \text{if } x < 0 \\ x & \text{if } x \geq 0 \end{cases}$

(b) $f(x) = \begin{cases} x \sin \dfrac{1}{x} & \text{if } x \neq 0 \\ 0 & \text{if } x = 0 \end{cases}$

(assume the general properties of the sine function are known)

(c) $f(x) = \begin{cases} x^2 & \text{if } x \neq 0 \\ 1 & \text{if } x = 0 \end{cases}$

(d) $f(x) = \begin{cases} 0 & \text{if } x \text{ is not rational} \\ \dfrac{1}{q} & \text{if } x = \dfrac{p}{q}, \text{ where } p \text{ and } q \text{ are integers with no common divisors other than } \pm 1, \text{ and } q > 0. \end{cases}$

2. Let E, E' be metric spaces, $f: E \to E'$ a continuous function. Show that if S is a closed subset of E' then $f^{-1}(S)$ is a closed subset of E. Derive from this the results that if f is a continuous real-valued function on E then the sets $\{p \in E : f(p) \le 0\}$, $\{p \in E : f(p) \ge 0\}$, $\{p \in E : f(p) = 0\}$ are closed.

3. Let E, E' be metric spaces, $f: E \to E'$ a function, and suppose that S_1, S_2 are closed subsets of E such that $E = S_1 \cup S_2$. Show that if the restrictions of f to S_1 and to S_2 are continuous then f is continuous.

4. Let U, V be (open or closed) intervals in $\mathbf{R}$, and let $f: U \to V$ be a function which is strictly increasing (i.e., if $x, y \in U$ and $x < y$ then $f(x) < f(y)$) and onto. Prove that f and f^{-1} are continuous.

5. Let E, E' be metric spaces, $f: E \to E'$ a function, and let $p \in E$. Define the *oscillation of f at p* to be

$$\text{g.l.b.}\ \{a \in \mathbf{R} : \text{there exists an open ball in } E \text{ of center } p \text{ such that for any } x, y \text{ in this ball we have } d'(f(x), f(y)) \le a\}$$

if the bracketed set is not empty; if the set in brackets is empty we define the oscillation of f at p to be the symbol $+\infty$. Prove that f is continuous at p if and only if the oscillation of f at p is zero, and that for any real number ϵ the set of points of E at which the oscillation of f is at least ϵ is closed.

6. Let E be a metric space, S a subset of E, and let $f: E \to \mathbf{R}$ be the function which takes the value 1 at each point of S and 0 at each point of $\complement S$. Prove that the set of points of E at which f is not continuous is precisely the boundary of S (cf. Prob. 17, Chap. III).

7. Let U be an open interval in $\mathbf{R}$, let $a \in U$, let E' be a metric space, and let $f: U - \{a\} \to E'$ be a function. Define

$$\lim_{x \to a+} f(x) = \lim_{x \to a} f_+(x),$$

where f_+ is the restriction of f to $U \cap \{x \in \mathbf{R} : x \ge a\}$, and

$$\lim_{x \to a-} f(x) = \lim_{x \to a} f_-(x),$$

where f_- is the restriction of f to $U \cap \{x \in \mathbf{R} : x \le a\}$, if these limits exist. Prove that $\lim_{x \to a} f$ exists if and only if $\lim_{x \to a+} f$ and $\lim_{x \to a-} f$ exist and are equal.

8. Let $U = \{x \in \mathbf{R} : x > a\}$, for some positive real number a, and let f be a real-valued function on U. Define

$$\lim_{x \to +\infty} f(x) = \lim_{y \to 0} g(y),$$

where $g: (0, 1/a) \to \mathbf{R}$ is given by $g(y) = f(1/y)$, if this latter limit exists. Prove that $\lim_{x \to +\infty} f(x)$ exists if and only if, given any $\epsilon > 0$, there exists a number N such that if $x, y \in \mathbf{R}$ and $x, y > N$ then $|f(x) - f(y)| < \epsilon$.

9. (a) Prove that $\sqrt{x}$ is continuous on $\{x \in \mathbf{R} : x \ge 0\}$.

   (b) Evaluate $\lim_{x \to 1} \dfrac{x - 1}{\sqrt{x} - 1}$.

   (c) Evaluate $\lim_{x \to +\infty} \dfrac{x}{2x^2 + 1}$ (cf. Prob. 8).

10. Discuss the continuity of the function $f\colon E^2 \to \mathbf{R}$ if

(a) $f(x, y) = \begin{cases} \dfrac{1}{x^2 + y^2} & \text{if } (x, y) \neq (0, 0) \\ 0 & \text{if } (x, y) = (0, 0) \end{cases}$

(b) $f(x, y) = \begin{cases} \dfrac{xy}{x^2 + y^2} & \text{if } (x, y) \neq (0, 0) \\ 0 & \text{if } (x, y) = (0, 0) \end{cases}$

(c) $f(x, y) = \begin{cases} \dfrac{xy^2}{x^2 + y^2} & \text{if } (x, y) \neq (0, 0) \\ 0 & \text{if } (x, y) = (0, 0). \end{cases}$

11. Give a proof of the first proposition of § 3 (on the continuity of sums, products, etc. of continuous functions) that is based directly on the definition of continuity.

12. Prove the analog of the first proposition of § 3 for complex-valued functions on a metric space (cf. Probs. 20, 21 of Chap. III).

13. Write down the details of the following alternate proof that a continuous real-valued function f on a compact metric space E is bounded and attains a maximum: If f is not bounded, then for $n = 1, 2, 3, \ldots$ there is a point $p_n \in E$ such that $|f(p_n)| > n$, and a contradiction arises from the existence of a convergent subsequence of $p_1, p_2, p_3, \ldots$. Thus f is bounded and we can find a sequence of points $q_1, q_2, q_3, \ldots$ of E such that $\lim_{n\to\infty} f(q_n) = \text{l.u.b. } \{f(p) : p \in E\}$.

    A maximum will be attained by f at the limit of a convergent subsequence of $q_1, q_2, q_3, \ldots$.

14. (a) Prove that if S is a nonempty compact subset of a metric space E and $p_0 \in E$ then $\min \{d(p_0, p) : p \in S\}$ exists ("distance from p_0 to S").
    (b) Prove that if S is a nonempty closed subset of E^n and $p_0 \in E^n$ then $\min \{d(p_0, p) : p \in S\}$ exists.

15. Prove that for any nonempty compact metric space E, $\max \{d(p, q) : p, q \in E\}$ exists ("diameter of E"). (*Hint:* Start with a sequence of pairs of points $\{(p_n, q_n)\}_{n=1,2,3,\ldots}$ of E such that

$$\lim_{n\to\infty} d(p_n, q_n) = \text{l.u.b. } \{d(p, q) : p, q \in E\}$$

    and pass to convergent subsequences.)

16. Let E, E' be metric spaces, $f\colon E \to E'$ a continuous function. Prove that if E is compact and f is one-one onto then $f^{-1}\colon E' \to E$ is continuous. (*Hint:* f sends closed sets onto closed sets, therefore open sets onto open sets.)

17. Is the function x^2 uniformly continuous on $\mathbf{R}$? The function $\sqrt{|x|}$? Why?

18. Prove that for any metric space E, the identity function on E is uniformly continuous.

19. Prove that for any metric space E and any $p_0 \in E$, the real-valued function sending any p into $d(p_0, p)$ is uniformly continuous.

20. State precisely and prove: A uniformly continuous function of a uniformly continuous function is uniformly continuous.

21. Let S be a subset of the metric space E with the property that each point of $\mathcal{C}S$ is a cluster point of S (one then calls S *dense* in E). Let E' be a complete metric space and $f: S \to E'$ a uniformly continuous function. Prove that f can be extended to a continuous function from E into E' in one and only one way, and that this extended function is also uniformly continuous.

22. Let V, V' be normed vector spaces (cf. Prob. 22, Chap. III) and $f: V \to V'$ a linear transformation. Prove the following statements.
    (a) If f is continuous at one point it is continuous everywhere, and in fact is uniformly continuous.
    (b) f is continuous if and only if the set $\{\|f(x)\|/\|x\| : x \in V, x \neq 0\}$ is bounded.
    (c) f is continuous if V is finite dimensional. (*Hint:* Use a basis of V.)
    (d) The set of all infinite sequences of real numbers with only a finite number of nonzero terms is a normed vector space if we define
    $$(x_1, x_2, x_3, \ldots) + (y_1, y_2, y_3, \ldots) = (x_1 + y_1, x_2 + y_2, x_3 + y_3, \ldots)$$
    $$c(x_1, x_2, x_3, \ldots) = (cx_1, cx_2, cx_3, \ldots)$$
    $$\|(x_1, x_2, x_3, \ldots)\| = \max\{|x_1|, |x_2|, |x_3|, \ldots\},$$
    and the map sending $(x_1, x_2, x_3, \ldots)$ into $(x_1, 2x_2, 3x_3, \ldots)$ is a one-one linear transformation of this normed vector space onto itself that is not continuous.

23. Use Problem 22 to prove that if V is a finite dimensional vector space over $\mathbf{R}$ and $\|\ \|_1$, $\|\ \|_2$ are two norm functions on V (i.e., real-valued functions such that $(V, \|\ \|_1)$ and $(V, \|\ \|_2)$ are normed vector spaces), then there exist positive real numbers m, M such that $m \leq \|x\|_1/\|x\|_2 \leq M$ for all nonzero $x \in V$. Deduce that any finite dimensional normed vector space is complete (as a metric space).

24. Give another proof of the intermediate value theorem by completing the following argument: If f is a continuous real-valued function on the closed interval $[a, b]$ in $\mathbf{R}$ and $f(a) < \gamma < f(b)$, then
$$f(\text{l.u.b. } \{x \in [a, b] : f(x) \leq \gamma\}) = \gamma.$$

25. Give a proof of the intermediate value theorem using uniform continuity. (*Hint:* Using the notation of this theorem, uniform continuity implies that, given any $\epsilon > 0$, if we divide $[a, b]$ into a sufficiently large number of subintervals of equal length then for at least one of the division points p we shall have $|f(p) - \gamma| < \epsilon$. Choose a sequence of p's corresponding to a sequence of ϵ's approaching zero, then a suitable subsequence.)

26. Let $a, b \in \mathbf{R}$, $a < b$, and let f be a continuous real-valued function on $[a, b]$. Prove that if f is one-one then $f([a, b])$ is $[f(a), f(b)]$ or $[f(b), f(a)]$, whichever expression makes sense.

27. Show that if $f: \mathbf{R} \to \mathbf{R}$ is a polynomial function of odd degree, then $f(\mathbf{R}) = \mathbf{R}$.

28. Show that any open or closed interval in E^n is connected.

29. A metric space E is said to be *arcwise connected* if, given any $p, q \in E$, there is a continuous function $f: [0, 1] \to E$ such that $f(0) = p$, $f(1) = q$. Show that
    (a) an arcwise connected metric space is connected
    (b) any connected open subset of E^n is arcwise connected.

30. Prove that a continuous real-valued function on a closed interval in E^2 cannot be one-one.

31. (A space-filling curve)
    (a) Show that the subset S of $[0, 1]$ consisting of all numbers having decimal expansions of the form
    $$.a_1b_1c_10a_2b_2c_20a_3b_3c_30\ \ldots$$
    (each a_n, b_n, c_n being one of the integers $0, 1, \ldots, 9$) is closed.
    (b) Show that the real-valued functions $\varphi_1, \varphi_2, \varphi_3$ on S which send
    $$.a_1b_1c_10a_2b_2c_20a_3b_3c_30\ \ldots$$
    into the real numbers with decimal expansions
    $$.a_1a_2a_3\ldots,\quad .b_1b_2b_3\ldots,\quad .c_1c_2c_3\ldots$$
    respectively are continuous. (Note that each number in S has a unique decimal expansion, so that $\varphi_1, \varphi_2, \varphi_3$, are well-defined.)
    (c) Show that there are unique continuous real-valued functions f_1, f_2, f_3 on $[0, 1]$ whose restrictions to S are $\varphi_1, \varphi_2, \varphi_3$ respectively, which equal 0 at 1, and which are linear on each open interval in $\mathcal{C}S$ (cf. Prob. 5, Chap. III).
    (d) Prove that the function $f: [0, 1] \to E^3$ defined by $f(x) = (f_1(x), f_2(x), f_3(x))$ for all $x \in [0, 1]$ is a continuous map of $[0, 1]$ onto the unit cube $\{(x_1, x_2, x_3) \in E^3 : x_1, x_2, x_3 \in [0, 1]\}$.

32. Show that the sequence of functions
$$\sqrt{x},\ \sqrt{x + \sqrt{x}},\ \sqrt{x + \sqrt{x + \sqrt{x}}},\ \ldots$$
on $\{x \in \mathbf{R} : x \geq 0\}$ is convergent and find the limit function.

33. (a) Show that the sequence of functions $x, x^2, x^3, \ldots$ converges uniformly on $[0, a]$ for any $a \in (0, 1)$, but not on $[0, 1]$.
    (b) Show that the sequence of functions $x(1 - x), x^2(1 - x), x^3(1 - x), \ldots$ converges uniformly on $[0, 1]$.

34. Is the sequence of functions $f_1, f_2, f_3, \ldots$ on $[0, 1]$ uniformly convergent if
$$f_n(x) = \frac{x}{1 + nx^2}?\quad \text{If } f_n(x) = \frac{nx}{1 + nx^2}?\quad \text{If } f_n(x) = \frac{nx}{1 + n^2x^2}?$$

35. Show that if the function $f: \mathbf{R} \to \mathbf{R}$ is uniformly continuous, then the sequence of functions $f(x + 1), f\left(x + \frac{1}{2}\right), f\left(x + \frac{1}{3}\right), \ldots$ is uniformly convergent.

36. Does the sequence of functions $x, \frac{x}{2}, \frac{x}{3}, \frac{x}{4}, \ldots$ converge uniformly on $\mathbf{R}$?

37. Let $f_1, f_2, f_3, \ldots$ and $g_1, g_2, g_3, \ldots$ be uniformly convergent sequences of real-valued functions on a metric space E. Show that the sequence $f_1 + g_1, f_2 + g_2, f_3 + g_3, \ldots$ is uniformly convergent. How about $f_1 g_1, f_2 g_2, f_3 g_3, \ldots$?

38. Prove that the limit of a uniformly convergent sequence of bounded functions (from one metric space into another) is bounded.

39. Give an example of a convergent sequence of continuous real-valued functions on $[0, 1]$ whose limit function fails to be continuous at an infinite number of points.

40. Let $a, b \in \mathbf{R}$, $a < b$, and for $n = 1, 2, 3, \ldots$ let $f_n: [a, b] \to \mathbf{R}$ be an increasing function (i.e., $f_n(x) \leq f_n(y)$ if $x \leq y$). Prove that if the sequence $f_1, f_2, f_3, \ldots$ converges to f then f is increasing, and that if f is continuous then the convergence is uniform.

41. Let $f, f_1, f_2, f_3, \ldots$ be continuous real-valued functions on the compact metric space E, with $f = \lim_{n\to\infty} f_n$. Prove that if $f_1(p) \leq f_2(p) \leq f_3(p) \leq \ldots$ for all $p \in E$ then the sequence $f_1, f_2, f_3, \ldots$ converges uniformly.

42. Show that the closed ball in $C([0, 1])$ of center 0 and radius 1 is not compact. (*Hint:* Consider the sequence of functions $x, x^2, x^3, \ldots$.)

43. If E is a compact metric space and $p_0 \in E$ we get a real-valued function F on $C(E)$ by setting $F(f) = f(p_0)$ for all $f \in C(E)$. Prove that F is uniformly continuous.

44. Generalize Problem 43 as follows: If E and E' are compact metric spaces and $\varphi: E \to E'$ is a continuous function, map $C(E')$ into $C(E)$ by sending each $f \in C(E')$ into $f \circ \varphi \in C(E)$. Prove that this map is uniformly continuous.

45. Let E be a compact metric space. Show that $C(E)$ is a complete normed vector space (cf. Prob. 22, Chap. III) if we add its elements in the usual way, multiply them by real numbers in the usual way, and take $\|f\| = \max\{|f(p)| : p \in E\}$ for all $f \in C(E)$. Show that the map of Problem 44 is a continuous linear transformation.

46. Prove the analog of the last theorem of the chapter when E is not compact but with a restriction to bounded continuous functions, the distance between two such functions f and g being taken as

$$\text{l.u.b.}\,\{d'(f(p), g(p)) : p \in E\}.$$

Do the same thing for bounded functions from E to E' that are not necessarily continuous. What is the relation between the two metric spaces so obtained?

CHAPTER V

Differentiation

The subject of this chapter is one-variable differential calculus. The essential items, and even their development, are familiar from elementary calculus. This ground can be covered with speed and precision since all the difficult work has been done in the preceding chapter.

§ 1. THE DEFINITION OF DERIVATIVE.

Definition. Let f be a real-valued function on an open subset U of $\mathbf{R}$. Let $x_0 \in U$. We say that f is *differentiable at* x_0 if

$$\lim_{x \to x_0} \frac{f(x) - f(x_0)}{x - x_0}$$

exists. If it exists, this limit, often denoted $f'(x_0)$, is called the *derivative of f at x_0*.

We remark to begin with that the notion of limit used here is exactly that of the preceding chapter, since x_0 is a cluster point of the metric space U and we are considering a function from the complement $\mathcal{C}\{x_0\}$ of $\{x_0\}$ in U into the metric space $\mathbf{R}$ (namely the function which associates to each $x \in \mathcal{C}\{x_0\}$ the element $(f(x) - f(x_0))/(x - x_0)$ of $\mathbf{R}$). As always, if the limit exists it is unique. Thus $f'(x_0)$, if it exists, is necessarily unique.

A clearly equivalent definition of $f'(x_0)$ is given by

$$f'(x_0) = \lim_{h \to 0} \frac{f(x_0 + h) - f(x_0)}{h}.$$

Here h is understood to vary in some open ball in $\mathbf{R}$ of center 0.

The equation

$$\lim_{x \to x_0} \frac{f(x) - f(x_0)}{x - x_0} = f'(x_0)$$

is equivalent to the existence, for each $\epsilon > 0$, of a number $\delta > 0$ such that

$$\left| \frac{f(x) - f(x_0)}{x - x_0} - f'(x_0) \right| \leq \epsilon$$

whenever $x \in U$, $x \neq x_0$, and $|x - x_0| < \delta$. The last inequality is equivalent to

$$|f(x) - f(x_0) - f'(x_0)(x - x_0)| \leq \epsilon |x - x_0|,$$

which also holds if $x = x_0$. Note that if δ is taken small enough and $|x - x_0| < \delta$ then it is automatically true that $x \in U$, since U is open. Thus we can say, somewhat more briefly, that the equation

$$\lim_{x \to x_0} \frac{f(x) - f(x_0)}{x - x_0} = f'(x_0)$$

is equivalent to the existence, for each $\epsilon > 0$, of a number $\delta > 0$ such that

$$|f(x) - f(x_0) - f'(x_0)(x - x_0)| \leq \epsilon |x - x_0|$$

whenever $|x - x_0| < \delta$.

Recall that a function $\varphi: \mathbf{R} \to \mathbf{R}$ is called linear if there exist numbers $c, k \in \mathbf{R}$ such that $\varphi(x) = cx + k$ for all $x \in \mathbf{R}$. We then have $\varphi(x) = \varphi(x_0) + c(x - x_0)$ for any $x_0 \in \mathbf{R}$. The real-valued function on $\mathbf{R}$ sending any x into $f(x_0) + f'(x_0)(x - x_0)$ is linear. It follows that *f is differentiable at x_0 if and only if f can be closely approximated near x_0 by a linear function* in the sense that there exists a linear function that differs from f by a very small fraction of $|x - x_0|$ if x is sufficiently near x_0; worded precisely, this condition is that there exist a linear function φ such that for any $\epsilon > 0$ there exists a $\delta > 0$ such that

$$|f(x) - \varphi(x)| \leq \epsilon |x - x_0|$$

whenever $|x - x_0| < \delta$.

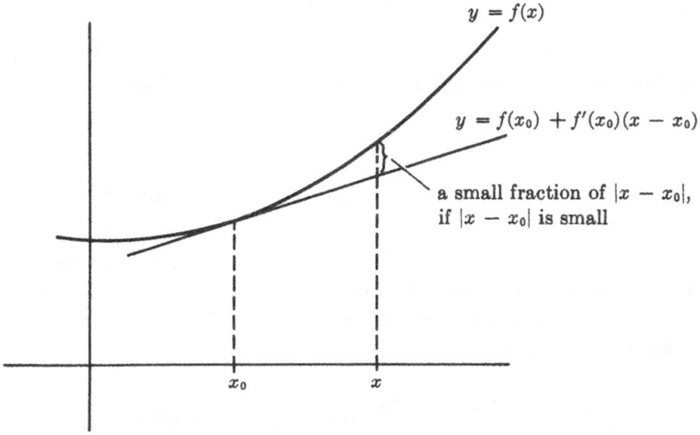

FIGURE 25. Graph of a function that is differentiable at x_0. Near x_0 the graph is very close to a certain straight line (the "tangent line at $x = x_0$"), in the sense indicated.

Proposition. *Let U be an open subset of $\mathbf{R}$, $f: U \to \mathbf{R}$. If f is differentiable at $x_0 \in U$ then f is continuous at x_0.*

Pick any $\epsilon_0 > 0$ and then a suitable number $\delta_0 > 0$ such that

$$|f(x) - f(x_0) - f'(x_0)(x - x_0)| \leq \epsilon_0 |x - x_0|$$

whenever $|x - x_0| < \delta_0$. Then if $|x - x_0| < \delta_0$ we have

$$\begin{aligned} |f(x) - f(x_0)| &\leq |f(x) - f(x_0) - f'(x_0)(x - x_0)| + |f'(x_0)(x - x_0)| \\ &\leq (\epsilon_0 + |f'(x_0)|)\,|x - x_0|. \end{aligned}$$

If, for any $\epsilon > 0$, we choose $\delta = \min\{\delta_0, \epsilon/(\epsilon_0 + |f'(x_0)|)\}$, we have $|f(x) - f(x_0)| < \epsilon$ whenever $|x - x_0| < \delta$, proving the proposition.

Definition. Let f be a real-valued function on an open subset U of $\mathbf{R}$. If $f'(x_0)$ exists for all $x_0 \in U$ then f is called *differentiable on* U (or just *differentiable*). The function f', often denoted df/dx, is called the *derivative of* f.

The notation df/dx (or $df(x)/dx$) has many obvious defects, but at least we usually know what is meant.

A differentiable function is necessarily continuous, but a continuous function is not necessarily differentiable. For example the absolute value function is continuous on all of $\mathbf{R}$ but it is not differentiable at 0, since if $x \neq 0$

$$\frac{|x| - |0|}{x - 0} = \frac{|x|}{x} = \begin{cases} 1 & \text{if } x > 0 \\ -1 & \text{if } x < 0. \end{cases}$$

No limit can exist as x approaches 0 since any open ball in $\mathbf{R}$ of center 0 contains both numbers greater than zero and numbers less than zero.

§ 2. RULES OF DIFFERENTIATION.

In very simple cases it is easy to differentiate (that is, compute derivatives) directly from the definition. For example, if f is a constant function, that is if $f(x) = c$ for all $x \in \mathbf{R}$, where c is some fixed real number, then for any $x_0 \in \mathbf{R}$ we have

$$f'(x_0) = \lim_{x \to x_0} \frac{f(x) - f(x_0)}{x - x_0} = \lim_{x \to x_0} \frac{c - c}{x - x_0} = \lim_{x \to x_0} 0 = 0.$$

If g is the identity function, that is if $g(x) = x$ for all $x \in \mathbf{R}$, then for any $x_0 \in \mathbf{R}$ we have

$$g'(x_0) = \lim_{x \to x_0} \frac{g(x) - g(x_0)}{x - x_0} = \lim_{x \to x_0} \frac{x - x_0}{x - x_0} = \lim_{x \to x_0} 1 = 1.$$

These results are usually written

$$\frac{dc}{dx} = 0, \quad \frac{dx}{dx} = 1.$$

For more complicated functions, differentiation by direct recourse to the definition is impractical, so special rules are developed. The following proposition makes the differentiation of rational functions almost mechanical. The formulas for differentiating exponential, logarithmic and trigonometric functions will have to wait until the next two chapters, where these functions are given adequate definitions.

Proposition. *Let f and g be real-valued functions on an open subset U of* **R**. *If f and g are differentiable at the point $x_0 \in U$, then so are $f + g$, $f - g$, fg, and, if $g(x_0) \neq 0$, f/g. Their derivatives at x_0 are given by the formulas*

$$\begin{aligned}
(f+g)'(x_0) &= f'(x_0) + g'(x_0)\\
(f-g)'(x_0) &= f'(x_0) - g'(x_0)\\
(fg)'(x_0) &= f(x_0)g'(x_0) + g(x_0)f'(x_0)\\
\left(\frac{f}{g}\right)'(x_0) &= \frac{g(x_0)f'(x_0) - f(x_0)g'(x_0)}{(g(x_0))^2}.
\end{aligned}$$

The proof, to be given shortly, is by direct computation. The limit formulas of the corollary on page 76 are used repeatedly. The continuity of f and g at p_0 is also used, in the form of the statements

$$\lim_{x\to x_0} f(x) = f(x_0), \quad \lim_{x\to x_0} g(x) = g(x_0).$$

In the case of the function f/g, the assumption that $g(x_0) \neq 0$ insures $g(x) \neq 0$ for all x in some open ball of center x_0 (by the continuity of g at x_0 and the result on page 75), so that it is permissible to restrict U to a smaller open set containing x_0 on which g is nowhere zero, banishing all concerns about possible zero denominators. Now that we have given all the reasons for the validity of the formal proofs, here are the formal proofs themselves:

$$\begin{aligned}
(f+g)'(x_0) &= \lim_{x\to x_0} \frac{f(x) + g(x) - f(x_0) - g(x_0)}{x - x_0}\\
&= \lim_{x\to x_0} \frac{f(x) - f(x_0)}{x - x_0} + \lim_{x\to x_0} \frac{g(x) - g(x_0)}{x - x_0}\\
&= f'(x_0) + g'(x_0).
\end{aligned}$$

The proof for $(f - g)'(x_0)$ is the same; just replace each g by $-g$.

$$\begin{aligned}
(fg)'(x_0) &= \lim_{x\to x_0} \frac{f(x)g(x) - f(x_0)g(x_0)}{x - x_0}\\
&= \lim_{x\to x_0} \left(f(x)\frac{g(x) - g(x_0)}{x - x_0} + g(x_0)\frac{f(x) - f(x_0)}{x - x_0}\right)\\
&= \lim_{x\to x_0} f(x) \cdot \lim_{x\to x_0} \frac{g(x) - g(x_0)}{x - x_0} + g(x_0) \cdot \lim_{x\to x_0} \frac{f(x) - f(x_0)}{x - x_0}\\
&= f(x_0)g'(x_0) + g(x_0)f'(x_0).
\end{aligned}$$

To find $(f/g)'(x_0)$, it is a little easier to first find $(1/g)'(x_0)$:

$$\begin{aligned}
(1/g)'(x_0) &= \lim_{x\to x_0} \frac{(1/g(x)) - (1/g(x_0))}{x - x_0} = \lim_{x\to x_0} \frac{g(x_0) - g(x)}{(x - x_0)g(x_0)g(x)}\\
&= -\frac{\displaystyle\lim_{x\to x_0} \frac{g(x) - g(x_0)}{x - x_0}}{g(x_0) \displaystyle\lim_{x\to x_0} g(x)} = -\frac{g'(x_0)}{(g(x_0))^2}.
\end{aligned}$$

Therefore our final step is

$$(f/g)'(x_0) = (f \cdot (1/g))'(x_0) = f(x_0) \cdot (1/g)'(x_0) + (1/g(x_0)) \cdot f'(x_0)$$
$$= -\frac{f(x_0)g'(x_0)}{(g(x_0))^2} + \frac{f'(x_0)}{g(x_0)} = \frac{g(x_0)f'(x_0) - f(x_0)g'(x_0)}{(g(x_0))^2}.$$

Corollary 1. *If f is a real-valued function on an open subset of* **R** *and $c \in \mathbf{R}$, then*

$$\frac{dcf}{dx} = c\frac{df}{dx}.$$

This means, of course, that for any $x_0 \in \mathbf{R}$ at which f is differentiable, $(cf)'(x_0)$ exists and equals $cf'(x_0)$. This follows from the formula for differentiating a product, together with the known result that the derivative of a constant function is zero.

Corollary 2. *For any integer n, $dx^n/dx = nx^{n-1}$.*

It is understood that if $n \leq 0$ then the function x^n is defined only on the nonzero real numbers. The result is known if $n = 0$ or 1. If n is a positive integer greater than one we repeatedly apply the formula for differentiating a product, as follows:

$$\frac{dx^2}{dx} = \frac{d}{dx}(x \cdot x) = x\frac{dx}{dx} + x\frac{dx}{dx} = x \cdot 1 + x = 2x$$
$$\frac{dx^3}{dx} = \frac{d}{dx}(x \cdot x^2) = x\frac{dx^2}{dx} + x^2\frac{dx}{dx} = x \cdot 2x + x^2 = 3x^2$$
$$\frac{dx^4}{dx} = \frac{d}{dx}(x \cdot x^3) = x\frac{dx^3}{dx} + x^3\frac{dx}{dx} = x \cdot 3x^2 + x^3 = 4x^3$$
$$\frac{dx^5}{dx} = \frac{d}{dx}(x \cdot x^4) = x\frac{dx^4}{dx} + x^4\frac{dx}{dx} = x \cdot 4x^3 + x^4 = 5x^4.$$

Clearly this process can be continued indefinitely. Each computation works out as above, giving at each stage the formula $dx^n/dx = nx^{n-1}$. This proves the result for $n \geq 0$. If $n < 0$, we set $n = -m$, so that $m > 0$, and complete the proof with the computation

$$\frac{dx^n}{dx} = \frac{d}{dx}\left(\frac{1}{x^m}\right) = \frac{x^m \cdot \dfrac{d1}{dx} - 1 \cdot \dfrac{dx^m}{dx}}{(x^m)^2} = \frac{-mx^{m-1}}{x^{2m}} = -mx^{-m-1} = nx^{n-1}.$$

The next result is the so-called "chain rule," or rule for differentiating a function of a function. Informally stated, if $u = u(y)$ and $y = y(x)$, so that $u = u(y(x))$, then

$$\frac{du}{dx} = \frac{du}{dy} \cdot \frac{dy}{dx}.$$

Proposition. *Let U and V be open subsets of* $\mathbf{R}$, *and let $f\colon U \to V$, $g\colon V \to \mathbf{R}$ be functions. Let $x_0 \in U$ be such that $f'(x_0)$ and $g'(f(x_0))$ exist. Then $(g \circ f)'(x_0)$ exists and*

$$(g \circ f)'(x_0) = g'(f(x_0))f'(x_0).$$

For any fixed y_0 for which $g'(y_0)$ exists, set

$$A(y, y_0) = \begin{cases} \dfrac{g(y) - g(y_0)}{y - y_0} & \text{if } y \in V,\, y \neq y_0 \\ g'(y_0) & \text{if } y = y_0. \end{cases}$$

Then

$$g(y) - g(y_0) = A(y, y_0)(y - y_0)$$

for all $y \in V$. Also

$$\lim_{y \to y_0} A(y, y_0) = g'(y_0) = A(y_0, y_0),$$

so that $A(y, y_0)$ is continuous at y_0. Now set $y_0 = f(x_0)$, $y = f(x)$. Since a continuous function of a continuous function is continuous,

$$\lim_{x \to x_0} A(f(x), f(x_0)) = g'(f(x_0)).$$

Hence

$$\begin{aligned}(g \circ f)'(x_0) &= \lim_{x \to x_0} \frac{g(f(x)) - g(f(x_0))}{x - x_0} = \lim_{x \to x_0} \frac{A(f(x), f(x_0))(f(x) - f(x_0))}{x - x_0} \\ &= \lim_{x \to x_0} A(f(x), f(x_0)) \cdot \lim_{x \to x_0} \frac{f(x) - f(x_0)}{x - x_0} = g'(f(x_0))f'(x_0).\end{aligned}$$

§3. THE MEAN VALUE THEOREM.

Proposition. *Let f be a real-valued function on an open subset U of* $\mathbf{R}$ *that attains a maximum or a minimum at the point $x_0 \in U$. Then if f is differentiable at x_0, $f'(x_0) = 0$.*

If $f'(x_0) \neq 0$, there exists a real number $\delta > 0$ such that if $x \neq x_0$ and $|x - x_0| < \delta$ then

$$\left| \frac{f(x) - f(x_0)}{x - x_0} - f'(x_0) \right| < \frac{|f'(x_0)|}{2},$$

that is,

$$f'(x_0) - \frac{|f'(x_0)|}{2} < \frac{f(x) - f(x_0)}{x - x_0} < f'(x_0) + \frac{|f'(x_0)|}{2}.$$

Since $|f'(x_0)|$ equals either $f'(x_0)$ or $-f'(x_0)$, each of the two extreme terms of the last inequality is necessarily either $f'(x_0)/2$ or $3f'(x_0)/2$, both of which have the same sign as $f'(x_0)$. Thus if $x \neq x_0$ and $|x - x_0| < \delta$, then $(f(x) - f(x_0))/(x - x_0)$ has a constant sign. But since f attains a maximum or a minimum at x_0, $f(x) - f(x_0)$ is always nonpositive or always nonnegative for all $x \in U$, hence always negative or always positive if $x \neq x_0$ and $|x - x_0| < \delta$. On the other hand the denominator $x - x_0$ can be either positive or negative. Therefore we can find an $x \neq x_0$ such that $|x - x_0| < \delta$ and $(f(x) - f(x_0))/(x - x_0)$ is either positive or negative, whichever we wish. This contradiction proves that the assumption that $f'(x_0) \neq 0$ is false.

***Lemma* (*Rolle's theorem*).** *Let $a, b \in \mathbf{R}$, $a < b$, and let f be a continuous real-valued function on $[a, b]$ that is differentiable on (a, b) and such that $f(a) = f(b) = 0$. Then there exists a number $c \in (a, b)$ such that $f'(c) = 0$.*

For since $[a, b]$ is compact, f must attain a maximum at at least one point of $[a, b]$, and also a minimum. If both maximum and minimum are attained at the end points a, b then since $f(a) = f(b) = 0$ we must have $f(x) = 0$ for all $x \in [a, b]$, so that $f'(c) = 0$ for all $c \in (a, b)$. In the contrary case, f attains a maximum or a minimum at some point $c \in (a, b)$, and the previous result gives $f'(c) = 0$.

A slight generalization of Rolle's theorem is the mean value theorem given below; Rolle's theorem is the special case where $f(a) = f(b) = 0$. The mean value theorem is illustrated in Figure 26.

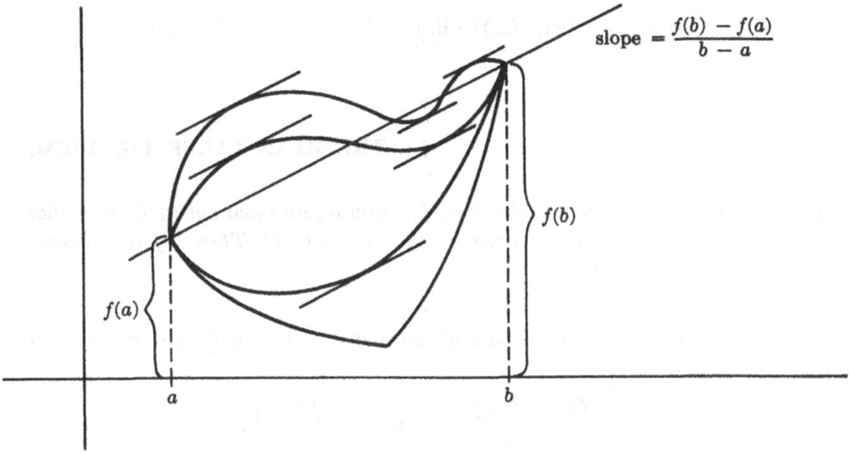

FIGURE 26. The geometric sense of the mean value theorem: the graph of a differentiable function has at least one tangent parallel to any chord. This is illustrated for several functions. The bottom curve shows how the theorem fails if differentiability is missing at one point.

Theorem (Mean value theorem). *Let $a, b \in \mathbf{R}$, $a < b$, and let f be a continuous real-valued function on $[a, b]$ that is differentiable on (a, b). Then there exists a number $c \in (a, b)$ such that*

$$f(b) - f(a) = (b - a)f'(c).$$

To prove this, define a new function $F\colon [a, b] \to \mathbf{R}$ by

$$F(x) = f(x) - f(a) - \frac{f(b) - f(a)}{b - a} \cdot (x - a)$$

for all $x \in [a, b]$. (Geometrically $F(x)$ is the vertical distance between the graph of f over $[a, b]$ and the line segment through the end points of this graph.) Then F is continuous on $[a, b]$, differentiable on (a, b), and $F(a) = F(b) = 0$. By Rolle's theorem, there exists a $c \in (a, b)$ such that $F'(c) = 0$. Thus

$$F'(c) = f'(c) - \frac{f(b) - f(a)}{b - a} = 0,$$

proving the result.

Corollary 1. *If a real-valued function f on an open interval in $\mathbf{R}$ has derivative zero at each point, then f is constant.*

We have to show that for any points a, b in the open interval we have $f(a) = f(b)$. Without loss of generality suppose $a < b$. For some $c \in (a, b)$ we have $f(b) - f(a) = (b - a)f'(c) = 0$. Thus indeed $f(a) = f(b)$.

Corollary 2. *If f and g are real-valued functions on an open interval in $\mathbf{R}$ which have the same derivative at each point, then f and g differ by a constant.*

For $(f - g)' = f' - g' = 0$, so $f - g$ is constant.

Definition. A real-valued function on a subset U of $\mathbf{R}$ is called

$$\left.\begin{array}{l} \textit{increasing} \\ \textit{strictly increasing} \\ \textit{decreasing} \\ \textit{strictly decreasing} \end{array}\right\} \quad \begin{array}{l} \text{if, whenever } a, b \in U \\ \text{and } a < b, \text{ we have} \end{array} \quad \left\{\begin{array}{l} f(a) \le f(b) \\ f(a) < f(b) \\ f(a) \ge f(b) \\ f(a) > f(b). \end{array}\right.$$

Corollary 3. *If f is a real-valued function on an open interval in $\mathbf{R}$ that has a positive (negative) derivative at each point, then f is strictly increasing (strictly decreasing).*

For if $a < b$, then $f(b) - f(a) = (b - a)f'(c)$ has the same sign as $f'(c)$.

§ 4. TAYLOR'S THEOREM.

Let U be an open subset of $\mathbf{R}$, $f: U \to \mathbf{R}$ a differentiable function. If the function $f': U \to \mathbf{R}$ is differentiable, we say that f is *twice differentiable* and call $(f')'$ the *second derivative of* f, writing $(f')'$ as f'' or $f^{(2)}$. If $f^{(2)}$ is differentiable, we say that f is *three times differentiable* and call $(f^{(2)})'$ the *third derivative of* f, writing $(f^{(2)})'$ as f''' or $f^{(3)}$. Similarly for functions that are 4, 5, 6, ... times differentiable. For any integer $n > 1$ and any $x_0 \in U$ we say that f is *n times differentiable at* x_0 if the restriction of f to some open ball of center x_0 is $(n-1)$ times differentiable and $(f^{(n-1)})'(x_0)$ exists; we then write $(f^{(n-1)})'(x_0) = f^{(n)}(x_0)$. Thus f is n times differentiable, for a given positive integer n, if and only if it is n times differentiable at each point of U. The nth derivative $f^{(n)}$ of f is often denoted

$$f^{(n)} = \frac{d^n f}{dx^n}.$$

In the rest of this section n will be one of the nonnegative integers $0, 1, 2, \ldots$. For convenience the *zeroth derivative of a function f* is defined to be $f^{(0)} = f$. Recall that

$$n! = 1 \cdot 2 \cdot 3 \cdots n$$

if $n = 1, 2, 3, \ldots$, so that

$$(n+1)! = (n+1) \cdot n!.$$

In order that the last equation also hold if $n = 0$, we define $0! = 1$.

Lemma. *Let U be an open interval in $\mathbf{R}$ and let the function $f: U \to \mathbf{R}$ be $(n+1)$ times differentiable. If for any $a, b \in U$ we define $R_n(b, a) \in \mathbf{R}$ by the equation*

$$f(b) = f(a) + \frac{f'(a)(b-a)}{1!} + \frac{f''(a)(b-a)^2}{2!} + \cdots + \frac{f^{(n)}(a)(b-a)^n}{n!} + R_n(b, a),$$

then

$$\frac{d}{dx} R_n(b, x) = -\frac{f^{(n+1)}(x)(b-x)^n}{n!}.$$

For any $x \in U$ we have

$$f(b) = f(x) + f'(x)\frac{(b-x)}{1!} + f''(x)\frac{(b-x)^2}{2!} + \cdots + f^{(n)}(x)\frac{(b-x)^n}{n!} + R_n(b, x).$$

For fixed b, each term in this equation is the value at x of a real-valued function on U. Each of these functions except the last is differentiable, hence also the last. Differentiating both sides of the equation we obtain

$$\begin{aligned}
0 &= f'(x) + \left(f'(x)\frac{d}{dx}\frac{(b-x)}{1!} + f''(x)\frac{(b-x)}{1!}\right) \\
&\quad + \left(f''(x)\frac{d}{dx}\frac{(b-x)^2}{2!} + f'''(x)\frac{(b-x)^2}{2!}\right) + \cdots \\
&\qquad + \left(f^{(n)}(x)\frac{d}{dx}\frac{(b-x)^n}{n!} + f^{(n+1)}(x)\frac{(b-x)^n}{n!}\right) + \frac{d}{dx}R_n(b, x) \\
&= f'(x) + \left(-f'(x) + f''(x)\frac{(b-x)}{1!}\right) \\
&\quad + \left(-f''(x)\frac{(b-x)}{1!} + f'''(x)\frac{(b-x)^2}{2!}\right) + \cdots \\
&\qquad + \left(-f^{(n)}(x)\frac{(b-x)^{n-1}}{(n-1)!} + f^{(n+1)}(x)\frac{(b-x)^n}{n!}\right) + \frac{d}{dx}R_n(b, x) \\
&= f^{(n+1)}(x)\frac{(b-x)^n}{n!} + \frac{d}{dx}R_n(b, x).
\end{aligned}$$

Theorem (Taylor's theorem). *Let U be an open interval in $\mathbf{R}$ and let the function $f: U \to \mathbf{R}$ be $(n+1)$ times differentiable. Then for any $a, b \in U$ we have*

$$\begin{aligned}
f(b) = f(a) &+ \frac{f'(a)}{1!}(b-a) + \frac{f''(a)}{2!}(b-a)^2 + \cdots \\
&+ \frac{f^{(n)}(a)}{n!}(b-a)^n + \frac{f^{(n+1)}(c)}{(n+1)!}(b-a)^{n+1},
\end{aligned}$$

where c is some number between a and b (or, if a and b are equal, $c = a$).

This is trivial if $a = b$, so assume that $a \neq b$. We need to show that

$$R_n(b, a) = \frac{f^{(n+1)}(c)}{(n+1)!}(b-a)^{n+1}$$

for some c between a and b. Since $a \neq b$ there is a unique real number K such that

$$R_n(b, a) = K\frac{(b-a)^{n+1}}{(n+1)!}.$$

The function $\varphi: U \to \mathbf{R}$ defined by

$$\varphi(x) = R_n(b, x) - K\frac{(b-x)^{n+1}}{(n+1)!}$$

for all $x \in U$ is differentiable. Furthermore $\varphi(a) = \varphi(b) = 0$. Thus the restriction of φ to the interval $[a, b]$ (or to the interval $[b, a]$ if $a > b$) satisfies the conditions of Rolle's theorem. Hence $\varphi'(c) = 0$ for some c between a and b. Since

$$\varphi'(x) = -f^{(n+1)}(x)\frac{(b-x)^n}{n!} + K\frac{(b-x)^n}{n!}$$

we have $K = f^{(n+1)}(c)$. Thus

$$R_n(b, a) = \frac{f^{(n+1)}(c)}{(n+1)!}(b-a)^{n+1},$$

as was to be proved.

Note that the case $n = 0$ of this theorem is essentially the mean value theorem. There is a little more generality here in that it is not assumed that $a < b$, but the assumption that f is differentiable on an open interval containing a and b is considerably more stringent than the analogous condition in the mean value theorem, where f was assumed differentiable only *between* a and b. However it is not difficult to get a somewhat more long-winded statement of Taylor's theorem which is an authentic generalization of the mean value theorem (see subsequent Problem 15).

The term "Taylor's theorem" we have attached to the above result is a convenient misnomer. Taylor's original statement was much weaker.

PROBLEMS

1. Discuss the differentiability of the function $f: \mathbf{R} \to \mathbf{R}$ if

(a) $f(x) = \begin{cases} x \sin \frac{1}{x} & \text{if } x \neq 0 \\ 0 & \text{if } x = 0 \end{cases}$

(assume the general properties of the sine function are known)

(b) $f(x) = \begin{cases} x^2 \sin \frac{1}{x} & \text{if } x \neq 0 \\ 0 & \text{if } x = 0 \end{cases}$

(c) $f(x) = \sqrt{|x|}$.

2. Let the real-valued function f on the open subset U of $\mathbf{R}$ be differentiable at the point $x_0 \in U$.

(a) Prove that $f'(x_0) = \lim_{h \to 0} \frac{f(x_0 + h) - f(x_0 - h)}{2h}$.

(b) If $\alpha, \beta \in \mathbf{R}$, compute $\lim_{h \to 0} \frac{f(x_0 + \alpha h) - f(x_0 + \beta h)}{h}$.

3. Here is a "proof" of the chain rule:

$$(g \circ f)'(x_0) = \lim_{x \to x_0} \frac{g(f(x)) - g(f(x_0))}{x - x_0} = \lim_{x \to x_0} \left(\frac{g(f(x)) - g(f(x_0))}{f(x) - f(x_0)} \cdot \frac{f(x) - f(x_0)}{x - x_0} \right)$$

$$= \lim_{x \to x_0} \frac{g(f(x)) - g(f(x_0))}{f(x) - f(x_0)} \cdot \lim_{x \to x_0} \frac{f(x) - f(x_0)}{x - x_0}$$

$$= g'(f(x_0)) f'(x_0).$$

(a) What is wrong with this "proof"?
(b) Alter this slightly into a correct proof.

4. Prove that if f is a differentiable real-valued function on an open interval in $\mathbf{R}$ then f is increasing (decreasing) if and only if f' is nonnegative (nonpositive) at each point of the interval.

5. Assuming the elementary properties of the trigonometric functions, show that $\tan x - x$ is strictly increasing on $(0, \pi/2)$, while the function $\frac{\sin x}{x}$ is strictly decreasing.

6. Prove that a differentiable real-valued function on $\mathbf{R}$ with bounded derivative is uniformly continuous.

7. Let $a, b \in \mathbf{R}$, $a < b$, and let f be a differentiable real-valued function on an open subset of $\mathbf{R}$ that contains $[a, b]$. Show that if γ is any real number between $f'(a)$ and $f'(b)$ then there exists a number $c \in (a, b)$ such that $\gamma = f'(c)$. (*Hint:* Combine the mean value theorem with the intermediate value theorem for the function $(f(x_1) - f(x_2))/(x_1 - x_2)$ on the set $\{(x_1, x_2) \in E^2 : a \leq x_1 < x_2 \leq b\}$.)

8. Let $a, b \in \mathbf{R}$, $a < b$, and let f, g be continuous real-valued functions on $[a, b]$ that are differentiable on (a, b). Prove that there exists a number $c \in (a, b)$ such that

$$f'(c)(g(b) - g(a)) = g'(c)(f(b) - f(a)).$$

(*Hint:* Consider the function

$$F(x) = (f(x) - f(a))(g(b) - g(a)) - (g(x) - g(a))(f(b) - f(a)).)$$

9. Use Problem 8 (Cauchy mean value theorem) to prove the following versions of L'Hospital's rule:

(a) Let U be an open interval in $\mathbf{R}$ and let f and g be differentiable real-valued functions on U, with g and g' nowhere zero on U. Let a be an extremity of U. Suppose that $\lim_{x \to a} f(x) = \lim_{x \to a} g(x) = 0$. Then

$$\lim_{x \to a} \frac{f(x)}{g(x)} = \lim_{x \to a} \frac{f'(x)}{g'(x)}$$

if the right-hand limit exists.

(b) Same as (a), except that it is assumed that

$$\lim_{x \to a} \frac{1}{f(x)} = \lim_{x \to a} \frac{1}{g(x)} = 0.$$

(c) Same as (a), except that $U = \{x \in \mathbf{R} : x > \alpha\}$ for some $\alpha \in \mathbf{R}$ and a is replaced by the symbol $+\infty$ (cf. Prob. 8, Chap. IV).

(d) Same as (c), except that it is assumed that

$$\lim_{x\to+\infty}\frac{1}{f(x)} = \lim_{x\to+\infty}\frac{1}{g(x)} = 0.$$

10. State precisely and prove: An n times differentiable function of an n times differentiable function is n times differentiable.

11. Let f be a real-valued function on an open subset U of $\mathbf{R}$ that is twice differentiable at $x_0 \in U$. Show that if $f'(x_0) = 0$ and $f''(x_0) < 0$ $\bigl(f''(x_0) > 0\bigr)$ then the restriction of f to some open ball of center x_0 attains a maximum (minimum) at x_0.

12. A real-valued function on an open interval in $\mathbf{R}$ is called *convex* if no point on the line segment between any two points of its graph lies below the graph. If the function is differentiable, this condition is known to be equivalent to the condition that no point of the graph lie below any point of any tangent to the graph. If the function is twice differentiable the condition is known to be equivalent to the second derivative of the function being nonnegative at all points. State these conditions in precise analytic terms and prove them.

13. Use Problem 9(a) to show that if f is a real-valued function on an open subset U of $\mathbf{R}$ that is n times differentiable at the point $x_0 \in U$ then

$$\lim_{h\to 0}\frac{f(x_0+h) - f(x_0) - f'(x_0)\dfrac{h}{1!} - f''(x_0)\dfrac{h^2}{2!} - \cdots - f^{(n-1)}(x_0)\dfrac{h^{n-1}}{(n-1)!}}{h^n} = \frac{f^{(n)}(x_0)}{n!}.$$

14. Use Taylor's theorem to prove the "binomial theorem" for positive integral exponent n:

$$(a+x)^n = a^n + na^{n-1}x + \frac{n(n-1)}{2}a^{n-2}x^2 + \frac{n(n-1)(n-2)}{2\cdot 3}a^{n-3}x^3 + \cdots + x^n.$$

15. Show that Taylor's theorem may be strengthened as follows: Let f be a continuous real-valued function on the closed interval in $\mathbf{R}$ of extremities a and b that is $(n+1)$ times differentiable on the open interval with these same extremities and suppose that $\lim_{x\to a} f'(x), \lim_{x\to a} f''(x), \ldots, \lim_{x\to a} f^{(n)}(x)$ exist and that $f', f'', \ldots, f^{(n)}$ are bounded. Then

$$f(b) = f(a) + \left(\lim_{x\to a} f'(x)\right)\frac{(b-a)}{1!} + \cdots + \left(\lim_{x\to a} f^{(n)}(x)\right)\frac{(b-a)^n}{n!} + f^{(n+1)}(c)\frac{(b-a)^{n+1}}{(n+1)!}$$

for some c between a and b.

CHAPTER VI

Riemann Integration

We discuss in this chapter the definition and basic properties of the Riemann integral for real-valued functions of one real variable. The integration of functions of several real variables will be discussed in the last chapter, together with some finer points of the one-variable case. Here we are concerned only with the simplest results, up to the integrability of a continuous function and the fundamental theorem of calculus. The details of the proofs will be the only essentially new material for most students. In the last section we apply our results by giving a rigorous treatment of the logarithmic and exponential functions.

§ 1. DEFINITIONS AND EXAMPLES.

Definition. Let $a, b \in \mathbf{R}$, $a < b$. By a *partition of the closed interval* $[a, b]$ is meant a finite sequence of numbers $x_0, x_1, \ldots, x_N$ such that $a = x_0 < x_1 < \cdots < x_N = b$. The *width* of this partition is defined to be

$$\max \{x_i - x_{i-1} : i = 1, 2, \ldots, N\}.$$

If f is a real-valued function on $[a, b]$, by a *Riemann sum for f corresponding to the given partition* is meant a sum

$$\sum_{i=1}^{N} f(x_i')(x_i - x_{i-1}),$$

where $x_{i-1} \leq x_i' \leq x_i$ for each $i = 1, 2, \ldots, N$.

Thus, given any function f: $[a, b] \rightarrow \mathbf{R}$ and a partition $x_0, x_1, \ldots, x_N$ of $[a, b]$, there are lots of Riemann sums for f corresponding to this partition, depending on the choice of $x_1', x_2', \ldots, x_N'$. In the special case where $f(x) \geq 0$ for each $x \in [a, b]$, each Riemann sum can be considered an approximation for the "area under the curve $y = f(x)$ between a and b", that is, the "area bounded by the x-axis, the graph of f, and the lines $x = a$ and $x = b$", as illustrated in Figure 27. However this geometric interpretation must not be overworked for at least two reasons. First of all we do not want to restrict ourselves to functions that are positive. Second, our arguments must have validity independent of geometric intuition. But the geometric interpretation does make the following definition reasonable.

Definition. Let $a, b \in \mathbf{R}$, $a < b$, and let f be a real-valued function on $[a, b]$. We say that f is *Riemann integrable on* $[a, b]$ if there exists a number $A \in \mathbf{R}$ such that, for any $\epsilon > 0$, there exists a $\delta > 0$ such that $|S - A| < \epsilon$ whenever S is a Riemann sum for f corresponding to any partition of $[a, b]$ of width less than δ. In this case A is called the *Riemann integral of f between a and b* and it is denoted $\int_a^b f(x)dx$.

It makes sense to speak of *the* Riemann integral of f between a and b since A is unique, by the usual argument: If A, A' are Riemann integrals of f between a and b then given any $\epsilon > 0$ there exists a $\delta > 0$ such that $|S - A|$, $|S - A'| < \epsilon$ whenever S is a Riemann sum for f corresponding to any partition of $[a, b]$ of width less than δ. There are partitions of $[a, b]$ of width less than any prescribed positive number since, for example, the partition by N equal subdivisions (with $x_i = a + i(b - a)/N$ for $i = 0$,

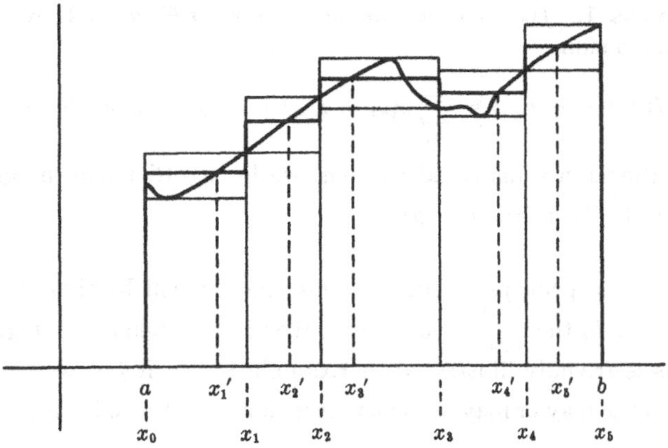

FIGURE 27. Area under a curve approximated by a Riemann sum. The indicated choice of $x_1', x_2', \ldots, x_5'$ gives a certain Riemann sum corresponding to the partition $x_0, x_1, \ldots, x_5$, and this sum can be considered an approximation of the area under the curve. The maximum (minimum) value of the Riemann sums for f corresponding to the given partition is given by the sum of the areas of the tallest (shortest) rectangles in the figure of bases $[x_0, x_1]$, $[x_1, x_2]$, $\ldots$, $[x_4, x_5]$, and the "true" "area under the curve" must lie between these latter extremes, as does our original Riemann sum. Thus the error in making our original approximation to the area under the curve is at most the total of the differences in area between the tallest and the shortest rectangles. It seems reasonable that if we divide $[a, b]$ into more and more pieces of widths approaching zero, then all our Riemann sums will approach a certain definite limit, the true "area under the curve". (Of course the only way to make this rigorous is to use this or another procedure to *define* the notion "area under a curve". For a specific curve the latter notion need not exist, just as limits do not always exist.)

$1, \ldots, N)$ has width $(b - a)/N$, which is small if N is large. Hence we can actually find a Riemann sum S for f corresponding to a partition of $[a, b]$ of width less than δ, so that the two inequalities $|S - A| < \epsilon$, $|S - A'| < \epsilon$ hold. Hence $|A - A'| < 2\epsilon$. Since ϵ was an arbitrary positive number we must have $|A - A'| = 0$, or $A = A'$.

Note the use of x in $\int_a^b f(x)dx$ as a "dummy variable"; we could equally well have written $\int_a^b f(t)dt$, or $\int_a^b f(u)du$.

We follow the usual convention of saying that f is, or is not, Riemann integrable on $[a, b]$, and in the former case writing $\int_a^b f(x)dx$, even if f is a function defined on a larger set than $[a, b]$, by implicitly replacing f by its restriction to $[a, b]$.

EXAMPLE 1. $f(x) = c$, a constant, for all $x \in [a, b]$. Here we have any Riemann sum

$$\sum_{i=1}^{N} f(x_i')(x_i - x_{i-1}) = \sum_{i=1}^{N} c(x_i - x_{i-1}) = c(x_N - x_0) = c(b - a).$$

Since all Riemann sums equal $c(b - a)$ we have f Riemann integrable on $[a, b]$, with $\int_a^b f(x)dx = c(b - a)$.

One of the principal results of this chapter will be that if f is continuous on $[a, b]$ then $\int_a^b f(x)dx$ exists, that is f is Riemann integrable on $[a, b]$; this is trivially illustrated in Example 1. But if f is not continuous then $\int_a^b f(x)dx$ may or may not exist, as is shown by the following examples.

EXAMPLE 2. Let ξ be a fixed point of $[a, b]$, let $c \in \mathbf{R}$, and set

$$f(x) = \begin{cases} 0 & \text{if } x \neq \xi \\ c & \text{if } x = \xi. \end{cases}$$

For any Riemann sum S corresponding to a partition of $[a, b]$ of width less than δ we have $|S| < 2|c|\delta$ (the coefficient 2 appearing since ξ may be one of the partition points x_i and we may in this case have both x_i' and x_{i+1}' equal to c.) So clearly $\int_a^b f(x)dx = 0$.

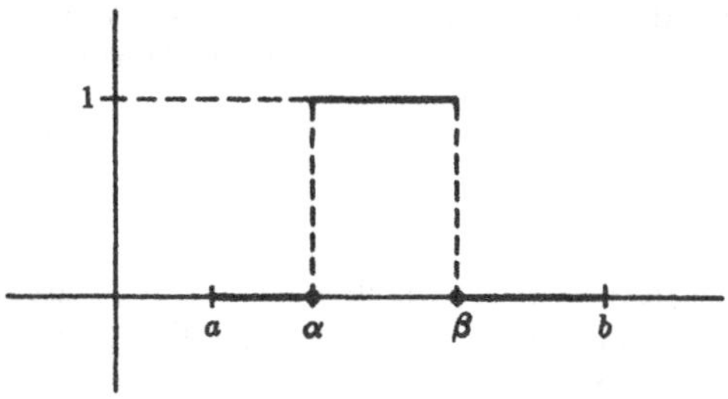

FIGURE 28. Graph of the function of Example 3.

EXAMPLE 3. Let $\alpha, \beta \in [a, b]$ with $\alpha < \beta$. Let $f\colon [a, b] \to \mathbf{R}$ be defined by

$$f(x) = \begin{cases} 1 & \text{if } x \in (\alpha, \beta) \\ 0 & \text{if } x \in [a, b],\ x \notin (\alpha, \beta). \end{cases}$$

Let $x_0, x_1, \ldots, x_N$ be a partition of $[a, b]$ of width less than δ and consider a Riemann sum for f corresponding to this partition, say

$$S = \sum_{i=1}^{N} f(x_i')(x_i - x_{i-1}),$$

where $x_{i-1} \leq x_i' \leq x_i$ for $i = 1, 2, \ldots, N$. Since $f(x_i')$ is 1 or 0 according as the point x_i' is in the open interval (α, β) or not, we have

$$S = \sum{}^{*} (x_i - x_{i-1}),$$

the asterisk indicating that we include in the sum only those i for which $x_i' \in (\alpha, \beta)$. Now choose p, q from among $1, 2, \ldots, N$ such that

$$x_{p-1} \leq \alpha < x_p, \quad x_{q-1} < \beta \leq x_q.$$

Then $x_i' \in (\alpha, \beta)$ if $p + 1 \leq i \leq q - 1$ and $x_i' \notin (\alpha, \beta)$ if $i < p$ or $i > q$. Therefore

$$\sum_{p+1 \leq i \leq q-1} (x_i - x_{i-1}) \leq S \leq \sum_{p \leq i \leq q} (x_i - x_{i-1}).$$

By the choice of p and q, $\beta - \alpha \leq x_q - x_{p-1} < (q - p + 1)\delta$, so that if δ is sufficiently small we must have $p + 1 \leq q - 1$, in which case the last inequality simplifies to

$$x_{q-1} - x_p \leq S \leq x_q - x_{p-1}.$$

Therefore

$$(x_{q-1} - \beta) - (x_p - \alpha) \leq S - (\beta - \alpha) \leq (x_q - \beta) - (x_{p-1} - \alpha).$$

Since the partition has width less than δ, each of the quantities $x_{q-1} - \beta$, $x_p - \alpha$, $x_q - \beta$, $x_{p-1} - \alpha$ is of absolute value less than δ. Therefore

$$|S - (\beta - \alpha)| < 2\delta.$$

Since δ was an arbitrarily small positive number we conclude that f is Riemann integrable on $[a, b]$ and that $\int_a^b f(x)dx = \beta - \alpha$.

EXAMPLE 4. Define $f: [a, b] \to \mathbf{R}$ by setting $f(x) = 1$ if x is rational, otherwise $f(x) = 0$. (This is the restriction to $[a, b]$ of Example 6, page 70.) Any interval in $\mathbf{R}$ is known to contain both points that are rational and points that are not. Hence for any partition $x_0, x_1, \ldots, x_N$ of $[a, b]$ we can choose the x_i''s to be either all rational, or all not, in which case the Riemann sums are respectively $b - a$ and 0. That is, $b - a$ and 0 are Riemann sums for f corresponding to any partition of $[a, b]$, no matter what the width. It is clear that f is not Riemann integrable on $[a, b]$.

In the future, for brevity, we shall say that a function is *integrable* on a closed interval, rather than *Riemann integrable,* and speak of its *integral* instead of its *Riemann integral.* It should be borne in mind however that there are other integration processes than that of Riemann, and for these other integration processes our results may or may not be true. For example the most commonly used integral after that of Riemann is that of Lebesgue. A given real-valued function on $[a, b]$ may or may not be Lebesgue integrable. If it is then its Lebesgue integral is a certain real number. If a function is Riemann integrable then it is also Lebesgue integrable and the two integrals are the same (hence can be denoted by the same symbol $\int_a^b f(x)dx$). But many functions that are not Riemann integrable are Lebesgue integrable, so the Lebesgue integral can be of greater use. For example, the function of Example 4 above is Lebesgue integrable; as a matter of fact its Lebesgue integral is zero, in line with the fact that in some sense the points of the interval $[a, b]$ that are rational are relatively few in comparison with those that are not. We repeat for emphasis that from now on integrable means Riemann integrable, integral means Riemann integral.

§ 2. LINEARITY AND ORDER PROPERTIES OF THE INTEGRAL.

Proposition. *Riemann integration has the following properties:*

(*1*) *If f and g are integrable real-valued functions on the interval $[a, b]$ then $f + g$ is integrable on $[a, b]$ and*

$$\int_a^b (f(x) + g(x))dx = \int_a^b f(x)dx + \int_a^b g(x)dx.$$

(*2*) *If f is an integrable real-valued function on the interval $[a, b]$ and $c \in \mathbf{R}$ then cf is integrable on $[a, b]$ and*

$$\int_a^b cf(x)dx = c\int_a^b f(x)dx.$$

These facts are easily proved by looking at the various Riemann sums, as follows. Given any $\epsilon > 0$ there are numbers $\delta_1, \delta_2 > 0$ such that if S_1, S_2 are any Riemann sums for f, g respectively corresponding to partitions of $[a, b]$ of widths less than δ_1, δ_2 respectively, then

$$\left|S_1 - \int_a^b f(x)dx\right| < \frac{\epsilon}{2}, \quad \left|S_2 - \int_a^b g(x)dx\right| < \frac{\epsilon}{2}.$$

Hence if $x_0, x_1, \ldots, x_N$ is any partition of $[a, b]$ of width less than $\min\{\delta_1, \delta_2\}$ and if $x_1', \ldots, x_N'$ are such that $x_{i-1} \le x_i' \le x_i$ for $i = 1, \ldots, N$, then

$$\begin{aligned}\Big|\sum_{i=1}^N (f(x_i') + g(x_i'))(x_i - x_{i-1}) - \Big(\int_a^b f(x)dx + \int_a^b g(x)dx\Big)\Big| \\ = \Big|\Big(\sum_{i=1}^N f(x_i')(x_i - x_{i-1}) - \int_a^b f(x)dx\Big) \\ + \Big(\sum_{i=1}^N g(x_i')(x_i - x_{i-1}) - \int_a^b g(x)dx\Big)\Big|\end{aligned}$$

$$\leq \left| \sum_{i=1}^{N} f(x_i')(x_i - x_{i-1}) - \int_a^b f(x)dx \right| + \left| \sum_{i=1}^{N} g(x_i')(x_i - x_{i-1}) - \int_a^b g(x)dx \right|$$

$$< \frac{\epsilon}{2} + \frac{\epsilon}{2} = \epsilon.$$

This proves part (1). For part (2), given any $\epsilon > 0$ there is a number $\delta > 0$ such that for any Riemann sum S for f corresponding to any partition of $[a, b]$ of width less than δ we have $\left| S - \int_a^b f(x)dx \right| < \epsilon/|c|$ (it is permissible to restrict our attention to the case $c \neq 0$ if we note that the case $c = 0$ is a triviality). Then if $x_0, x_1, \ldots, x_N$ is any partition of $[a, b]$ of width less than δ and $x_i' \in [x_{i-1}, x_i]$ for $i = 1, \ldots, N$ we have

$$\left| \sum_{i=1}^{N} cf(x_i')(x_i - x_{i-1}) - c\int_a^b f(x)dx \right| = |c| \cdot \left| \sum_{i=1}^{N} f(x_i')(x_i - x_{i-1}) - \int_a^b f(x)dx \right| < |c| \cdot \frac{\epsilon}{|c|} = \epsilon,$$

finishing the proof.

An immediate consequence of the proposition is that (under the hypotheses of part (1))

$$\int_a^b (f(x) - g(x))dx = \int_a^b f(x)dx - \int_a^b g(x)dx.$$

This comes from applying part (1) to the functions f and $-g$, the latter being integrable by part (2), with $c = -1$.

Proposition. *If f is an integrable real-valued function on the interval $[a, b]$ and $f(x) \geq 0$ for all $x \in [a, b]$, then*

$$\int_a^b f(x)dx \geq 0.$$

For if we are given any $\epsilon > 0$ we may find a Riemann sum S for f on $[a, b]$ such that $\left| S - \int_a^b f(x)dx \right| < \epsilon$. Then $\int_a^b f(x)dx \geq S - \epsilon$. Clearly $S \geq 0$, so that $\int_a^b f(x)dx \geq -\epsilon$. This being true for all $\epsilon > 0$, we have $\int_a^b f(x)dx \geq 0$.

Corollary 1. *If f and g are integrable real-valued functions on the interval $[a, b]$ and $f(x) \leq g(x)$ for all $x \in [a, b]$, then*

$$\int_a^b f(x)dx \leq \int_a^b g(x)dx.$$

For $\int_a^b g(x)dx - \int_a^b f(x)dx = \int_a^b (g(x) - f(x))dx \geq 0$.

Corollary 2. *If f is an integrable real-valued function on the interval $[a, b]$ and $m, M \in \mathbf{R}$ are such that $m \leq f(x) \leq M$ for all $x \in [a, b]$, then*

$$m(b-a) \leq \int_a^b f(x)dx \leq M(b-a).$$

For $\int_a^b m dx \leq \int_a^b f(x)dx \leq \int_a^b M dx$, and we know that for any constant c we have $\int_a^b c dx = c(b-a)$.

§ 3. EXISTENCE OF THE INTEGRAL.

Lemma 1. *A real-valued function f on the interval $[a, b]$ is integrable on $[a, b]$ if and only if, given any $\epsilon > 0$, there exists a number $\delta > 0$ such that $|S_1 - S_2| < \epsilon$ whenever S_1 and S_2 are Riemann sums for f corresponding to partitions of $[a, b]$ of width less than δ.*

First suppose f integrable on $[a, b]$. Then given any $\epsilon > 0$ there is a $\delta > 0$ such that $\left|S - \int_a^b f(x)dx\right| < \epsilon/2$ whenever S is a Riemann sum for f corresponding to a partition of $[a, b]$ of width less than δ. If S_1 and S_2 are two such Riemann sums then

$$\begin{aligned} |S_1 - S_2| &= \left|\left(S_1 - \int_a^b f(x)dx\right) - \left(S_2 - \int_a^b f(x)dx\right)\right| \\ &\leq \left|S_1 - \int_a^b f(x)dx\right| + \left|S_2 - \int_a^b f(x)dx\right| < \frac{\epsilon}{2} + \frac{\epsilon}{2} = \epsilon. \end{aligned}$$

This proves the "only if" part of the lemma.

Conversely, assume the hypothesis of the "if" part of the lemma. For $n = 1, 2, 3, \ldots$ choose any partition of $[a, b]$ of width less than $1/n$ and a Riemann sum $S^{(n)}$ for f corresponding to this partition. Then $S^{(1)}, S^{(2)}, S^{(3)}, \ldots$ is a Cauchy sequence of real numbers. (For, by assumption, for every $\epsilon > 0$ there exists a $\delta > 0$ such that $|S_1 - S_2| < \epsilon$ whenever S_1 and S_2 are Riemann sums for f corresponding to partitions of $[a, b]$ of width less than δ, so if we choose an integer N such that $1/N < \delta$ then we have $|S^{(n)} - S^{(m)}| < \epsilon$ whenever $n, m > N$.) Since $\mathbf{R}$ is complete, the sequence $S^{(1)}, S^{(2)}, S^{(3)}, \ldots$ converges. Let its limit be A. Given any $\epsilon > 0$ now choose $\delta > 0$ such that $|S_1 - S_2| < \epsilon/2$ whenever S_1 and S_2 are Riemann sums for f corresponding to partitions of $[a, b]$ of width less than δ, and choose an integer N such that $|S^{(N)} - A| < \epsilon/2$ and $1/N < \delta$. Then for any Riemann sum S for f corresponding to a partition of $[a, b]$ of width less than δ we have

$$\begin{aligned} |S - A| = |(S - S^{(N)}) + (S^{(N)} - A)| &\leq |S - S^{(N)}| + |S^{(N)} - A| \\ &< \frac{\epsilon}{2} + \frac{\epsilon}{2} = \epsilon. \end{aligned}$$

Thus f is integrable on $[a, b]$.

At this point it would be easy to use Lemma 1 to give a direct proof of the main result of this section, the integrability of continuous functions. However we postpone this result because it is natural to wonder what integrable functions are like in general, and the next lemma and proposition will give us a better intuitive understanding of the situation.

Definition. A real-valued function f on the interval $[a, b]$ is called a *step function* if there exists a partition $x_0, x_1, \ldots, x_N$ of $[a, b]$ such that f is constant on each open subinterval $(x_0, x_1), (x_1, x_2), \ldots, (x_{N-1}, x_N)$.

For example the functions of Examples 1, 2, and 3 of § 1 are step functions. A step function of more general appearance is indicated in Figure 29.

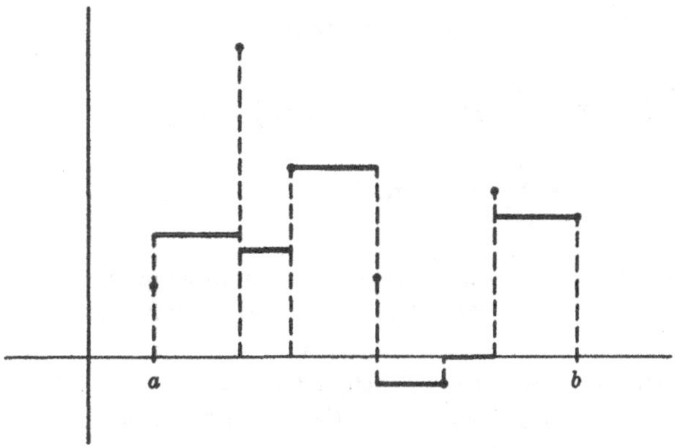

FIGURE 29. Graph of a step function on $[a, b]$.

Lemma 2. *A step function is integrable. In particular, if $x_0, x_1, \ldots, x_N$ is a partition of the interval $[a, b]$, if $c_1, \ldots, c_N \in \mathbf{R}$ and if f: $[a, b] \to \mathbf{R}$ is such that $f(x) = c_i$ if $x_{i-1} < x < x_i$ for $i = 1, \ldots, N$, then*

$$\int_a^b f(x)dx = \sum_{i=1}^{N} c_i(x_i - x_{i-1}).$$

Note that the values of $f(x_0), f(x_1), \ldots, f(x_N)$ have no effect on the integral. It is convenient to place this lemma here, but its proof could have been given much earlier, immediately after the definition of integral in § 1. However it is most simple to base the proof on Examples 2 and 3 of § 1 and the first proposition of § 2, as follows:

For $i = 1, 2, \ldots, N$ define $\varphi_i\colon [a, b] \to \mathbf{R}$ by

$$\varphi_i(x) = \begin{cases} 1 & \text{if } x \in (x_{i-1}, x_i) \\ 0 & \text{if } x \in [a, b],\ x \notin (x_{i-1}, x_i). \end{cases}$$

Then $f - \sum_{i=1}^{N} c_i\varphi_i$ is a function on $[a, b]$ that takes on the value zero at all points except possibly the points $x_0, x_1, \ldots, x_N$; hence it is the sum of a finite number of functions of the type of Example 2 of § 1. By Example 2, together with the linearity of the integral, we have

$$\int_a^b \left(f(x) - \sum_{i=1}^{N} c_i\varphi_i(x)\right)dx = 0.$$

But each φ_i is a function of the type of Example 3 of § 1, hence is integrable, with $\int_a^b \varphi_i = x_i - x_{i-1}$. Again using the linearity of the integral we get $f = \left(f - \sum_{i=1}^{N} c_i\varphi_i\right) + \sum_{i=1}^{N} c_i\varphi_i$ integrable and

$$\begin{aligned}\int_a^b f(x)dx &= \int_a^b \left(f(x) - \sum_{i=1}^{N} c_i\varphi_i(x)\right)dx + \sum_{i=1}^{N} c_i \int_a^b \varphi_i(x)dx \\ &= \sum_{i=1}^{N} c_i(x_i - x_{i-1}).\end{aligned}$$

For an illustration of the situation of the following proposition, where a function is sandwiched between two step functions, see Figure 27 (page 113).

Proposition. *The real-valued function f on the interval $[a, b]$ is integrable on $[a, b]$ if and only if, for each $\epsilon > 0$, there exist step functions f_1, f_2 on $[a, b]$ such that*

$$f_1(x) \le f(x) \le f_2(x) \quad \textit{for each } x \in [a, b]$$

and

$$\int_a^b (f_2(x) - f_1(x))dx < \epsilon.$$

We first prove that if the given condition holds then f is integrable. We use the criterion of Lemma 1. Given $\epsilon > 0$ we have to produce a $\delta > 0$ such that if S_1, S_2 are Riemann sums for f corresponding to partitions of $[a, b]$ of width less than δ then $|S_1 - S_2| < \epsilon$. Use the hypothesis to find step functions f_1, f_2 on $[a, b]$ such that

$$f_1(x) \le f(x) \le f_2(x) \quad \text{for all } x \in [a, b]$$

and

$$\int_a^b (f_2(x) - f_1(x))dx < \frac{\epsilon}{3}.$$

Since f_1, f_2 are integrable on $[a, b]$ we can find a $\delta > 0$ such that any Riemann sum for f_1 (or f_2) corresponding to any partition of $[a, b]$ of width less than δ differs in absolute value from $\int_a^b f_1(x)dx$ $\left(\text{or } \int_a^b f_2(x)dx\right)$ by less than $\epsilon/3$. Now let S be any Riemann sum for f corresponding to a partition of $[a, b]$ of width less than δ; say that $x_0, x_1, \ldots, x_N$ is this partition and that $S = \sum_{i=1}^{N} f(x_i')(x_i - x_{i-1})$, where $x_{i-1} \leq x_i' \leq x_i$ for each $i = 1, \ldots, N$. Then since

$$f_1(x) \leq f(x) \leq f_2(x) \quad \text{for } x \in [a, b]$$

we have

$$\sum_{i=1}^{N} f_1(x_i')(x_i - x_{i-1}) \leq S \leq \sum_{i=1}^{N} f_2(x_i')(x_i - x_{i-1}).$$

By our choice of δ we have

$$\left| \sum_{i=1}^{N} f_1(x_i')(x_i - x_{i-1}) - \int_a^b f_1(x)dx \right| < \frac{\epsilon}{3}$$

and

$$\left| \sum_{i=1}^{N} f_2(x_i')(x_i - x_{i-1}) - \int_a^b f_2(x)dx \right| < \frac{\epsilon}{3},$$

implying

$$\int_a^b f_1(x)dx - \frac{\epsilon}{3} < S < \int_a^b f_2(x)dx + \frac{\epsilon}{3}.$$

Thus S belongs to a fixed open interval

$$\left(\int_a^b f_1(x)dx - \frac{\epsilon}{3}, \int_a^b f_2(x)dx + \frac{\epsilon}{3} \right)$$

of length

$$\frac{\epsilon}{3} + \int_a^b (f_2(x) - f_1(x))dx + \frac{\epsilon}{3} < \frac{\epsilon}{3} + \frac{\epsilon}{3} + \frac{\epsilon}{3} = \epsilon.$$

If S_1, S_2 are two Riemann sums for f corresponding to partitions of $[a, b]$ of width less than δ, then each S_1, S_2 will belong to the above interval, so that $|S_1 - S_2| < \epsilon$. This is what we wished to show, so half the proposition is proved.

To prove the remaining half of the proposition, we start with an integrable function f: $[a, b] \to \mathbf{R}$ and a fixed $\epsilon > 0$ and we have to produce step functions f_1, f_2 with the desired properties. Using Lemma 1, we can find a partition $x_0, x_1, \ldots, x_N$ of $[a, b]$ such that any two Riemann sums for f corresponding to this partition differ in absolute value by less than ϵ', where ϵ' is some arbitrary fixed positive number less than ϵ. That is, for arbitrary $x_i', x_i'' \in [x_{i-1}, x_i]$, $i = 1, \ldots, N$, we have

$$\left| \sum_{i=1}^{N} (f(x_i') - f(x_i''))(x_i - x_{i-1}) \right| < \epsilon'.$$

If we apply this inequality to the special case where, for some fixed index $j = 1, \ldots, N$, we have $x_i' = x_i''$ if $i \neq j$ and $x_j'' = x_j$, we get

$$|(f(x_j') - f(x_j))(x_j - x_{j-1})| < \epsilon',$$

implying

$$|f(x_j')| < \frac{\epsilon'}{x_j - x_{j-1}} + |f(x_j)|.$$

This last inequality holds for all $x_j' \in [x_{j-1}, x_j]$. Thus f is bounded on $[x_{j-1}, x_j]$. Therefore f is bounded on all of $[a, b]$. Thus for $i = 1, \ldots, N$ we can define

$$m_i = \text{g.l.b. } \{f(x_i') : x_i' \in [x_{i-1}, x_i]\}$$

and

$$M_i = \text{l.u.b. } \{f(x_i') : x_i' \in [x_{i-1}, x_i]\}$$

and we can define step functions f_1, f_2 on $[a, b]$ by

$$f_1(x) = \begin{cases} m_i & \text{if } x_{i-1} < x < x_i,\ i = 1, \ldots, N \\ \min\{m_1, \ldots, m_N\} & \text{if } x = x_i,\ i = 0, 1, \ldots, N \end{cases}$$

$$f_2(x) = \begin{cases} M_i & \text{if } x_{i-1} < x < x_i,\ i = 1, \ldots, N \\ \max\{M_1, \ldots, M_N\} & \text{if } x = x_i,\ i = 0, 1, \ldots, N. \end{cases}$$

Clearly $f_1(x) \leq f(x) \leq f_2(x)$ for all $x \in [a, b]$, and the proof will be complete if we can show that $\int_a^b (f_2(x) - f_1(x))dx < \epsilon$. To do this, for any real number $\eta > 0$ find specific $x_i', x_i'' \in [x_{i-1}, x_i]$, $i = 1, \ldots, N$, so that

$$f(x_i') < m_i + \eta, \quad f(x_i'') > M_i - \eta.$$

Then

$$\begin{aligned}\sum_{i=1}^{N} (f(x_i'') - f(x_i'))(x_i - x_{i-1}) &> \sum_{i=1}^{N} (M_i - m_i - 2\eta)(x_i - x_{i-1}) \\ &= \int_a^b (f_2(x) - f_1(x))dx - 2\eta(b - a).\end{aligned}$$

Since

$$\left|\sum_{i=1}^{N} (f(x_i') - f(x_i''))(x_i - x_{i-1})\right| < \epsilon'$$

we have

$$\int_a^b (f_2(x) - f_1(x))dx - 2\eta(b - a) < \epsilon',$$

or

$$\int_a^b (f_2(x) - f_1(x))dx < \epsilon' + 2\eta(b - a).$$

Since η was *any* positive number, we have

$$\int_a^b (f_2(x) - f_1(x))dx \leq \epsilon' < \epsilon,$$

and the proof is complete.

The following result, which occurred in the course of the proof of the proposition, is a trivial consequence of the proposition itself.

Corollary. *If the real-valued function f on $[a, b]$ is integrable on $[a, b]$, then it is bounded on $[a, b]$.*

Theorem. *If f is a continuous real-valued function on the interval $[a, b]$ then $\int_a^b f(x)dx$ exists.*

We shall prove this theorem by showing that the criterion of the preceding proposition obtains. Since f is uniformly continuous on $[a, b]$, given any $\epsilon > 0$ we can find a $\delta > 0$ such that whenever $x', x'' \in [a, b]$ and $|x' - x''| < \delta$ then $|f(x') - f(x'')| < \epsilon/(b - a)$. Choose any partition $x_0, x_1, \ldots, x_N$ of $[a, b]$ of width less than δ. For each $i = 1, \ldots, N$ choose $x_i', x_i'' \in [x_{i-1}, x_i]$ such that the restriction of f to $[x_{i-1}, x_i]$ attains a minimum at x_i' and a maximum at x_i''. Define step functions f_1, f_2 on $[a, b]$ by

$$f_1(x) = \begin{cases} f(x_i') & \text{if } x_{i-1} < x < x_i,\ i = 1, \ldots, N \\ f(x) & \text{if } x = x_i,\ i = 0, 1, \ldots, N. \end{cases}$$

$$f_2(x) = \begin{cases} f(x_i'') & \text{if } x_{i-1} < x < x_i,\ i = 1, \ldots, N \\ f(x) & \text{if } x = x_i,\ i = 0, 1, \ldots, N. \end{cases}$$

Then $f_1(x) \leq f(x) \leq f_2(x)$ for all $x \in [a, b]$. Furthermore for each $i = 1, \ldots, N$ we have $|x_i' - x_i''| \leq x_i - x_{i-1} < \delta$, so that $|f(x_i') - f(x_i'')| < \epsilon/(b - a)$ and therefore $f_2(x) - f_1(x) < \epsilon/(b - a)$ for all $x \in [a, b]$. Therefore

$$\int_a^b (f_2(x) - f_1(x))dx \leq \max \{f_2(x) - f_1(x) : x \in [a, b]\} \cdot (b - a)$$

$$< \frac{\epsilon}{b - a} \cdot (b - a) = \epsilon.$$

Thus the criterion of the last proposition is satisfied.

§ 4. THE FUNDAMENTAL THEOREM OF CALCULUS.

Proposition. *Let $a, b, c \in \mathbf{R}$, $a < b < c$, and let f be a real-valued function on $[a, c]$. Then f is integrable on $[a, c]$ if and only if it is integrable on both $[a, b]$ and $[b, c]$, in which case*

$$\int_a^b f(x)dx + \int_b^c f(x)dx = \int_a^c f(x)dx.$$

It is convenient to use the proposition of the preceding section. If f is integrable on both $[a, b]$ and $[b, c]$, then for any $\epsilon > 0$ we can find step functions h_1 and h_2 on $[a, b]$ and k_1 and k_2 on $[b, c]$ such that

$$h_1(x) \leq f(x) \leq h_2(x) \quad \text{for each } x \in [a, b]$$

$$k_1(x) \leq f(x) \leq k_2(x) \quad \text{for each } x \in [b, c]$$

with $\int_a^b (h_2(x) - h_1(x))dx$, $\int_b^c (k_2(x) - k_1(x))dx$ each less than $\epsilon/2$.

Define functions f_1, f_2 on $[a, c]$ by

$$f_1(x) = \begin{cases} h_1(x) & \text{if } a \leq x \leq b \\ k_1(x) & \text{if } b < x \leq c \end{cases}$$

$$f_2(x) = \begin{cases} h_2(x) & \text{if } a \leq x \leq b \\ k_2(x) & \text{if } b < x \leq c. \end{cases}$$

Then f_1, f_2 are step functions,

$$f_1(x) \leq f(x) \leq f_2(x) \quad \text{for all } x \in [a, c],$$

and

$$\int_a^b (f_2(x) - f_1(x))dx < \frac{\epsilon}{2}, \quad \int_b^c (f_2(x) - f_1(x))dx < \frac{\epsilon}{2}.$$

Now $f_2 - f_1$ is a step function on $[a, c]$ and the proposition is clearly true for step functions, so that

$$\int_a^c (f_2(x) - f_1(x))dx = \int_a^b (f_2(x) - f_1(x))dx + \int_b^c (f_2(x) - f_1(x))dx.$$

Therefore

$$\int_a^c (f_2(x) - f_1(x))dx < \frac{\epsilon}{2} + \frac{\epsilon}{2} = \epsilon.$$

This shows that f is integrable on $[a, c]$. Conversely, if f is integrable on $[a, c]$ then for any $\epsilon > 0$ there are step functions f_1, f_2 on $[a, c]$ such that

$$f_1(x) \leq f(x) \leq f_2(x) \text{ for all } x \in [a, c]$$

and

$$\int_a^c (f_2(x) - f_1(x))dx < \epsilon.$$

Since

$$\int_a^b (f_2(x) - f_1(x))dx \geq 0, \quad \int_b^c (f_2(x) - f_1(x))dx \geq 0$$

and

$$\int_a^c (f_2(x) - f_1(x))dx = \int_a^b (f_2(x) - f_1(x))dx + \int_b^c (f_2(x) - f_1(x))dx$$

we have

$$\int_a^b (f_2(x) - f_1(x))dx < \epsilon, \quad \int_b^c (f_2(x) - f_1(x))dx < \epsilon.$$

Thus, again by the proposition of the last section, f is integrable on both $[a, b]$ and $[b, c]$. To complete the proof, suppose f integrable on $[a, b]$, $[b, c]$ and $[a, c]$. Given $\epsilon > 0$ we can find $\delta > 0$ such that any Riemann sum for f corresponding to partitions of $[a, b]$, $[b, c]$ and $[a, c]$ of width less than δ differ in absolute value from $\int_a^b f(x)dx$, $\int_b^c f(x)dx$ and $\int_a^c f(x)dx$ respectively by less than $\epsilon/3$. Take partitions of $[a, b]$ and of $[b, c]$ of width less than δ and Riemann sums S_1, S_2 for f corresponding to these partitions. Then $S_1 + S_2$ is a Riemann sum for f corresponding to a partition of $[a, c]$ of width less than δ, and we have

$$\left|S_1 - \int_a^b f(x)dx\right| < \frac{\epsilon}{3}, \quad \left|S_2 - \int_b^c f(x)dx\right| < \frac{\epsilon}{3},$$
$$\left|S_1 + S_2 - \int_a^c f(x)dx\right| < \frac{\epsilon}{3}.$$

Therefore

$$\begin{aligned}\left|\int_a^b f(x)dx + \int_b^c f(x)dx - \int_a^c f(x)dx\right| &\\ \leq \left|\int_a^b f(x)dx - S_1\right| + \left|\int_b^c f(x)dx - S_2\right| &\\ + \left|S_1 + S_2 - \int_a^c f(x)dx\right| < \frac{\epsilon}{3} + \frac{\epsilon}{3} + \frac{\epsilon}{3} = \epsilon.\end{aligned}$$

Since ϵ was any positive number, we have

$$\int_a^b f(x)dx + \int_b^c f(x)dx = \int_a^c f(x)dx.$$

Definition. If f is an integrable real-valued function on the interval $[a, b]$, we set

$$\int_b^a f(x)dx = -\int_a^b f(x)dx$$

and, for any $c \in [a, b]$,

$$\int_c^c f(x)dx = 0.$$

Corollary. *If f is a real-valued function on an interval in* $\mathbf{R}$ *which contains the points a, b, c and if two of the quantities $\int_a^b f(x)dx$, $\int_b^c f(x)dx$, $\int_a^c f(x)dx$ exist, then the third exists and*

$$\int_a^b f(x)dx + \int_b^c f(x)dx = \int_a^c f(x)dx.$$

The special cases $a = b$, $b = c$, and $a = c$ are all trivial to verify, so we may assume the three numbers a, b, c distinct. The points a, b, c determine a certain closed interval in $\mathbf{R}$ that is expressed as the union of two closed subintervals, namely the interval $[\min\{a, b, c\}, \max\{a, b, c\}]$ and the two subintervals determined by that point among a, b, c which is between the other two. The existence of any of $\int_a^b f(x)dx$, $\int_b^c f(x)dx$, $\int_a^c f(x)dx$ is equivalent to the existence of the corresponding $\int_b^a f(x)dx$, $\int_c^b f(x)dx$, $\int_c^a f(x)dx$, so the proposition tells us immediately that the existence of two of the integrals in question implies that of the third. Thus we may assume that all the integrals in question exist and it remains to prove the equality, which we may take to be in the equivalent but more symmetric form

$$\int_a^b f(x)dx + \int_b^c f(x)dx + \int_c^a f(x)dx = 0.$$

To prove this last equality, we note that it does not alter its sense under the cyclic permutations of a, b, c which send a, b, c into b, c, a respectively, or into c, a, b respectively. Hence we may assume without loss of generality that b is between a and c. Thus we are reduced to the two special cases $a < b < c$ and $a > b > c$. The truth of the equality in the first case follows directly from the proposition, while the second case is the same as the first, but with a change of signs.

A further consequence of the proposition is that if a real-valued function f is integrable on a closed interval in $\mathbf{R}$ then $\int_a^b f(x)dx$ exists for all a, b in this closed interval. We remark that if $|f(x)| \leq M$ for all x in the closed interval then for any a, b in the interval

$$\left| \int_a^b f(x)dx \right| \leq M\,|b - a|.$$

This is trivial if $a = b$, a consequence of the fact that $-M \leq f(x) \leq M$ for all x in the interval and the last corollary of § 2 if $a < b$, and a consequence of the last case and symmetry of sign if $a > b$.

***Theorem* (*Fundamental theorem of calculus*).** *Let f be a continuous real-valued function on an open interval U in $\mathbf{R}$ and let $a \in U$. Let the function F on U be defined by $F(x) = \int_a^x f(t)dt$ for all $x \in U$. Then F is differentiable and $F' = f$.*

Since f is continuous, $F(x) = \int_a^x f(t)dt$ is defined for all $x \in U$. We have to show that for any fixed $x_0 \in U$

$$\lim_{x \to x_0} \frac{F(x) - F(x_0)}{x - x_0} = f(x_0).$$

For any $x \in U$, $x \neq x_0$, we have

$$\left|\frac{F(x) - F(x_0)}{x - x_0} - f(x_0)\right| = \left|\frac{\int_a^x f(t)dt - \int_a^{x_0} f(t)dt}{x - x_0} - f(x_0)\right|$$
$$= \left|\frac{\int_{x_0}^x f(t)dt}{x - x_0} - \frac{\int_{x_0}^x f(x_0)dt}{x - x_0}\right| = \left|\frac{\int_{x_0}^x (f(t) - f(x_0))dt}{x - x_0}\right|.$$

Since f is continuous at x_0, given any $\epsilon > 0$ we can find a $\delta > 0$ such that $|f(x) - f(x_0)| < \epsilon$ if $x \in U$ and $|x - x_0| < \delta$. Thus if $x \in U$, $|x - x_0| < \delta$ and $x \neq x_0$ then for any t in the closed interval of extremities x_0 and x we have $|f(t) - f(x_0)| < \epsilon$, so that

$$\left|\int_{x_0}^x (f(t) - f(x_0))dt\right| \leq \epsilon|x - x_0|.$$

(We have used the remark immediately preceding the statement of the theorem.) Therefore if $x \in U$, $|x - x_0| < \delta$ and $x \neq x_0$ we have

$$\left|\frac{F(x) - F(x_0)}{x - x_0} - f(x_0)\right| \leq \frac{\epsilon|x - x_0|}{|x - x_0|} = \epsilon.$$

This proves the desired limit statement.

Corollary 1. *If f is a continuous real-valued function on an open interval in $\mathbf{R}$, then there exists a real-valued function F on the same interval whose derivative is f.*

For if a is any fixed point in the interval, $F(x) = \int_a^x f(t)dt$ will do.

We recall that if $F' = f$ then F is called an *antiderivative* or *primitive* of f. Corollary 1 says that any continuous real-valued function on an open interval has an antiderivative. If F is an antiderivative of f on an open interval, so is $F + c$ for any constant $c \in \mathbf{R}$. Furthermore, any antiderivative of f must have the form $F + c$: for if G is another antiderivative of f, then $(G - F)' = G' - F' = 0$, so $G - F$ must be constant, by Corollary 1 of the mean value theorem.

Corollary 2. *If the real-valued function F on an open interval U in $\mathbf{R}$ has the continuous derivative f and $a, b \in U$, then*

$$\int_a^b f(t)dt = F(b) - F(a).$$

Since $\frac{d}{dx}\left(\int_a^x f(t)dt - F(x)\right) = f(x) - f(x) = 0$, $\int_a^x f(t)dt - F(x)$ is constant. Thus $\int_a^x f(t)dt = F(x) + c$, for some $c \in \mathbf{R}$. In particular $0 = \int_a^a f(t)dt = F(a) + c$, so that $c = -F(a)$. Therefore $\int_a^x f(t)dt = F(x) - F(a)$. Hence $\int_a^b f(t)dt = F(b) - F(a)$.

This corollary is a powerful tool for the computation of integrals. For example, if n is a positive integer then the function $x^{n+1}/(n+1)$ has the continuous derivative x^n, so that

$$\int_a^b x^n\,dx = (b^{n+1} - a^{n+1})/(n+1).$$

***Corollary 3* (*Change of variable theorem*).** *Let U, V be open intervals in $\mathbf{R}$, $\varphi\colon U \to V$ a differentiable function with continuous derivative, and $f\colon V \to \mathbf{R}$ a continuous function. Then for any $a, b \in U$*

$$\int_{\varphi(a)}^{\varphi(b)} f(v)dv = \int_a^b f(\varphi(u))\varphi'(u)du.$$

Let $F\colon V \to \mathbf{R}$ be the function defined by $F(y) = \int_{\varphi(a)}^{y} f(v)dv$ for all $y \in V$. Then F is differentiable and $F' = f$. The function $G\colon U \to \mathbf{R}$ defined by $G(x) = \int_{\varphi(a)}^{\varphi(x)} f(v)dv$ is the composite $G = F \circ \varphi$ of two differentiable functions, hence is itself differentiable. By the chain rule we have $G'(x) = F'(\varphi(x))\varphi'(x) = f(\varphi(x))\varphi'(x)$ for all $x \in U$. Hence $G(x) = \int_a^x f(\varphi(u))\varphi'(u)du + c$, for some constant $c \in \mathbf{R}$. Setting $x = a$ we get $c = 0$, so that $G(x) = \int_a^x f(\varphi(u))\varphi'(u)du$. This last equation holds for all $x \in U$. Setting $x = b$ gives us Corollary 3.

§ 5. THE LOGARITHMIC AND EXPONENTIAL FUNCTIONS.

In this section we develop in a rigorous fashion the familiar properties of some of the functions which are dealt with in elementary calculus.

Definition. If $x \in \mathbf{R}$, $x > 0$, then $\log x = \int_1^x \frac{dt}{t}$.

Proposition. *The function* $\log\colon \{x \in \mathbf{R} : x > 0\} \to \mathbf{R}$ *is differentiable with $d\log x/dx = 1/x$, it is strictly increasing, assumes all values in $\mathbf{R}$, and satisfies the rules*

$$\begin{aligned} \log xy &= \log x + \log y && \text{if } x, y > 0 \\ \log \frac{x}{y} &= \log x - \log y && \text{if } x, y > 0 \\ \log x^n &= n \log x && \text{if } x > 0,\ n \text{ an integer.} \end{aligned}$$

The differentiability of log, together with the equation $d\log x/dx = 1/x$, comes from the fundamental theorem of calculus. Since $1/x > 0$ if $x > 0$, the derivative of log is always positive, so log is a strictly increasing function. If a is some fixed positive number and $y = ax$, the chain rule gives

$$\frac{d}{dx}\log y = \frac{1}{y}\frac{dy}{dx} = \frac{1}{ax}\cdot a = \frac{1}{x},$$

so that $d \log ax/dx = d \log x/dx$ and hence $\log ax = \log x + c$, for some $c \in \mathbf{R}$. Setting $x = 1$ gives $c = \log a$, so that $\log ax = \log a + \log x$. After changing notation we have

$$\log xy = \log x + \log y \quad \text{if } x, y > 0.$$

In the special case $x = 1/y$ this yields $\log 1/y = -\log y$, so that

$$\log \frac{x}{y} = \log x + \log \frac{1}{y} = \log x - \log y \quad \text{if } x, y > 0.$$

Clearly if $x > 0$ then

$$\begin{aligned}
&\log x^0 = \log 1 = 0 = 0 \cdot \log x \\
&\log x^1 = \log x = 1 \cdot \log x \\
&\log x^2 = \log x + \log x = 2 \log x \\
&\log x^3 = \log (x^2 \cdot x) = \log x^2 + \log x = 3 \log x \\
&\text{etc.}
\end{aligned}$$

so that

$$\log x^n = n \log x \quad \text{if } n = 0, 1, 2, 3, \ldots.$$

If $n = 0, 1, 2, 3, \ldots$, then $\log x^{-n} = \log 1/x^n = -\log x^n = (-n) \log x$, so that

$$\log x^n = n \log x \quad \text{if } x > 0, n \text{ any integer.}$$

Since $2 > 1$ we have $\log 2 > \log 1 = 0$. Since $\log 2^n = n \log 2$, given any $\gamma \in \mathbf{R}$ we can find integers n_1, n_2 such that

$$\log 2^{n_1} < \gamma < \log 2^{n_2}$$

(simply by taking $n_1 < \gamma/\log 2 < n_2$). By the intermediate value theorem there is some c between 2^{n_1} and 2^{n_2} such that $\log c = \gamma$. That is, the log function takes on all values. Everything desired has been proved.

Definition. exp is the inverse function of log, that is

$$\exp (x) = y \quad \text{means} \quad x = \log y.$$

This makes sense since the log function is one-one (being strictly increasing). We avoid the notation e^x for the moment to avoid confusion with our existing notation for powers.

Proposition. *The function* $\exp : \mathbf{R} \to \{x \in \mathbf{R} : x > 0\}$ *is differentiable, with* $d \exp (x)/dx = \exp (x)$. *It is strictly increasing, assumes all positive values, and satisfies the rules*

$$\begin{aligned}
\exp (x) \cdot \exp (y) &= \exp (x + y) \quad \textit{if } x, y \in \mathbf{R} \\
\frac{\exp (x)}{\exp (y)} &= \exp (x - y) \quad \textit{if } x, y \in \mathbf{R} \\
\exp (nx) &= (\exp (x))^n \quad \textit{if } x \in \mathbf{R}, n \textit{ an integer.}
\end{aligned}$$

To prove this, forget about differentiability for a moment. Then everything else follows immediately from the corresponding properties of the log function. (The easiest way to prove the equations is to check each to see if it gives a correct statement when log is applied to both sides. Everything works out, using the identity $\log(\exp(x)) = x$ and the corresponding formulas for log.) As for differentiability, we first prove that exp is continuous. We must show that for any $x_0 \in \mathbf{R}$ and any $\epsilon > 0$ there exists a $\delta > 0$ such that $|\exp(x) - \exp(x_0)| < \epsilon$ if $|x - x_0| < \delta$. We may assume that $\epsilon < \exp(x_0)$. Since exp is strictly increasing, if x is between $\log(\exp(x_0) - \epsilon)$ and $\log(\exp(x_0) + \epsilon)$ then $\exp(x)$ will be between $\exp(x_0) - \epsilon$ and $\exp(x_0) + \epsilon$. Hence we insure that $|\exp(x) - \exp(x_0)| < \epsilon$ whenever $|x - x_0| < \delta$ by choosing

$$\delta = \min\{x_0 - \log(\exp(x_0) - \epsilon), \log(\exp(x_0) + \epsilon) - x_0\}.$$

Thus exp is continuous. To prove exp differentiable and find its derivative, let $x_0 \in \mathbf{R}$ be fixed and write $\exp(x_0) = y_0$, $\exp(x) = y$. Then $\lim\limits_{x \to x_0} y = y_0$ and

$$\lim_{x \to x_0} \frac{\exp(x) - \exp(x_0)}{x - x_0} = \lim_{y \to y_0} \frac{y - y_0}{\log y - \log y_0} = \frac{1}{\lim\limits_{y \to y_0} \dfrac{\log y - \log y_0}{y - y_0}}$$

$$= \frac{1}{\dfrac{d \log y}{dy}(y_0)} = \frac{1}{\dfrac{1}{y_0}} = y_0 = \exp(x_0).$$

This ends the proof.

The symbol x^n has so far been defined only for integral values of n. In this case, if $x > 0$, we have $\log x^n = n \log x$, so that $x^n = \exp(n \log x)$. Hence the following definition is consistent with our existing notation.

Definition. If $x, n \in \mathbf{R}$, $x > 0$, then $x^n = \exp(n \log x)$.

Proposition. *For $x, y, n, m \in \mathbf{R}$, $x, y > 0$, we have*

$$x^n \cdot x^m = x^{n+m}$$

$$\frac{x^n}{x^m} = x^{n-m}$$

$$(x^n)^m = x^{nm}$$

$$(xy)^n = x^n y^n$$

$$\frac{d}{dx} x^n = n x^{n-1}.$$

The four algebraic identities follow immediately from the definition and previous results of this section. For example,

$$x^n \cdot x^m = \exp(n \log x) \cdot \exp(m \log x) = \exp(n \log x + m \log x)$$
$$= \exp((n + m) \log x) = x^{n+m}.$$

The proof of the last formula is an exercise in the chain rule:

$$\frac{d}{dx} x^n = \frac{d}{dx} \exp(n \log x) = \exp(n \log x) \frac{d}{dx}(n \log x) = x^n \cdot \frac{n}{x} = nx^{n-1}.$$

The rules for fractional exponents are of course contained in the last proposition. For example $x^{1/2} = \sqrt{x}$, since $(x^{1/2})^2 = x^1 = x$.

It is convenient to extend the definition of x^n slightly by setting $0^n = 0$ if $n > 0$, so that for any fixed positive n the function x^n is continuous for $x \geq 0$.

Definition. $e = \exp(1)$.

We immediately recover the standard notation for the exponential function: if $x \in \mathbf{R}$ then $e^x = \exp(x \log e) = \exp(x)$, since $\log e = 1$. Thus we may write the formulas of the proposition before last in their more convenient forms

$$e^x \cdot e^y = e^{x+y}, \quad \frac{d}{dx} e^x = e^x, \text{ etc.}$$

A rough approximation of e may be obtained by noting that for $1 \leq x \leq 2$ we have $1/2 \leq 1/x \leq 1$, so that $1/2 \leq \int_1^2 dx/x = \log 2 \leq 1$. As a matter of fact, we can get the slightly stronger relation

$$\frac{1}{2} < \log 2 < 1$$

by remarking that it is easy to find a larger step function than the constant $1/2$ that is less than or equal to $1/x$ for $1 \leq x \leq 2$, and a smaller one than the constant 1 that is greater than or equal to $1/x$ for $1 \leq x \leq 2$. From $1/2 < \log 2 < 1$ follows $1 < \log 2^2 < 2$. Hence $\log 2 < 1 < \log 4$, so

$$2 < e < 4.$$

It should be remarked that of course we could have obtained all the results of this section differently, *starting* with the exponential function. The argument (in outline) is as follows: For any fixed positive integer n the function x^n is continuous and strictly increasing for $x \geq 0$, assuming arbitrarily large values. Therefore by the intermediate value theorem any positive number has a unique positive n^{th} root. We define rational powers of a number $x > 0$ by setting

$$x^{m/n} = (\text{positive } n^{th} \text{ root of } x)^m$$

if m and n are integers with no common factor other than ± 1 and $n > 0$. We then prove the various rules of exponents, for *rational* exponents. If

$x > 0$ and $n \in \mathbf{R}$ is not rational, we can find a sequence $n_1, n_2, n_3, \ldots$ of rational numbers that converges to n; we then define x^n to be the limit of the sequence $x^{n_1}, x^{n_2}, x^{n_3}, \ldots$, first showing that this limit exists and is independent of the choice of the sequence $n_1, n_2, n_3, \ldots$. We then verify all the rules of exponents for arbitrary real exponents. Next we look at the function a^x, for some fixed $a > 0$, show by a suitable trick that it is differentiable, and show also that for a very special choice of base a, a choice denoted e, we get the magic formula $de^x/dx = e^x$. Finally we do the log function, which at this point is easy. Thus we end up with a more natural, but considerably longer, derivation of exactly the same results as before.

PROBLEMS

1. Compute $\int_0^1 x\,dx$ directly from the definition of the integral, assuming only that this integral exists.
2. Prove that $\int_0^1 f(x)dx = 0$ if $f(1/n) = 1$ for $n = 1, 2, 3, \ldots$ and $f(x) = 0$ for all other x.
3. Does $\int_0^1 f(x)dx$ exist if f is the function of Problem 1(d), Chap. IV?
4. Let $f: [a, b] \to \mathbf{R}$ and let $c \in \mathbf{R}$. Prove that if $\int_a^b f(x)dx$ exists then so does $\int_{a+c}^{b+c} f(x - c)dx$ and these two integrals are equal.
5. Prove that a continuous real-valued function on a closed interval in $\mathbf{R}$ is integrable, using only Lemma 1 of § 3 and uniform continuity.
6. Let $[a, b]$ be a closed interval in $\mathbf{R}$ and let V be a complete normed vector space (cf. Prob. 22, Chap. III).
   (a) Show that the definition of $\int_a^b f(x)dx$ for real-valued functions f on $[a, b]$ generalizes to functions f: $[a, b] \to V$.
   (b) Prove the analog of the criterion for integrability of Lemma 1 of § 3 for V-valued functions on $[a, b]$.
   (c) Using (b), prove that if f: $[a, b] \to V$ is continuous then $\int_a^b f(x)dx$ exists.
   (d) Prove that if f: $[a, b] \to V$ is continuous then
   $$\left\| \int_a^b f(x)dx \right\| \leq \int_a^b \| f(x) \| \, dx.$$
   (e) Prove that if V is finite-dimensional with basis $v_1, \ldots, v_n$, and if $f_1, \ldots, f_n$ are real-valued functions on $[a, b]$, then $\int_a^b (f_1(x)v_1 + \cdots + f_n(x)v_n)dx$ exists if and only if $\int_a^b f_1(x)dx, \ldots, \int_a^b f_n(x)dx$ exist, in which case
   $$\int_a^b (f_1(x)v_1 + \cdots + f_n(x)v_n)dx = \left(\int_a^b f_1(x)dx\right)v_1 + \cdots + \left(\int_a^b f_n(x)dx\right)v_n.$$
   (For part (e) you will need the result of Prob. 23, Chap. IV.)

7. Prove that if the real-valued function f on the interval $[a, b]$ is bounded and is continuous except at a finite number of points, then $\int_a^b f(x)dx$ exists.

8. Prove that if $f\colon [a, b] \to \mathbf{R}$ is increasing (or decreasing) then $\int_a^b f(x)dx$ exists.

9. Prove that if the real-valued function f on the interval $[a, b]$ is integrable on $[a, b]$ then so is $|f|$, and $\left|\int_a^b f(x)dx\right| \le \int_a^b |f(x)|dx$.

10. Prove that if the real-valued function f on the interval $[a, b]$ is integrable on $[a, b]$ then so is f^2. Using the identity $(f+g)^2 = f^2 + 2fg + g^2$, prove that the product of two integrable functions is integrable.

11. Prove that if f is a continuous real-valued function on the interval $[a, b]$ such that $f(x) \ge 0$ for all $x \in [a, b]$ and $f(x) > 0$ for some $x \in [a, b]$, then $\int_a^b f(x)dx > 0$.

12. Show that if f is a continuous real-valued function on the interval $[a, b]$ then $\int_a^b f(x)dx = f(\xi)(b-a)$ for some $\xi \in [a, b]$ (mean value theorem for integrals).

13. Show that if f is a continuous real-valued function on the interval $[a, b]$ and $f(x) \ge 0$ for all $x \in [a, b]$, then

$$\lim_{n\to\infty}\left(\int_a^b (f(x))^n dx\right)^{1/n} = \max\{f(x) : x \in [a, b]\}.$$

14. Show that if f is a continuous real-valued function on $\{x \in \mathbf{R} : x \ge 0\}$ and $\lim_{x\to+\infty} f(x) = c$ (cf. Prob. 8, Chap. IV), then

$$\lim_{x\to+\infty} \frac{1}{x}\int_0^x f(t)dt = c.$$

15. Let $[a, b]$ and $[c, d]$ be closed intervals in $\mathbf{R}$ and let f be a continuous real-valued function on $\{(x, y) \in E^2 : x \in [a, b], y \in [c, d]\}$. Show that the function $g\colon [c, d] \to \mathbf{R}$ defined by $g(y) = \int_a^b f(x, y)dx$ for all $y \in [c, d]$ is continuous.

16. Prove that the real-valued function on $C([a, b])$ which sends any f into $\int_a^b f(x)dx$ is uniformly continuous.

17. Prove that if u and v are real-valued functions on an open subset of $\mathbf{R}$ containing the interval $[a, b]$ and if u and v have continuous derivatives, then (integration by parts)

$$\int_a^b u(x)v'(x)dx = u(b)v(b) - u(a)v(a) - \int_a^b v(x)u'(x)dx.$$

18. Prove that

$$\int_0^{x/\sqrt{1+x^2}} \frac{dt}{\sqrt{1-t^2}} = \int_0^x \frac{du}{1+u^2} \quad \text{for all } x \in \mathbf{R}.$$

19. Show that if U is an open interval in $\mathbf{R}$ and the function $f: U \to \mathbf{R}$ has a continuous $(n+1)^{th}$ derivative on U, then for any $a, b \in U$ we have

$$f(b) = f(a) + \frac{f'(a)(b-a)}{1!} + \frac{f''(a)(b-a)^2}{2!} + \cdots + \frac{f^{(n)}(a)(b-a)^n}{n!} + \frac{\int_a^b (b-x)^n f^{(n+1)}(x)dx}{n!}.$$

20. A *curve in a metric space* E is a continuous function f from an interval $[a, b]$ in $\mathbf{R}$ into E; its *length* is

$$\text{l.u.b.} \left\{\sum_{i=1}^{N} d(f(x_{i-1}), f(x_i)) : x_0, x_1, \ldots, x_N \text{ is a partition of } [a, b]\right\},$$

if this l.u.b. exists. Prove that if $E = E^n$ and $f(x) = (f_1(x), \ldots, f_n(x))$, where $f_1, \ldots, f_n: [a, b] \to \mathbf{R}$ are continuous functions which have continuous derivatives on (a, b) that extend to continuous functions on $[a, b]$, then the curve has length and this length is

$$\int_a^b \sqrt{(f_1'(x))^2 + \cdots + (f_n'(x))^2}\, dx.$$

21. Compute
   (a) $\lim_{n\to\infty} \frac{1^k + 2^k + \cdots + n^k}{n^{k+1}}$, where $k \in \mathbf{R}, k > 0$
   (b) $\lim_{n\to\infty} \left(\frac{1}{n+1} + \frac{1}{n+2} + \cdots + \frac{1}{2n}\right)$.

22. Show that for $n = 1, 2, 3, \ldots$, the number

$$1 + \frac{1}{2} + \frac{1}{3} + \cdots + \frac{1}{n} - \log n$$

is positive, that it decreases as n increases, and hence that the sequence of these numbers converges to a limit between 0 and 1 (Euler's constant).

23. Prove that the only function $f: \mathbf{R} \to \mathbf{R}$ such that $f' = f$ and $f(0) = 1$ is given by $f(x) = e^x$.

24. Prove that
   (a) $\log(1 + x) \leq x$ for all $x > -1$, with equality if and only if $x = 0$
   (b) $e^x \geq 1 + x$ for all x, with equality if and only if $x = 0$
   (c) $\lim_{x\to 0} \frac{\log(1+x)}{x} = 1$
   (d) $\lim_{h\to 0} (1 + hx)^{1/h} = \lim_{n\to\infty} \left(1 + \frac{x}{n}\right)^n = e^x$
   (e) $\lim_{n\to\infty} n(x^{1/n} - 1) = \log x \quad$ if $x > 0$.

25. Show that for $x > e$ the function $\frac{\log x}{x}$ is strictly decreasing and that it gets arbitrarily close to zero, hence that
   (a) $\lim_{x\to+\infty} \frac{\log x}{x^\alpha} = 0 \quad$ if $\alpha > 0$
   (b) $\lim_{x\to 0} x^\alpha \log x = 0 \quad$ if $\alpha > 0$
   (c) $\lim_{x\to+\infty} \frac{x^n}{e^x} = 0 \quad$ for any $n \in \mathbf{R}$.

26. Define $f: \mathbf{R} \to \mathbf{R}$ by

$$f(x) = \begin{cases} e^{-1/x^2} & \text{if } x > 0 \\ 0 & \text{if } x \leq 0. \end{cases}$$

Prove that f has derivatives of all orders, with $f^{(n)}(0) = 0$ for all n.

27. If $a, b \in \mathbf{R}$, $a < b$, and $f: \{x \in \mathbf{R} : a < x \leq b\} \to \mathbf{R}$ is a continuous function, define the *improper integral* $\int_{a+}^{b} f(x)dx$ to be $\lim_{y \to a} \int_{y}^{b} f(x)dx$, if this limit exists. Show that if g is another continuous real-valued function on the set $\{x \in \mathbf{R} : a < x \leq b\}$ such that $|f(x)| \leq g(x)$ for all x in this set and $\int_{a+}^{b} g(x)dx$ exists, then $\int_{a+}^{b} f(x)dx$ exists. Hence show that $\int_{a+}^{b} f(x)dx$ exists if there exists an $\alpha < 1$ such that $(x - a)^{\alpha} f$ is bounded on (a, b).

28. If $a \in \mathbf{R}$ and $f: \{x \in \mathbf{R} : x \geq a\} \to \mathbf{R}$ is a continuous function, define the *improper integral* $\int_{a}^{+\infty} f(x)dx$ to be $\lim_{y \to +\infty} \int_{a}^{y} f(x)dx$, if this limit exists (cf. Prob. 8, Chap. IV). Show that if g is another continuous real-valued function on $\{x \in \mathbf{R} : x \geq a\}$, if $|f(x)| \leq g(x)$ for all $x \geq a$, and if $\int_{a}^{+\infty} g(x)dx$ exists, then $\int_{a}^{+\infty} f(x)dx$ exists. Hence show that $\int_{a}^{+\infty} f(x)dx$ exists if there exists an $\alpha > 1$ such that $x^{\alpha} f(x)$ is bounded.

CHAPTER VII

Interchange of Limit Operations

The various kinds of limiting processes we have studied (limit of a sequence of points in a metric space, limit of a function, differentiation, integration) do not always occur singly. In a given problem we may be called upon to take one kind of limit, then another kind. In such problems the order in which the operations are performed is naturally of importance. We have already treated such a problem in applying the two operations $\lim_{n\to\infty}$ and $\lim_{p\to p_0}$ to a sequence of functions $f_1, f_2, f_3, \ldots$ from one metric space into another, p_0 being a point of the first metric space. If $\lim_{p\to p_0} f_n(p)$ exists for each $n = 1, 2, 3, \ldots$ we get a sequence of points $\lim_{p\to p_0} f_1(p)$, $\lim_{p\to p_0} f_2(p)$, $\lim_{p\to p_0} f_3(p)$, $\ldots$ in the second metric space, and we may be able to take $\lim_{n\to\infty}$ of this sequence of points. On the other hand the limit function $\lim_{n\to\infty} f_n$ may exist and if it does we may be able to apply $\lim_{p\to p_0}$ to the limit function, again getting a point in the second metric space. However it may happen that all of these operations can be performed and we arrive at different answers in the two cases. In one extremely important case this cannot happen, for we have proved that if $f_1, f_2, f_3, \ldots$ is a uniformly convergent sequence of continuous functions then $\lim_{n\to\infty} f_n$ is also continuous, so that

$$\lim_{p\to p_0} \left((\lim_{n\to\infty} f_n)(p) \right) = (\lim_{n\to\infty} f_n)(p_0)$$

$$= \lim_{n\to\infty} f_n(p_0) = \lim_{n\to\infty} \left(\lim_{p\to p_0} f_n(p)\right).$$

In this chapter we prove a number of similar results for other pairs of limiting processes. No attempt will be made to be systematic; we only intend to provide some especially useful results. At the same time we take the opportunity to discuss the meanings of these results for infinite series, which is how infinite sequences usually arise in practice, developing the theory of infinite series sufficiently for the purposes of calculus. An exposition of the trigonometric functions is given as an easy application.

§ 1. INTEGRATION AND DIFFERENTIATION OF SEQUENCES OF FUNCTIONS.

If $f_1, f_2, f_3, \ldots$ is a sequence of Riemann integrable real-valued functions on a closed interval $[a, b]$ in $\mathbf{R}$ and $f_1, f_2, f_3, \ldots$ converges to the function f on $[a, b]$, can we assert that

$$\int_a^b f(x)dx = \lim_{n\to\infty} \int_a^b f_n(x)dx?$$

The following example shows that in general we cannot, not even if $f_1, f_2, f_3, \ldots$ are all continuous.

EXAMPLE. For $n = 1, 2, 3, \ldots$ let f_n: $[0, 1] \to \mathbf{R}$ be defined by Figure 30 (f_n can be defined analytically by setting $f_n(x) = 4n^2x$ for $0 \leq x \leq 1/2n$, $f_n(x) = 4n - 4n^2x$ for $1/2n < x \leq 1/n$, $f_n(x) = 0$ for $1/n < x \leq 1$). For each n, f_n is continuous and $\int_0^1 f_n(x)dx = 1$, so that $\lim_{n\to\infty} \int_0^1 f_n(x)dx = 1$. On the other hand, $f = \lim_{n\to\infty} f_n = 0$. (For clearly $f(0) = 0$ and if $x \neq 0$ then $f_n(x) = 0$ if $n > 1/x$.) Hence $\int_0^1 f(x)dx = 0 \neq \lim_{n\to\infty} \int_0^1 f_n(x)dx$.

If, however, $f_1, f_2, f_3, \ldots$ converges uniformly, there is no trouble:

Theorem. *Let $a, b \in \mathbf{R}$, $a < b$, and let $f_1, f_2, f_3, \ldots$ be a uniformly convergent sequence of continuous real-valued functions on $[a, b]$. Then*

$$\int_a^b (\lim_{n\to\infty} f_n(x))dx = \lim_{n\to\infty} \int_a^b f_n(x)dx.$$

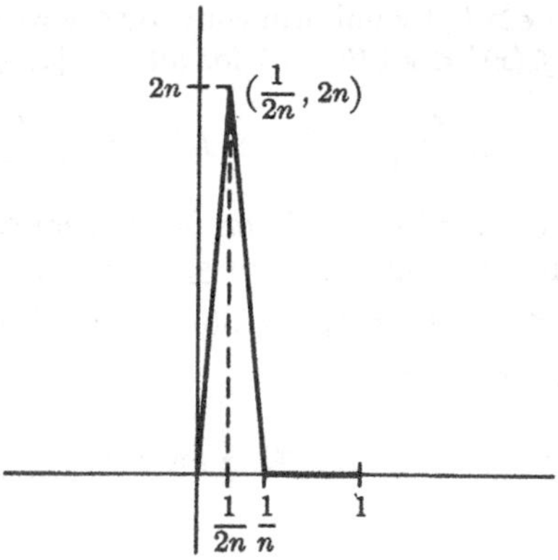

FIGURE 30. Graph of the function f_n of the example on p. 138.

Let $f = \lim_{n\to\infty} f_n$. Since each f_n is continuous and the convergence is uniform, f is continuous. In particular f is integrable on $[a, b]$. By the definition of uniform convergence, for any $\epsilon > 0$ there exists a positive integer N such that if $n > N$ then $|f(x) - f_n(x)| < \epsilon/(b - a)$ for all $x \in [a, b]$. We then have the inequalities

$$-\frac{\epsilon}{b-a} \leq f(x) - f_n(x) \leq \frac{\epsilon}{b-a}$$

for all $x \in [a, b]$, which imply

$$-\epsilon \leq \int_a^b (f(x) - f_n(x))dx \leq \epsilon$$

or

$$\left| \int_a^b f(x)dx - \int_a^b f_n(x)dx \right| \leq \epsilon.$$

This last inequality holds for all $n > N$, and therefore

$$\lim_{n\to\infty} \int_a^b f_n(x)dx = \int_a^b f(x)dx.$$

REMARK. The same theorem holds if we do not assume that each f_n is continuous, but merely Riemann integrable on $[a, b]$. Indeed the same proof will hold once it is shown that $f = \lim_{n\to\infty} f_n$ is integrable. This can be done easily using the criterion of the proposition of § 3 of the last chapter, as follows. Given any $\epsilon > 0$, by uniform convergence we can find an integer n such that $|f(x) - f_n(x)| < \epsilon/3(b - a)$ for all $x \in [a, b]$, so that

$$f_n(x) - \frac{\epsilon}{3(b-a)} \leq f(x) \leq f_n(x) + \frac{\epsilon}{3(b-a)}$$

for all $x \in [a, b]$. Since f_n is integrable on $[a, b]$ there exist step functions g_1, g_2 on $[a, b]$ such that $g_1(x) \leq f_n(x) \leq g_2(x)$ for all $x \in [a, b]$ and $\int_a^b (g_2(x) - g_1(x))dx < \epsilon/3$. Then $g_1 - \epsilon/3(b - a)$ and $g_2 + \epsilon/3(b - a)$ are step functions on $[a, b]$ such that

$$g_1(x) - \frac{\epsilon}{3(b-a)} \leq f(x) \leq g_2(x) + \frac{\epsilon}{3(b-a)}$$

for all $x \in [a, b]$ and

$$\int_a^b \left(\left(g_2(x) + \frac{\epsilon}{3(b-a)}\right) - \left(g_1(x) - \frac{\epsilon}{3(b-a)}\right)\right)dx < \epsilon.$$

By the proposition quoted, f is integrable on $[a, b]$.

To prove an analogous result for the differentiation of the limit of a sequence of differentiable functions one has to make slightly stronger assumptions.

Theorem. *Let $f_1, f_2, f_3, \ldots$ be a sequence of real-valued functions on an open interval U in $\mathbf{R}$, each having a continuous derivative. Suppose that the sequence $f_1', f_2', f_3', \ldots$ converges uniformly on U and that for some $a \in U$ the sequence $f_1(a), f_2(a), f_3(a), \ldots$ converges. Then* $\lim_{n\to\infty} f_n$ *exists, is differentiable, and*

$$(\lim_{n\to\infty} f_n)' = \lim_{n\to\infty} f_n'.$$

By the fundamental theorem of calculus we have

$$\int_a^x f_n'(t)dt = f_n(x) - f_n(a)$$

for any $x \in U$ and any $n = 1, 2, 3, \ldots$. Let $\lim_{n\to\infty} f_n' = g$. By the previous theorem $\lim_{n\to\infty} (f_n(x) - f_n(a))$ exists for any $x \in U$ and equals $\int_a^x g(t)dt$. Since $\lim_{n\to\infty} f_n(a)$ exists, so does $\lim_{n\to\infty} f_n(x)$. Setting $\lim_{n\to\infty} f_n(x) = f(x)$ we have

$$f(x) - f(a) = \int_a^x g(t)dt$$

for each $x \in U$. A second use of the fundamental theorem of calculus gives $f' = g$, which is what was to be proved.

§ 2. INFINITE SERIES.

If $a_1, a_2, a_3, \ldots$ is a sequence of real numbers, by the *infinite series*

$$a_1 + a_2 + a_3 + \cdots,$$

also denoted

$$\sum_{n=1}^{\infty} a_n\,,$$

we mean the sequence $a_1, a_1 + a_2, a_1 + a_2 + a_3, \ldots$. The terms of the latter sequence are called the *partial sums* of the series. If $A \in \mathbf{R}$, we say that the infinite series *converges to* A if the sequence of partial sums converges to A, that is if

$$\lim_{n\to\infty} (a_1 + a_2 + \cdots + a_n) = A.$$

If the series converges to A it is customary to call A the *sum of the series* (although this is not a sum at all, but a limit of sums) and it is customary to write

$$a_1 + a_2 + a_3 + \cdots = A$$

or

$$\sum_{n=1}^{\infty} a_n = A.$$

(This somewhat awkward convention, whereby we use the symbol $a_1 + a_2 + a_3 + \cdots$ to denote both the series and its sum, if the latter exists, rarely causes confusion, since it is usually clear from the context whether the series or its sum is meant.) If a series converges to some real number, the series is said to *converge*, or to be *convergent*; in the contrary case the series *diverges*, or is *divergent*.

Similarly, if $f_1, f_2, f_3, \ldots$ is a sequence of real-valued functions on a metric space E, by the infinite series

$$f_1 + f_2 + f_3 + \cdots$$

we mean the sequence of functions $f_1, f_1 + f_2, f_1 + f_2 + f_3, \ldots$. We say that the series *converges at* p, for a certain $p \in E$, if the series $f_1(p) + f_2(p) + f_3(p) + \cdots$ converges; otherwise the series $f_1 + f_2 + f_3 + \cdots$ is said to *diverge at* p. The series is said to *converge on* E (or, more simply, to *converge*) if it converges at each point of E; in this case there is a real-valued function f on E such that $f(p)$ is the sum of the series $f_1(p) + f_2(p) + f_3(p) + \cdots$ for each $p \in E$, and we of course write

$$f_1 + f_2 + f_3 + \cdots = f.$$

Finally, we say that $f_1 + f_2 + f_3 + \cdots$ *converges uniformly* on E if the sequence of partial sums $f_1, f_1 + f_2, f_1 + f_2 + f_3, \ldots$ converges uniformly on E.

EXAMPLE. If $a \in \mathbf{R}$, $|a| < 1$, then the "geometric" series $\sum_{n=0}^{\infty} a^n = 1 + a + a^2 + \cdots$ converges. In fact, since

$$(1 - a)(1 + a + a^2 + \cdots + a^{n-1}) = 1 - a^n$$

we have

$$1 + a + a^2 + \cdots + a^{n-1} = \frac{1 - a^n}{1 - a}$$

so that

$$\sum_{n=0}^{\infty} a^n = \lim_{n\to\infty} \frac{1 - a^n}{1 - a} = \frac{1}{1 - a}.$$

(We have here used the fact that $\lim_{n\to\infty} a^n = 0$ if $|a| < 1$. This was proved at the end of § 3 of Chapter III. Another proof is obtained by noting that $\log |a|^n = n \log |a|$ is negative and gets arbitrarily large in absolute value as n increases.) Letting x denote the identity function on $\mathbf{R}$, as usual, we get the analogous statement for the series of functions on the metric space $(-1, 1)$:

$$1 + x + x^2 + x^3 + \cdots = \frac{1}{1 - x}.$$

The elementary facts about infinite sequences of real numbers can be translated immediately into facts about infinite series of real numbers. For example, since an infinite sequence can have at most one limit, an infinite series can have at most one sum. As another example, since a sequence of real numbers converges if and only if it is a Cauchy sequence, a series of real numbers converges if and only if its sequence of partial sums is a Cauchy sequence. Since for $n > m$ the difference between the mth and n^{th} partial sums of the series $a_1 + a_2 + a_3 + \cdots$ is $a_{m+1} + a_{m+2} + \cdots + a_n$, we have the following result.

Proposition. *The series of real numbers $a_1 + a_2 + a_3 + \cdots$ converges if and only if, given any $\epsilon > 0$, there is a positive integer N such that if $n > m \geq N$ then*

$$|a_{m+1} + a_{m+2} + \cdots + a_n| < \epsilon.$$

The following two corollaries are immediate.

Corollary 1. *If the series of real numbers $a_1 + a_2 + a_3 + \cdots$ converges, then $\lim_{n\to\infty} a_n = 0$.*

Corollary 2. *If* $\sum_{n=1}^{\infty} a_n$ *and* $\sum_{n=1}^{\infty} b_n$ *are infinite series of real numbers such that* $a_n = b_n$ *whenever* n *is sufficiently large, then if one series converges so does the other.*

EXAMPLE 1. The geometric series $\sum_{n=1}^{\infty} a^n$ does not converge if $|a| \geq 1$ by the first corollary.

EXAMPLE 2. The "harmonic" series $\sum_{n=1}^{\infty} \frac{1}{n} = 1 + \frac{1}{2} + \frac{1}{3} + \cdots$ diverges. For whenever $n = 2m$ we have

$$a_{m+1} + a_{m+2} + \cdots + a_n = \frac{1}{m+1} + \frac{1}{m+2} + \cdots + \frac{1}{n}$$
$$\geq \frac{1}{n} + \frac{1}{n} + \cdots + \frac{1}{n} = m \cdot \frac{1}{n} = \frac{1}{2},$$

contrary to the condition of the proposition.

Proposition. *The following rules hold:*

(1) If $\sum_{n=1}^{\infty} a_n$ *and* $\sum_{n=1}^{\infty} b_n$ *are convergent series of real numbers, then the series* $\sum_{n=1}^{\infty} (a_n + b_n)$ *is also convergent and*

$$\sum_{n=1}^{\infty} (a_n + b_n) = \sum_{n=1}^{\infty} a_n + \sum_{n=1}^{\infty} b_n.$$

(2) If $\sum_{n=1}^{\infty} a_n$ *is a convergent series of real numbers and* $c \in \mathbf{R}$, *then* $\sum_{n=1}^{\infty} ca_n$ *is convergent and*

$$\sum_{n=1}^{\infty} ca_n = c \sum_{n=1}^{\infty} a_n.$$

The proposition is immediate from the third proposition of § 3 of Chapter III (page 48).

Proposition. *If* $a_1, a_2, a_3, \ldots$ *are nonnegative real numbers, then either the series* $\sum_{n=1}^{\infty} a_n$ *converges or it has arbitrarily large partial sums.*

For the sequence of partial sums is increasing, hence convergent in case it is bounded from above.

***Proposition* (*Comparison test*).** *If* $\sum_{n=1}^{\infty} a_n$ and $\sum_{n=1}^{\infty} b_n$ *are infinite series of real numbers such that* $|a_n| \leq b_n$ *for* $n = 1, 2, 3, \ldots$ *and* $\sum_{n=1}^{\infty} b_n$ *converges, then* $\sum_{n=1}^{\infty} a_n$ *converges.*

For let ϵ be any real number greater than zero. By the first proposition of this section there is a positive integer N such that if $n > m \geq N$ then

$$|b_{m+1} + b_{m+2} + \cdots + b_n| < \epsilon.$$

Thus if $n > m \geq N$ then

$$\begin{aligned} |a_{m+1} + a_{m+2} + \cdots + a_n| &\leq |a_{m+1}| + |a_{m+2}| + \cdots + |a_n| \\ &\leq b_{m+1} + b_{m+2} + \cdots + b_n < \epsilon. \end{aligned}$$

Thus, by the proposition just quoted, $\sum_{n=1}^{\infty} a_n$ is convergent.

Corollary. *Under the conditions of the proposition,*

$$\left| \sum_{n=1}^{\infty} a_n \right| \leq \sum_{n=1}^{\infty} b_n.$$

Each partial sum $a_1 + a_2 + \cdots + a_n$ is such that $|a_1 + a_2 + \cdots + a_n| \leq |a_1| + |a_2| + \cdots + |a_n| \leq b_1 + b_2 + \cdots + b_n \leq \sum_{n=1}^{\infty} b_n$, and since the closed interval $\left[-\sum_{n=1}^{\infty} b_n, \sum_{n=1}^{\infty} b_n \right]$ contains each partial sum $a_1 + a_2 + \cdots + a_n$ it also contains the limit $\sum_{n=1}^{\infty} a_n$ of these partial sums.

Definition. If $a_1, a_2, a_3, \ldots$ are real numbers, the series $\sum_{n=1}^{\infty} a_n$ is said to be *absolutely convergent*, or *converge absolutely*, if the series $\sum_{n=1}^{\infty} |a_n|$ is convergent.

According to the proposition before last, a series of real numbers $\sum_{n=1}^{\infty} a_n$ is absolutely convergent if and only if the set of partial sums of the series $\sum_{n=1}^{\infty} |a_n|$ is bounded. By the comparison test an absolutely convergent series is convergent. The comparison test is actually a test for absolute convergence, and the following "ratio test" is essentially a special case of it.

***Proposition* (*Ratio test*).** *If* $\sum_{n=1}^{\infty} a_n$ *is an infinite series of nonzero real numbers and if there exists a number* $\rho < 1$ *such that* $|a_{n+1}/a_n| \leq \rho$ *for all sufficiently large* n, *then the series converges absolutely. If* $|a_{n+1}/a_n| \geq 1$ *for all sufficiently large* n, *the series diverges.*

For the proof we may replace the phrases "for all sufficiently large n" by "for all n", since lopping off the first few terms of the series does not affect its convergence. Then if $|a_{n+1}/a_n| \leq \rho < 1$ we have

$$|a_{n+1}| \leq \rho|a_n| \leq \rho^2|a_{n-1}| \leq \cdots \leq \rho^n|a_1|,$$

so the absolute convergence of $\sum_{n=1}^{\infty} a_n$ follows from comparison with the geometric series $|a_1| + \rho|a_1| + \rho^2|a_1| + \cdots$. On the other hand, from the statement $|a_{n+1}/a_n| \geq 1$ comes the fact

$$|a_{n+1}| \geq |a_n| \geq |a_{n-1}| \geq \cdots \geq |a_1|,$$

so the sequence $a_1, a_2, a_3, \ldots$ does not have zero as its limit, proving that the series diverges.

Corollary. *If the series of real numbers* $\sum_{n=1}^{\infty} a_n$ *is such that* $\lim_{n\to\infty} |a_{n+1}/a_n|$ *exists and is less than (greater than) one then the series converges absolutely (diverges).*

For if the limit is less than 1 we may take the ρ of the proposition to be any number between this limit and 1, while if the limit is greater than one the proposition is directly applicable.

The series

$$1 - \frac{1}{2} + \frac{1}{3} - \frac{1}{4} + \cdots$$

is an example of a series of real numbers that is convergent without being absolutely convergent. The series is not absolutely convergent since the corresponding series of absolute values is $\sum_{n=1}^{\infty} 1/n$, the harmonic series, which is known to be divergent; the convergence of the series is a result of the following more general statement on "alternating" series.

Proposition. *Let* $a_1, a_2, a_3, \ldots$ *be a decreasing sequence of positive numbers converging to zero. Then the series*

$$\sum_{n=1}^{\infty} (-1)^{n-1} a_n = a_1 - a_2 + a_3 - a_4 + \cdots$$

converges to some positive number less than a_1.

Any partial sum of the above series can be written

$$(a_1 - a_2) + (a_3 - a_4) + \cdots + \begin{cases} a_{n-1} - a_n & \text{if } n \text{ is even} \\ (a_{n-2} - a_{n-1}) + a_n & \text{if } n \text{ is odd,} \end{cases}$$

which is a sum of nonnegative numbers, hence nonnegative, and this partial sum can also be written

$$a_1 - (a_2 - a_3) - (a_4 - a_5) - \cdots + \begin{cases} -(a_{n-1} - a_n) & \text{if } n \text{ is odd} \\ -(a_{n-2} - a_{n-1}) - a_n & \text{if } n \text{ is even,} \end{cases}$$

which is at most a_1. That is, each partial sum is located in $[0, a_1]$. If we delete the first few terms of an alternating series we are left with plus or minus another alternating series, so for any positive integers $n > m$ we have

$$\left| \sum_{i=m+1}^{n} (-1)^{i-1} a_i \right| = |a_{m+1} - a_{m+2} + a_{m+3} - \cdots \pm a_n| \leq a_{m+1}.$$

Since $\lim_{n\to\infty} a_n = 0$, the series $\sum_{n=1}^{\infty} (-1)^{n-1} a_n$ converges. Each partial sum lies in the closed interval $[0, a_1]$. The sum of the series is not zero since the partial sums $(a_1 - a_2) + (a_3 - a_4) + \cdots + (a_{2n-1} - a_{2n})$ are increasing and positive for large n, and the sum is less than a_1 since it equals $a_1 - (a_2 - a_3 + a_4 - a_5 + \cdots) < a_1$. Thus

$$\sum_{n=1}^{\infty} (-1)^{n-1} a_n \in (0, a_1).$$

Note that this result implies the seemingly stronger result that the difference between the sum of the series

$$a_1 - a_2 + a_3 - a_4 + \cdots$$

and its nth partial sum is less than a_{n+1} in absolute value, since this difference is again an alternating series.

The main properties of absolutely convergent series, proved in the next two propositions, are that their terms may be rearranged in any order or regrouped in any way without affecting the convergence or the sums of the series. This makes it possible to perform many kinds of manipulations with these series without concern about convergence problems, a fact that does not hold for series that are convergent but not absolutely convergent (cf. Problem 14).

Proposition. *Let $f\colon \{1, 2, 3, \ldots\} \to \{1, 2, 3, \ldots\}$ be a function that is one-one and onto. Then if $\sum_{n=1}^{\infty} a_n$ is an absolutely convergent series of real numbers, the series $\sum_{n=1}^{\infty} a_{f(n)}$ is also absolutely convergent and*

$$\sum_{n=1}^{\infty} a_{f(n)} = \sum_{n=1}^{\infty} a_n.$$

For any positive integer n, the numbers $f(1), f(2), \ldots, f(n)$ are a subset of $1, 2, \ldots, N$, for some N. Thus any partial sum of the series $\sum_{n=1}^{\infty} |a_{f(n)}|$ is less than or equal to a partial sum of the series $\sum_{n=1}^{\infty} |a_n|$. Since the latter series converges its partial sums are bounded, hence also the partial sums of the series $\sum_{n=1}^{\infty} |a_{f(n)}|$ are bounded. Thus the series $\sum_{n=1}^{\infty} a_{f(n)}$ is absolutely convergent. We know that

$$\sum_{n=1}^{\infty} a_{f(n)} - \sum_{n=1}^{\infty} a_n = \sum_{n=1}^{\infty} (a_{f(n)} - a_n)$$

and we shall complete the proof by showing that the latter sum is zero. For any $\epsilon > 0$ choose a positive integer N such that whenever $n > m \geq N$ we have $|a_{m+1}| + |a_{m+2}| + \cdots + |a_n| < \epsilon$. Then choose N' such that all the numbers $1, 2, \ldots, N$ are included among $f(1), f(2), \ldots, f(N')$. Clearly $N' \geq N$. If $n > N'$ we have

$$\sum_{i=1}^{n} (a_{f(i)} - a_i) = \sum_{i \in S_1} a_i - \sum_{j \in S_2} a_j,$$

where S_1 consists of those integers $f(1), f(2), \ldots, f(n)$ which do not occur among $1, 2, \ldots, n$, while S_2 consists of those integers $1, 2, \ldots, n$ which do not occur among $f(1), f(2), \ldots, f(n)$. Clearly S_1 and S_2 have no element in common and neither includes any of the numbers $1, 2, \ldots, N$, so that $S_1 \cup S_2 \subset \{N + 1, N + 2, \ldots, M\}$ for some M. Thus for $n > N'$ we have

$$\left| \sum_{i=1}^{n} (a_{f(i)} - a_i) \right| \leq \sum_{i \in S_1 \cup S_2} |a_i| \leq |a_{N+1}| + |a_{N+2}| + \cdots + |a_M| < \epsilon.$$

This proves that $\sum_{n=1}^{\infty} (a_{f(n)} - a_n) = 0$.

If S is a set and $\varphi: S \to \mathbf{R}$ a function then the expression $\sum_{s \in S} \varphi(s)$ is well-defined in case S is finite. This expression can sometimes be given a meaning, independent of any ordering of S, if S is infinite. In fact if S can be put in one-one correspondence with the natural numbers and if in so doing we obtain an absolutely convergent series then we can define $\sum_{s \in S} \varphi(s)$ to be the sum of that series. More precisely, if $f: \{1, 2, 3, \ldots\} \to S$ is a function that is one-one onto and if $\sum_{n=1}^{\infty} \varphi(f(n))$ is absolutely convergent, then we define $\sum_{s \in S} \varphi(s)$ to be $\sum_{n=1}^{\infty} \varphi(f(n))$ (which by the last proposition is independent of the choice of f).

Special cases of infinite sets S which can be put in one-one correspondence with $\{1, 2, 3, \ldots\}$ are

(1) any infinite subset of the natural numbers (for the elements of such a set can be written down in their natural order)
(2) the set of all ordered pairs $\{(n, m) : n, m = 1, 2, 3, \ldots\}$ of natural numbers (which can be written down in the order

$$(1,1), (1,2), (2,1), (1,3), (2,2), (3,1), (1,4), (2,3), (3,2), (4,1), \ldots)$$

(3) any infinite set of disjoint nonempty subsets of the natural numbers (which can be written down in the order of their smallest elements).

The following result says that the terms of an absolutely convergent series may be regrouped in any fashion without altering the absolute convergence or the sum.

Proposition. *Let $\sum_{n=1}^{\infty} a_n$ be an absolutely convergent series of real numbers and let $S_1, S_2, S_3, \ldots$ be a sequence (finite or infinite) of disjoint nonempty sets of natural numbers whose union $S_1 \cup S_2 \cup S_3 \cup \cdots$ is the entire set of natural numbers $\{1, 2, 3, \ldots\}$. Then for each i such that S_i is infinite the series $\sum_{n \in S_i} a_n$ is absolutely convergent, if the number of sets $S_1, S_2, S_3, \ldots$ is infinite then the series $\sum_{i=1}^{\infty} \left(\sum_{n \in S_i} a_n \right)$ is absolutely convergent, and in any case*

$$\sum_{i=1,2,3,\ldots} \left(\sum_{n \in S_i} a_n \right) = \sum_{n=1}^{\infty} a_n.$$

For any infinite subset S of $\{1, 2, 3, \ldots\}$, ordered in a sequence in any fashion, each partial sum of the series $\sum_{n \in S} |a_n|$ is less than or equal to some partial sum of the series $\sum_{n=1}^{\infty} |a_n|$. Since the partial sums of the latter series are bounded, so are the partial sums of $\sum_{n \in S} |a_n|$. Thus $\sum_{n \in S} a_n$ is absolutely convergent. Thus $\sum_{n \in S} a_n$ makes sense for any subset $S \subset \{1, 2, 3, \ldots\}$. We claim that

$$\sum_{n=1}^{\infty} a_n = \sum_{n \in S_1} a_n + \sum_{n \in S_2 \cup S_3 \cup \cdots} a_n.$$

This is clear if either S_1 or $S_2 \cup S_3 \cup \cdots$ is a finite set. On the other hand if both S_1 and $S_2 \cup S_3 \cup \cdots$ are infinite then we can order them into sequences and then use part (1) of the second proposition of this section to get the same result. Thus, by repeated application of this idea,

$$\sum_{n=1}^{\infty} a_n = \sum_{n\in S_1} a_n + \sum_{n\in S_2\cup S_3\cup\cdots} a_n = \sum_{n\in S_1} a_n + \sum_{n\in S_2} a_n + \sum_{n\in S_3\cup S_4\cup\cdots} a_n$$
$$= \cdots = \sum_{n\in S_1} a_n + \sum_{n\in S_2} a_n + \cdots + \sum_{n\in S_\nu} a_n + \sum_{n\in S_{\nu+1}\cup S_{\nu+2}\cup\cdots} a_n$$

in case there happen to be at least ν sets $S_1, S_2, S_3, \ldots$. We are done, except in the case where the number of sets $S_1, S_2, S_3, \ldots$ is infinite, where it remains to show that the series $\sum_{i=1}^{\infty}\left(\sum_{n\in S_i} a_n\right)$ is absolutely convergent and that its sum is $\sum_{n=1}^{\infty} a_n$. To prove that it converges to the sum $\sum_{n=1}^{\infty} a_n$ it suffices to show that

$$\lim_{\nu\to\infty} \sum_{n\in S_{\nu+1}\cup S_{\nu+2}\cup\cdots} a_n = 0.$$

To do this, for any $\epsilon > 0$ choose a positive integer N such that if $n > m \geq N$ then $|a_{m+1}| + |a_{m+2}| + \cdots + |a_n| < \epsilon$ and then choose N' such that $\{1, 2, \ldots, N\} \subset S_1 \cup S_2 \cup \cdots \cup S_{N'}$. If now $\nu > N'$ then the absolute value of any partial sum of the infinite series $\sum_{n\in S_{\nu+1}\cup S_{\nu+2}\cup\cdots} a_n$ (taking the terms of this series to be in any fixed order at all) is at most $\sum_{n\in S'} |a_n|$, where S' is some finite subset of $\{N+1, N+2, N+3, \ldots\}$, hence it is at most $|a_{N+1}| + |a_{N+2}| + \cdots + |a_M|$ for some $M > N$, hence is less than ϵ. Thus the above limit is indeed zero and $\sum_{i=1}^{\infty}\left(\sum_{n\in S_i} a_n\right)$ indeed converges to $\sum_{n=1}^{\infty} a_n$. Applying this to the absolutely convergent series $\sum_{n=1}^{\infty} |a_n|$, we see that $\sum_{i=1}^{\infty}\left(\sum_{n\in S_i} |a_n|\right)$ is convergent. Since $\left|\sum_{n\in S_i} a_n\right| \leq \sum_{n\in S_i} |a_n|$ for all $i = 1, 2, 3, \ldots$, the comparison test shows that $\sum_{i=1}^{\infty}\left(\sum_{n\in S_i} a_n\right)$ is absolutely convergent. This completes the proof.

For infinite series of real-valued functions on a metric space we have the following results, all immediate consequences of the definitions and results of preceding sections.

Proposition. *The infinite series $\sum_{n=1}^{\infty} f_n$ of real-valued functions on a metric space E converges uniformly if and only if, given any $\epsilon > 0$, there exists an integer N such that if $n > m \geq N$ then*

$$|f_{m+1}(p) + f_{m+2}(p) + \cdots + f_n(p)| < \epsilon$$

for all $p \in E$.

The infinite series $\sum_{n=1}^{\infty} f_n$ of real-valued functions on a metric space E is said to *converge absolutely* if the series $\sum_{n=1}^{\infty} f_n(p)$ is absolutely convergent for each $p \in E$.

Corollary. *If $\sum_{n=1}^{\infty} f_n$ is an infinite series of real-valued functions on a metric space E and $\sum_{n=1}^{\infty} a_n$ a convergent series of real numbers such that $|f_n(p)| \leq a_n$ for all $p \in E$ and all n, then $\sum_{n=1}^{\infty} f_n$ converges absolutely and uniformly.*

Proposition. *If $\sum_{n=1}^{\infty} f_n$ is a uniformly convergent series of continuous real-valued functions on a metric space E then its sum is a continuous function on E.*

Proposition. *If $a, b \in \mathbf{R}$, $a < b$, and $\sum_{n=1}^{\infty} f_n$ is a uniformly convergent series of continuous real-valued functions on $[a, b]$ then*

$$\int_a^b \left(\sum_{n=1}^{\infty} f_n \right)(x)dx = \sum_{n=1}^{\infty} \int_a^b f_n(x)dx.$$

Proposition. *Let $f_1, f_2, f_3, \ldots$ be a sequence of real-valued functions on an open interval U in $\mathbf{R}$, each having a continuous derivative. Suppose that the infinite series $\sum_{n=1}^{\infty} f_n'$ converges uniformly on U and that for some $a \in U$ the series $\sum_{n=1}^{\infty} f_n(a)$ converges. Then the series $\sum_{n=1}^{\infty} f_n$ converges to a differentiable function on U and*

$$\left(\sum_{n=1}^{\infty} f_n \right)' = \sum_{n=1}^{\infty} f_n'.$$

§ 3. POWER SERIES.

Let $a, c_0, c_1, c_2, \ldots$ be real numbers. The series of real-valued functions on $\mathbf{R}$

$$\sum_{n=0}^{\infty} c_n(x - a)^n = c_0 + c_1(x - a) + c_2(x - a)^2 + \cdots$$

is called a *power series* (in powers of $x - a$).

To avoid messy circumlocutions, one also calls the above expression a power series when x is not the identity function on $\mathbf{R}$ but rather some specific element of $\mathbf{R}$.

The first question about power series is for which $x \in \mathbf{R}$ the series converges. Here are three examples, all verified by the ratio test; the immediately following theorem asserts that these examples are typical.

EXAMPLE 1. $\sum_{n=0}^{\infty} \frac{x^n}{n!}$ converges for all $x \in \mathbf{R}$.

EXAMPLE 2. $\sum_{n=0}^{\infty} x^n$ converges if $|x| < 1$, diverges if $|x| \geq 1$.

EXAMPLE 3. $\sum_{n=0}^{\infty} n!\,x^n$ converges if $x = 0$, diverges for all other x.

Theorem. *For a given power series $\sum_{n=0}^{\infty} c_n(x-a)^n$ one of the following is true:*

(1) The series converges absolutely for all $x \in \mathbf{R}$.

(2) There exists a real number $r > 0$ such that the series converges absolutely for all $x \in \mathbf{R}$ such that $|x - a| < r$ and diverges for all x such that $|x - a| > r$.

(3) The series converges only if $x = a$.

Furthermore, for any $r_1 < r$ in case (2), or for an arbitrary $r_1 \in \mathbf{R}$ in case (1), the convergence is uniform for all x such that $|x - a| \leq r_1$.

For suppose that the series converges for $x = \xi$, for some $\xi \neq a$, and let $0 < b < |\xi - a|$. We shall show that $\sum_{n=0}^{\infty} c_n(x-a)^n$ converges absolutely and uniformly for all x such that $|x - a| \leq b$. To do this, note that since $\sum_{n=0}^{\infty} c_n(\xi - a)^n$ converges we have $\lim_{n\to\infty} c_n(\xi - a)^n = 0$, so that there exists a number M such that $|c_n(\xi - a)^n| \leq M$ for all n. If $|x - a| \leq b$ then

$$|c_n(x-a)^n| = |c_n(\xi - a)^n| \cdot \left|\frac{x-a}{\xi - a}\right|^n \leq M \cdot \left|\frac{b}{\xi - a}\right|^n.$$

But $\sum_{n=0}^{\infty} M|b/(\xi - a)|^n$ is a geometric series with ratio $|b/(\xi - a)| < 1$, so by comparison with this series $\sum_{n=0}^{\infty} c_n(x-a)^n$ converges absolutely and uniformly for all x such that $|x - a| \leq b$. Now consider the set S of all

$\xi \in \mathbf{R}$ such that the series $\sum_{n=0}^{\infty} c_n(\xi - a)^n$ converges. It may happen that $S = \{a\}$, which is possibility (3) above. It may happen that the set S is unbounded, in which case for every $r_1 \in \mathbf{R}$ there exists a $\xi \in S$ such that $r_1 < |\xi - a|$ and what we have already shown proves that we are in case (1) above. The last possibility is that S is bounded, $S \neq \{a\}$. Here we set $r = \text{l.u.b.}\ \{|\xi - a| : \xi \in S\}$. Then $r > 0$, the series diverges if $|x - a| > r$, and for any $r_1 < r$ there is a $\xi \in S$ such that $r_1 < |\xi - a|$, so that the series converges absolutely and uniformly for all x such that $|x - a| \leq r_1$. Since r_1 was any number less than r this proves that case (2) obtains.

The number r of case (2) is called the *radius of convergence* of the given power series, the interval $(a - r, a + r)$ the *interval of convergence*. In cases (1), (3) we also use the expression "radius of convergence", meaning by this the symbol ∞ or the number zero respectively.

If a power series has radius of convergence $r \neq 0, \infty$, it may or may not converge at the extremities $a - r, a + r$ of the interval of convergence; for example the power series

$$x^2 - \frac{x^4}{2} + \frac{x^6}{3} - \frac{x^8}{4} + \cdots,$$

$$x - \frac{x^2}{2} + \frac{x^3}{3} - \frac{x^4}{4} + \cdots,$$

and

$$x + x^2 + x^3 + \cdots$$

all have interval of convergence $(-1, 1)$ and the first converges at both extremities, the second at one but not the other, and the third at neither extremity.

Lemma. *Let* $\sum_{n=0}^{\infty} c_n(x - a)^n$ *be a power series with radius of convergence* r *(possibly* $r = 0$ *or* $r = \infty$*). Then the series*

$$\sum_{n=0}^{\infty} nc_n(x - a)^{n-1}$$

and

$$\sum_{n=0}^{\infty} \frac{c_n}{n+1}(x - a)^{n+1}$$

also have radius of convergence r.

We shall first show that if the series $\sum_{n=0}^{\infty} c_n(x - a)^n$ converges for $x = \xi \neq a$, then the other two series converge for all x such that $|x - a| <$

$|\xi - a|$. As in the proof of the last theorem, since $\sum_{n=0}^{\infty} c_n(\xi - a)^n$ converges the terms of this series approach zero, hence are bounded, so there exists a number M such that $|c_n(\xi - a)^n| \le M$ for all n. Thus

$$|n\, c_n(x - a)^{n-1}| = \frac{n\,|c_n(\xi - a)^n|}{|\xi - a|}\left|\frac{x - a}{\xi - a}\right|^{n-1} \le \frac{nM}{|\xi - a|}\left|\frac{x - a}{\xi - a}\right|^{n-1}$$

and similarly

$$\left|\frac{c_n}{n+1}(x - a)^{n+1}\right| \le \frac{M\,|\xi - a|}{n+1}\left|\frac{x - a}{\xi - a}\right|^{n+1}.$$

To show that the series $\sum_{n=0}^{\infty} nc_n(x - a)^{n-1}$ and $\sum_{n=0}^{\infty} \frac{c_n}{n+1}(x - a)^{n+1}$ converge if $|x - a| < |\xi - a|$ it therefore suffices to show that the series

$$\sum_{n=0}^{\infty} \frac{nM}{|\xi - a|}\left|\frac{x - a}{\xi - a}\right|^{n-1}$$

and

$$\sum_{n=0}^{\infty} \frac{M\,|\xi - a|}{n+1}\left|\frac{x - a}{\xi - a}\right|^{n+1}$$

converge, which is easily accomplished using the ratio test.

Thus if $\sum_{n=0}^{\infty} c_n(x - a)^n$ has radius of convergence r, then either of the two series obtained from this one by "differentiating term by term" or "integrating term by term" has radius of convergence at least r. But the original series $\sum_{n=0}^{\infty} c_n(x - a)^n$ can be obtained from the series $\sum_{n=0}^{\infty} nc_n(x - a)^{n-1}$ by integrating term by term (except for the term for $n = 0$) and from the series $\sum_{n=0}^{\infty} \frac{c_n}{n+1}(x - a)^{n+1}$ by differentiating term by term, so the previous argument applies in reverse, showing that the radius of convergence of the original series is at least that of either of the two others. Thus all three series have the same radius of convergence r.

Theorem. *If the power series $\sum_{n=0}^{\infty} c_n(x - a)^n$ has radius of convergence $r > 0$ (possibly $r = \infty$) then the function f on $(a - r, a + r)$ (or on $\mathbf{R}$, if $r = \infty$) given by*

$$f(x) = \sum_{n=0}^{\infty} c_n(x - a)^n$$

is differentiable. Furthermore for any $x \in (a - r, a + r)$ (or $x \in \mathbf{R}$, if $r = \infty$) we have

$$f'(x) = \sum_{n=0}^{\infty} nc_n(x - a)^{n-1} \quad \text{and} \quad \int_a^x f(t)dt = \sum_{n=0}^{\infty} \frac{c_n}{n+1}(x - a)^{n+1}.$$

By the lemma the three series involved have the same radius of convergence r. Pick any positive number $r_1 < r$. Then each series converges uniformly on $[a - r_1, a + r_1]$ by the last theorem. By the last proposition of the last section we have

$$\left(\sum_{n=0}^{\infty} c_n(x-a)^n\right)' = \sum_{n=0}^{\infty} \left(c_n(x-a)^n\right)' = \sum_{n=0}^{\infty} nc_n(x-a)^{n-1}$$

on $(a - r_1, a + r_1)$. Similarly the result on term-by-term integration for $x \in [a - r_1, a + r_1]$ follows from the immediate predecessor of the quoted proposition. Since r_1 was any positive number less than r, these same results are true on $(a - r, a + r)$.

Let $a, c_0, c_1, c_2, \ldots$ be real numbers. We say that a real-valued function f on an open subset of $\mathbf{R}$ *has the power series expansion* $\sum_{n=0}^{\infty} c_n(x-a)^n$ *there* if

$$f(x) = \sum_{n=0}^{\infty} c_n(x-a)^n$$

for all x in the open subset. In this case f' exists on the open subset and has a power series expansion there, and similarly for f'', f''', etc. In fact, from

$$f(x) = c_0 + c_1(x-a) + c_2(x-a)^2 + c_3(x-a)^3 + c_4(x-a)^4 + \cdots$$

follows

$$f'(x) = c_1 + 2c_2(x-a) + 3c_3(x-a)^2 + 4\,c_4(x-a)^3 + \cdots,$$

$$f''(x) = 2c_2 + 2\cdot 3c_3(x-a) + 3\cdot 4\,c_4(x-a)^2 + \cdots,$$

$$f'''(x) = 2\cdot 3c_3 + 2\cdot 3\cdot 4\,c_4(x-a) + \cdots,$$

$$\cdots$$

$$f^{(n)}(x) = n!c_n + \cdots.$$

For $x = a$ (assuming this point to be in the open set on which f is defined) we get $f(a) = c_0$, $f'(a) = c_1$, $f''(a) = 2c_2$, $f'''(a) = 2\cdot 3c_3, \ldots, f^{(n)}(a) = n!c_n$. We restate these results as follows.

Corollary. *If the function f has the power series expansion $\sum_{n=0}^{\infty} c_n(x-a)^n$ on an open subset of $\mathbf{R}$ that contains a, then f has continuous derivatives of all orders on this open subset and $c_n = f^{(n)}(a)/n!$ for all n. In particular, if f has a power series expansion in powers of $x - a$ then this power series is unique.*

For any real-valued function f defined on an open subset of $\mathbf{R}$ that contains a and possessing derivatives of all orders at a, the power series

$$\sum_{n=0}^{\infty} \frac{f^{(n)}(a)}{n!}(x-a)^n$$

is called the *Taylor series of f at the point a*.

EXAMPLE. If $|x| < 1$ then $1/(1+x) = 1 - x + x^2 - x^3 + \cdots$. Therefore

$$\log(1+x) = \int_0^x \frac{dt}{1+t} = x - \frac{x^2}{2} + \frac{x^3}{3} - \frac{x^4}{4} + \cdots \quad \text{if } |x| < 1.$$

Is this power series representation of $\log(1+x)$ valid for other values of x? Certainly not for $|x| > 1$, for then the series diverges (and furthermore $\log(1+x)$ is not defined for $x < -1$). Certainly not if $x = -1$, for the same reasons. But if $x = 1$ the series converges. Does it converge to $\log 2$? The answer is *yes*, that is, it is true that

$$\log 2 = 1 - \frac{1}{2} + \frac{1}{3} - \frac{1}{4} + \cdots,$$

but this statement needs proof. Since a uniformly convergent series of continuous functions has a continuous sum and the function $\log(1+x)$ is continuous at $x = 1$, it suffices to show that the series $\sum_{n=1}^{\infty} (-1)^{n-1}x^n/n$ is uniformly convergent for $x \in [0, 1]$. This is true since the sum of any number of consecutive terms starting with the nth has absolute value at most $x^n/n \leq 1/n$, since for $0 < x \leq 1$ we have an alternating series.

Suppose now that f is a real-valued function on an open interval in $\mathbf{R}$ containing a and that f has derivatives of all orders. When does f have a power series expansion in powers of $x - a$? That is, when is it true that $f(x) = \sum_{n=0}^{\infty} f^{(n)}(a)(x-a)^n/n!$? Reverting to a previous notation (end of Chapter V),

$$f(x) = f(a) + \frac{f'(a)(x-a)}{1!} + \cdots + \frac{f^{(n)}(a)(x-a)^n}{n!} + R_n(x, a),$$

we see that we have $f(x) = \sum_{n=0}^{\infty} f^{(n)}(a)(x-a)^n/n!$ for any particular x if and only if $\lim_{n\to\infty} R_n(x, a) = 0$. This can be a useful criterion, since Taylor's theorem gives us some practical information on $R_n(x, a)$.

EXAMPLE. The Taylor series of e^x at the point 0 is $\sum_{n=0}^{\infty} x^n/n!$. This series converges for all $x \in \mathbf{R}$. By Taylor's theorem we can write $R_n(x, 0) = e^{\xi}x^{n+1}/(n+1)!$, where ξ is some number between 0 and x. Since e^x is an increasing function we have

$$|R_n(x,0)| \leq \frac{e^{|x|}|x|^{n+1}}{(n+1)!}.$$

Since $\sum_{n=0}^{\infty} x^n/n!$ converges, $\lim_{n\to\infty} x^{n+1}/(n+1)! = 0$. Thus $\lim_{n\to\infty} R_n(x, 0) = 0$ and therefore

$$e^x = \sum_{n=0}^{\infty} \frac{x^n}{n!}$$

for all $x \in \mathbf{R}$.

§ 4. THE TRIGONOMETRIC FUNCTIONS.

We want to define the trigonometric functions and derive their standard properties in a rigorous manner. The usual geometric way of doing this, using angles and arc length, relies on intuition, but it is possible to make this method entirely logical. However it is much simpler to use an alternate approach. We shall confine the discussion to the sine and cosine functions, since all the other trigonometric functions, as well as their inverses, may be got from these.

We look for real-valued functions on $\mathbf{R}$ that are everywhere twice differentiable and satisfy the differential equation

$$f'' = -f.$$

If such a function f exists, from the equation $f'' = -f$ we deduce $f''' = -f'$, so that f is three times differentiable, then we get $f^{(4)} = -f'' = f$, so that f is four times differentiable. From $f^{(4)} = f$ we get $f^{(5)} = f'$, $f^{(6)} = f'' = -f$, $f^{(7)} = -f'$, $f^{(8)} = -f'' = f$, etc. Thus f has derivatives of all orders and its Taylor series at the point 0 is

$$f(0) + f'(0)x - \frac{f(0)}{2!}x^2 - \frac{f'(0)}{3!}x^3 + \frac{f(0)}{4!}x^4 + \cdots.$$

For any particular $x \neq 0$, Taylor's theorem gives us the estimate

$$R_n(x,0) = \frac{f^{(n+1)}(\xi)}{(n+1)!}x^{n+1}$$

for some ξ between 0 and x. Letting M be an upper bound for $|f(\xi)|$, $|f'(\xi)|$ for ξ ranging over the closed interval with extremities 0 and x, we have

$$|R_n(x,0)| \leq \frac{M|x|^{n+1}}{(n+1)!}.$$

Thus $\lim_{n\to\infty} R_n(x, 0) = 0$. As a consequence, if a function f with the desired properties exists, it is equal to its Taylor series for all $x \in \mathbf{R}$.

For any $c_1, c_2 \in \mathbf{R}$ the series

$$c_1 + c_2 x - \frac{c_1 x^2}{2!} - \frac{c_2 x^3}{3!} + \frac{c_1 x^4}{4!} + \cdots$$

converges for all $x \in \mathbf{R}$, as will be seen by comparison with the series $(|c_1| + |c_2|) \sum_{n=0}^{\infty} |x|^n/n!$. Taking $c_1 = 0$, $c_2 = 1$ we are led to define the sine function sin by

$$\sin x = x - \frac{x^3}{3!} + \frac{x^5}{5!} - \frac{x^7}{7!} + \cdots$$

and taking $c_1 = 1$, $c_2 = 0$ we define the cosine function cos by

$$\cos x = 1 - \frac{x^2}{2!} + \frac{x^4}{4!} - \frac{x^6}{6!} + \cdots.$$

The functions sin and cos are defined on all of $\mathbf{R}$. Differentiating their series term by term gives

$$\frac{d}{dx}\sin x = \cos x, \quad \frac{d}{dx}\cos x = -\sin x.$$

Thus sin and cos both satisfy the equation $f'' = -f$ and any function $f: \mathbf{R} \to \mathbf{R}$ that satisfies this equation must be of the form

$$f(x) = c_1 \cos x + c_2 \sin x$$

for certain constants $c_1, c_2 \in \mathbf{R}$.

It follows immediately from the series expansions that

$$\sin 0 = 0, \qquad \cos 0 = 1$$

$$\sin(-x) = -\sin x, \quad \cos(-x) = \cos x.$$

Also

$$\frac{d}{dx}(\sin^2 x + \cos^2 x) = 2 \sin x \frac{d}{dx} \sin x + 2 \cos x \frac{d}{dx} \cos x = 0,$$

so that $\sin^2 x + \cos^2 x$ is constant. Since $\sin^2 0 + \cos^2 0 = 1$ we get

$$\sin^2 x + \cos^2 x = 1.$$

To derive the familiar addition formulas, fix some $\alpha \in \mathbf{R}$. Then

$$\frac{d}{dx}\sin(x + \alpha) = \cos(x + \alpha)\frac{d}{dx}(x + \alpha) = \cos(x + \alpha)$$

and

$$\frac{d}{dx}\cos(x + \alpha) = -\sin(x + \alpha)\frac{d}{dx}(x + \alpha) = -\sin(x + \alpha),$$

so that

$$\frac{d^2 \sin(x+\alpha)}{dx^2} = -\sin(x+\alpha).$$

Hence we can write

$$\sin(x+\alpha) = c_1 \cos x + c_2 \sin x$$

for certain $c_1, c_2 \in \mathbf{R}$. Differentiating gives

$$\cos(x+\alpha) = -c_1 \sin x + c_2 \cos x.$$

Setting $x = 0$ in the last two equations gives $c_1 = \sin \alpha$, $c_2 = \cos \alpha$, so that

$$\sin(x+\alpha) = \sin x \cos \alpha + \cos x \sin \alpha$$

$$\cos(x+\alpha) = \cos x \cos \alpha - \sin x \sin \alpha.$$

To derive the periodicity properties of sin and cos reason as follows: $\sin x > 0$ if $x \in (0, 2)$, since then all the expressions

$$x - \frac{x^3}{3!}, \frac{x^5}{5!} - \frac{x^7}{7!}, \frac{x^9}{9!} - \frac{x^{11}}{11!}, \cdots$$

are positive. Since $d \cos x/dx = -\sin x$, $\cos x$ is a decreasing function on $(0, 2)$. Now

$$\cos 1 = 1 - \frac{1}{2!} + \frac{1}{4!} - \frac{1}{6!} + \cdots > 0,$$

while

$$\cos 2 = 1 - \frac{2^2}{2!} + \frac{2^4}{4!} - \left(\frac{2^6}{6!} - \frac{2^8}{8!}\right) - \cdots < 1 - \frac{2^2}{2!} + \frac{2^4}{4!} < 0.$$

It follows that $\cos x$ is zero at some unique point of the interval $(1, 2)$. This unique point we denote $\pi/2$ (this is a *definition* of π; note that at the moment we have only the rough approximation $2 < \pi < 4$). We deduce that on the interval $[0, \pi/2]$ $\cos x$ decreases from 1 to 0. Since the derivative of $\sin x$ is $\cos x$, which is positive if $x \in (0, \pi/2)$, $\sin x$ increases on this interval. Using the facts that $\sin 0 = 0$ and $\sin^2 x + \cos^2 x = 1$, we see that $\sin x$ increases from 0 to 1 on the interval $[0, \pi/2]$. The addition formulas then give

$$\sin\left(x + \frac{\pi}{2}\right) = \cos x, \quad \cos\left(x + \frac{\pi}{2}\right) = -\sin x.$$

Repeated application of these give $\sin(x+\pi) = \sin((x+\pi/2)+\pi/2) = \cos(x+\pi/2) = -\sin x$, $\sin(x+2\pi) = \sin((x+\pi)+\pi) = -\sin(x+\pi) = \sin x$. Similarly, or by differentiating the last formula, we get $\cos(x+2\pi) = \cos x$. In other words, sin and cos have period 2π.

§ 5. DIFFERENTIATION UNDER THE INTEGRAL SIGN.

The result we give here is only the simplest of a number of similar results, but it is also the most useful and its proof is illustrative of the others. The notion of partial derivative enters, but only for convenience of notation. None of the properties of partial differentiation to be developed in Chapter IX will be used here, only the definition, which is essentially a one-variable matter.

Let U be a subset of E^2 with the property that for each $x \in \mathbf{R}$ the subset of $\mathbf{R}$ given by $\{y \in \mathbf{R} : (x, y) \in U\}$ is open. That is, U is the union of open subsets of vertical lines in the plane E^2. Then if f is a real-valued function on U and $(x_0, y_0) \in U$, by

$$\frac{\partial f}{\partial y}(x_0, y_0)$$

we denote the derivative at y_0 of the function sending y into $f(x_0, y)$, provided this derivative exists; that is

$$\frac{\partial f}{\partial y}(x_0, y_0) = \lim_{y \to y_0} \frac{f(x_0, y) - f(x_0, y_0)}{y - y_0},$$

if this limit exists. If $\frac{\partial f}{\partial y}(x_0, y_0)$ exists for all $(x_0, y_0) \in U$ we have a real-valued function on U whose value at each $(x_0, y_0) \in U$ is $\frac{\partial f}{\partial y}(x_0, y_0)$, and we of course denote this function by $\frac{\partial f}{\partial y}$.

Theorem. *Let $a, b, c, d \in \mathbf{R}$, $a < b$, $c < d$, and let f be a continuous real-valued function on the subset of E^2 given by*

$$\{(x, y) \in E^2 : a \leq x \leq b, c < y < d\}.$$

Suppose that $\frac{\partial f}{\partial y}$ exists and is continuous on this set. Then the function $F: (c, d) \to \mathbf{R}$ defined by

$$F(y) = \int_a^b f(x, y)dx$$

is differentiable and

$$F'(y) = \int_a^b \frac{\partial f}{\partial y}(x, y)dx$$

for all $y \in (c, d)$.

For a fixed $y \in (c, d)$, both f and $\partial f/\partial y$ are continuous functions of x for $x \in [a, b]$, so both integrals in question exist. We have to show that

$F'(y_0)$ exists and is as indicated for each $y_0 \in (c, d)$. So let $y_0 \in (c, d)$ be fixed. Choose numbers c', d' such that $c < c' < y_0 < d' < d$. Then the set

$$S = \{(x, y) \in E^2 : x \in [a, b], y \in [c', d']\}$$

is compact, so that the continuous function $\partial f/\partial y$ is uniformly continuous on S. Given any $\epsilon > 0$ choose $\delta > 0$ such that if $(x, y) \in S$, $(x_1, y_1) \in S$ and $\sqrt{(x - x_1)^2 + (y - y_1)^2} < \delta$ then

$$\left| \frac{\partial f}{\partial y}(x, y) - \frac{\partial f}{\partial y}(x_1, y_1) \right| < \frac{\epsilon}{b - a}.$$

We may assume that $\delta < \min\{y_0 - c', d' - y_0\}$. Then if $y \in \mathbf{R}$ and $|y - y_0| < \delta$ we have $(x, y) \in S$ for any $x \in [a, b]$. If in addition to $|y - y_0| < \delta$ we have $y \neq y_0$ then

$$\begin{aligned} &\left| \frac{F(y) - F(y_0)}{y - y_0} - \int_a^b \frac{\partial f}{\partial y}(x, y_0)dx \right| \\ &\qquad = \left| \int_a^b \left(\frac{f(x, y) - f(x, y_0)}{y - y_0} - \frac{\partial f}{\partial y}(x, y_0) \right) dx \right| \\ &\qquad = \left| \int_a^b \left(\frac{\partial f}{\partial y}(x, \eta) - \frac{\partial f}{\partial y}(x, y_0) \right) dx \right| \end{aligned}$$

where η (which depends on both x and y) is always between y and y_0. (We have used the mean value theorem.) But $\sqrt{(x - x)^2 + (\eta - y_0)^2} = |\eta - y_0| < |y - y_0| < \delta$, so that

$$\left| \frac{\partial f}{\partial y}(x, \eta) - \frac{\partial f}{\partial y}(x, y_0) \right| < \frac{\epsilon}{b - a}.$$

Thus

$$\left| \frac{F(y) - F(y_0)}{y - y_0} - \int_a^b \frac{\partial f}{\partial y}(x, y_0)dx \right| \leq \epsilon$$

if $|y - y_0| < \delta$, $y \neq y_0$. Therefore

$$F'(y_0) = \lim_{y \to y_0} \frac{F(y) - F(y_0)}{y - y_0} = \int_a^b \frac{\partial f}{\partial y}(x, y_0)dx,$$

as was to be shown.

PROBLEMS

1. Find a sequence of continuous functions $f_n: \mathbf{R} \to \mathbf{R}$ such that $\lim_{x \to 0} \lim_{n \to \infty} f_n(x)$ and $\lim_{n \to \infty} \lim_{x \to 0} f_n(x)$ exist and are unequal.

2. If $f: E^2 - \{(0, 0)\} \to \mathbf{R}$, three limits we can consider are $\lim_{y\to 0} \lim_{x\to 0} f(x, y)$, $\lim_{x\to 0} \lim_{y\to 0} f(x, y)$, and $\lim_{(x,y)\to(0,0)} f(x, y)$. Compute these limits, if they exist, for $f(x, y) = \dfrac{xy}{x^2 + y^2}$, and for $f(x, y) = \dfrac{x^2 - y^2}{x^2 + y^2}$.
3. Find a sequence of continuous functions $f_n: [0, 1] \to \mathbf{R}$ that converges to the zero function and such that the sequence $\int_0^1 f_1(x)dx, \int_0^1 f_2(x)dx, \int_0^1 f_3(x)dx, \ldots$ increases without bound.
4. Find a uniformly convergent sequence of differentiable functions $f_n: (0, 1) \to \mathbf{R}$ such that the sequence $f_1', f_2', f_3', \ldots$ does not converge.
5. Construct a convergent sequence of Riemann integrable real-valued functions on $[0, 1]$ whose limit function is not Riemann integrable.
6. Prove the following fact, implicitly used several times in the text: For any positive integer m, a series of real numbers $\sum_{n=1}^{\infty} a_n$ is convergent if and only if $\sum_{n=1}^{\infty} a_{m+n}$ is convergent, and in that case
$$\sum_{n=1}^{\infty} a_n = a_1 + a_2 + \cdots + a_m + \sum_{n=1}^{\infty} a_{m+n}.$$
7. Show that if $a_1 + a_2 + a_3 + \cdots$ is a convergent series of real numbers and $\nu_1, \nu_2, \nu_3, \ldots$ is a subsequence of the sequence $1, 2, 3, \ldots$, then
$$(a_1 + a_2 + \cdots + a_{\nu_1}) + (a_{\nu_1+1} + \cdots + a_{\nu_2}) + (a_{\nu_2+1} + \cdots + a_{\nu_3}) + \cdots = \sum_{n=1}^{\infty} a_n.$$
8. Let $a_1, a_2, a_3, \ldots$ be a decreasing sequence of positive numbers. Show that
   (a) if $a_1 + a_2 + a_3 + \cdots$ converges then $\lim_{n\to\infty} na_n = 0$
   (b) $a_1 + a_2 + a_3 + \cdots$ converges if and only if $a_1 + 2a_2 + 4a_4 + 8a_8 + \cdots$ converges.
9. (Integral test). Let $f: \{x \in \mathbf{R} : x \geq 1\} \to \mathbf{R}$ be a decreasing positive-valued function. Prove that $\sum_{n=1}^{\infty} f(n)$ converges if and only if $\lim_{n\to\infty} \int_1^n f(x)dx$ exists. (*Hint:* Draw a diagram.)
10. Use the preceding problem to tell for which $p > 0$ the following series converge:
$$\sum_{n=1}^{\infty} \frac{1}{n^p}, \quad \sum_{n=2}^{\infty} \frac{1}{n(\log n)^p}, \quad \sum_{n=3}^{\infty} \frac{1}{n \log n\, (\log \log n)^p}.$$
11. Show the convergence of the series
$$\sum_{n=1}^{\infty} \left(\frac{1}{n} - \frac{1}{n + x}\right)$$
of real-valued functions on $\mathbf{R} - \{-1, -2, -3, \ldots\}$.
12. Show that if $a_1 + a_2 + a_3 + \cdots$ is an absolutely convergent series of real numbers, then $a_1^2 + a_2^2 + a_3^2 + \cdots$ converges.

13. (Root test). Let $\sum_{n=1}^{\infty} a_n$ be a series of real numbers. Show that if there exists a number $\rho < 1$ such that $\sqrt[n]{|a_n|} \leq \rho$ for all sufficiently large n, then the series is absolutely convergent.

14. Prove that a series of real numbers which is convergent but not absolutely convergent can have its terms rearranged in such a way that the new series converges to any preassigned real number, or such that the partial sums of the new series become arbitrarily large, or become arbitrarily small.

15. Prove that if $\sum_{n=1}^{\infty} a_n$ and $\sum_{n=1}^{\infty} b_n$ are absolutely convergent series of real numbers then the series $\sum_{n,m=1}^{\infty} a_n b_m$ is also absolutely convergent, and

$$\sum_{n,m=1}^{\infty} a_n b_m = \left(\sum_{n=1}^{\infty} a_n\right)\left(\sum_{n=1}^{\infty} b_n\right).$$

16. Let $a_1, a_2, a_3, \ldots$ be a sequence of nonnegative real numbers, let $S_1, S_2, S_3, \ldots$ be a sequence (finite or infinite) of disjoint nonempty sets of natural numbers whose union is $\{1, 2, 3, \ldots\}$, and suppose that for each i such that S_i is infinite the series $\sum_{n \in S_i} a_n$ converges and that if the number of sets $S_1, S_2, S_3, \ldots$ is infinite then the series $\sum_{i=1}^{\infty} \left(\sum_{n \in S_i} a_n\right)$ converges. Prove that the series $\sum_{n=1}^{\infty} a_n$ converges.

17. Let V be a complete normed vector space (Prob. 22, Chap. III). The definitions of an infinite series of real numbers and the convergence and sum of such a series generalize immediately to series of elements of V.
   (a) Verify the analog for series of elements of V of the convergence criterion of the first proposition of § 2.
   (b) Define the notion of absolute convergence for series of elements of V and verify the rearranging and regrouping properties of absolutely convergent series of elements of V.
   (c) Define the notion of uniform convergence for a series of V-valued functions on a metric space and prove that the sum of a uniformly convergent series of continuous V-valued functions is continuous.

18. Let $c_0, c_1, c_2, \ldots \in \mathbf{R}$. Prove that if $\lim_{n\to\infty} |c_n/c_{n+1}|$ exists, it is equal to the radius of convergence of the power series $\sum_{n=0}^{\infty} c_n x^n$.

19. Let $c_0, c_1, c_2, \ldots \in \mathbf{R}$. Prove that the radius of convergence of the power series $\sum_{n=0}^{\infty} c_n x^n$ is $1/\limsup_{n\to\infty} \sqrt[n]{|c_n|}$. (Cf. Prob. 18, Chap. III for the definition of lim sup; the quoted expression is to be interpreted as 0 if the lim sup does not exist and as ∞ if the lim sup is 0.)

20. Find the radii of convergence of the following power series:

   (a) $\sum_{n=1}^{\infty} n(\log n)\, x^n$

(b) $\sum_{n=2}^{\infty} (\log n)^{\log n} x^n$

(c) $\sum_{n=1}^{\infty} \frac{x^n}{n\sqrt{n}}$

(d) $\sum_{n=1}^{\infty} \frac{x^n}{(\sqrt{n})^n}$

(e) $\sum_{n=1}^{\infty} \frac{n^n x^n}{n!}$.

21. Show that a power series $\sum_{n=0}^{\infty} c_n x^n$ has the same radius of convergence as $\sum_{n=0}^{\infty} c_{n+m} x^n$, for any positive integer m.

22. Let $a, c_0, c_1, c_2, \ldots \in \mathbf{R}$, with at least one of $c_0, c_1, c_2, \ldots$ nonzero, and let the power series $\sum_{n=0}^{\infty} c_n(x-a)^n$ have positive radius of convergence r. Show that there exists a positive number $\delta < r$ such that the sum of the series is nonzero for every real number x such that $0 < |x - a| < \delta$.

23. Let $a, c_0, c_1, c_2, \ldots \in \mathbf{R}$ and let the power series $\sum_{n=0}^{\infty} c_n(x-a)^n$ have radius of convergence $r > 0$ and converge to $f(x)$ if $|x - a| < r$. Show that if $b \in \mathbf{R}$ and $|b - a| < r$, then there exists a power series in powers of $x - b$ which converges to $f(x)$ whenever $|x - b| < r - |b - a|$.

(*Hint:* Expand out $\sum_{n=0}^{\infty} c_n\,((x-b) + (b-a))^n$ by the binomial theorem.)

24. Let $a \in \mathbf{R}$, $a \neq 0, 1, 2, \ldots$. Show that the "binomial series"

$$1 + ax + \frac{a(a-1)}{2!}x^2 + \frac{a(a-1)(a-2)}{3!}x^3 + \cdots$$

has radius of convergence 1. Let $f(x)$ be the sum of this series on its interval of convergence. Show that $(1 + x) f'(x) = af(x)$, and hence that $f(x) = (1 + x)^a$ for $|x| < 1$.

25. Show that the series $\sum_{n,m=1}^{\infty} \frac{1}{(n+m)!}$ is absolutely convergent and find its sum.

26. Use Problem 15 and the binomial theorem to show that

$$\left(\sum_{n=0}^{\infty} \frac{x^n}{n!}\right)\left(\sum_{n=0}^{\infty} \frac{y^n}{n!}\right) = \sum_{n=0}^{\infty} \frac{(x+y)^n}{n!}$$

for all $x, y \in \mathbf{R}$. Hence give an alternate development of the theory of the exponential function e^x.

27. Find a real-valued function on $\mathbf{R}$ possessing derivatives of all orders whose Taylor series at a certain point converges to the function only at that point. (*Hint:* Start with Prob. 26, Chap. VI.)

28. Define the functions tan, cot, sec, csc in terms of sin and cos and compute their derivatives.

29. Show that the functions sin and tan (cf. Prob. 28) are each increasing on $(-\pi/2, \pi/2)$. Hence define the functions $\sin^{-1}$ and $\tan^{-1}$ (on $(-1, 1)$ and **R** respectively), prove them differentiable, and compute their derivatives.

30. Starting with the formula for $d \tan^{-1} x/dx$ (cf. Prob. 29), give the reasons justifying the argument that for $|x| < 1$ we have

$$\tan^{-1} x = \int_0^x \frac{dt}{1+t^2} = \int_0^x (1 - t^2 + t^4 - \cdots)\, dt = x - \frac{x^3}{3} + \frac{x^5}{5} - \cdots,$$

and therefore

$$\frac{\pi}{4} = 1 - \frac{1}{3} + \frac{1}{5} - \frac{1}{7} + \cdots.$$

31. Starting with the formula for $d \sin^{-1} x/dx$ (cf. Prob. 29), make use of the binomial series (cf. Prob. 24) to find the Taylor series for $\sin^{-1}$ at the point 0.

32. The definitions of an infinite series of real numbers and the convergence and sum of such a series extend verbatim to series of complex numbers (cf. Prob. 20, Chap. III). Verify that the convergence criterion of the first proposition of § 2, the notion of absolute convergence, and the rearranging and regrouping properties of absolutely convergent series hold for series of complex numbers, and that the notion of uniform convergence extends to series of complex-valued functions on a metric space, as well as the theorem that the sum of a uniformly convergent series of continuous functions is continuous. (There is no need to prove any of this if you have done Problem 17.) The notion of real power series extends to power series with complex coefficients in powers of a complex variable. Verify that the first theorem of § 3 generalizes almost verbatim to such complex power series.

33. (a) Verify that the complex power series

$$1 + \frac{z}{1!} + \frac{z^2}{2!} + \frac{z^3}{3!} + \cdots$$

$$1 - \frac{z^2}{2!} + \frac{z^4}{4!} - \frac{z^6}{6!} + \cdots$$

$$z - \frac{z^3}{3!} + \frac{z^5}{5!} - \cdots$$

converge for all $z \in \mathbf{C}$ (cf. Prob. 32).

(b) Denoting the sums of the series of part (a) by e^z, $\cos z$, and $\sin z$ respectively (which agrees with our previous conventions if $z \in \mathbf{R}$), prove that

$$e^{z_1} \cdot e^{z_2} = e^{z_1+z_2}$$

for all $z_1, z_2 \in \mathbf{C}$ (cf. Prob. 26) and that

$$e^{iz} = \cos z + i \sin z$$

for all $z \in \mathbf{C}$.

(c) Verify that

$$\cos z = \frac{e^{iz} + e^{-iz}}{2}$$

$$\sin z = \frac{e^{iz} - e^{-iz}}{2i}$$

$$\cos^2 z + \sin^2 z = 1$$

for all $z \in \mathbf{C}$, and that the usual equations hold for $\cos(z_1 + z_2)$ and $\sin(z_1 + z_2)$.

(d) Prove that any complex number $z \neq 0$ can be written $z = e^{\zeta}$ for some $\zeta \in \mathbf{C}$.

(e) Prove that for any $z \in \mathbf{C}$ and any positive integer n we can write $z = w^n$ for some $w \in \mathbf{C}$.

34. The "fundamental theorem of algebra" states that for any positive integer n and any $a_1, a_2, \ldots, a_n \in \mathbf{C}$ there exists at least one $\zeta \in \mathbf{C}$ such that

$$\zeta^n + a_1 \zeta^{n-1} + a_2 \zeta^{n-2} + \cdots + a_n = 0.$$

Expand the following outline into a proof of this theorem.

(a) Let $f: \mathbf{C} \to \mathbf{C}$ be defined by $f(z) = z^n + a_1 z^{n-1} + \cdots + a_n$. Then $|f(z)|$ is large if $|z|$ is large (cf. Prob. 20, Chap. III).

(b) Since the function $|f(z)|$ is continuous (cf. Prob. 12, Chap. IV), it therefore attains a minimum at some point $\zeta \in \mathbf{C}$.

(c) We can write

$$f(z) = f(\zeta) + a(z - \zeta)^m (1 + (z - \zeta) g(z))$$

where m is a positive integer, $a \in \mathbf{C}$, $a \neq 0$, and $g(z)$ is a polynomial in z.

(d) Choose $\alpha \in \mathbf{C}$ such that $\alpha^m = -f(\zeta)/a$ (cf. Problem 33(e)). Then if $f(\zeta) \neq 0$ we have $|f(\zeta + t\alpha)| < |f(\zeta)|$ for any sufficiently small positive real number t, which is a contradiction. Thus $f(\zeta) = 0$.

35. Let $[a, b]$ and $[c, d]$ be closed intervals in $\mathbf{R}$ and let f be a continuous real-valued function on $\{(x, y) \in E^2 : x \in [a, b], y \in [c, d]\}$. By Prob. 15, Chap. VI, $\int_a^b f(x, y)dx$ is continuous in y and $\int_c^d f(x, y)dy$ is continuous in x, so that

$$\int_c^d \left(\int_a^b f(x, y)dx \right) dy \quad \text{and} \quad \int_a^b \left(\int_c^d f(x, y)dy \right) dx$$

exist. Prove that these integrals are equal by computing d/dt of

$$\int_c^d \left(\int_a^t f(x, y)dx \right) dy \quad \text{and} \quad \int_a^t \left(\int_c^d f(x, y)dy \right) dx$$

for $t \in (a, b)$.

36. Let f be a real-valued function on an open subset of E^2. Prove that if $\frac{\partial}{\partial x}\left(\frac{\partial f}{\partial y}\right)$ and $\frac{\partial}{\partial y}\left(\frac{\partial f}{\partial x}\right)$ exist and are continuous then they are equal. ($\partial/\partial y$ has been defined in the text; the definition of $\partial/\partial x$ is analogous). (*Hint:* Use Problem 35 to show that if the set $\{(x, y) \in E^2 : x \in [a, b], y \in [c, d]\}$ is entirely contained in the set on which f is defined, then

$$\int_a^b \left(\int_c^d \frac{\partial}{\partial x}\left(\frac{\partial f}{\partial y}\right) dy \right) dx = \int_a^b \left(\int_c^d \frac{\partial}{\partial y}\left(\frac{\partial f}{\partial x}\right) dy \right) dx.)$$

37. Let $a, b, c \in \mathbf{R}$, $b < c$, and let f be a continuous real-valued function on the set $\{(x, y) \in E^2 : x \geq a, y \in [b, c]\}$. Let $F: [b, c] \to \mathbf{R}$ be another function. We say that $\int_a^{+\infty} f(x, y)\, dx$ *converges uniformly to* $F(y)$ *on* $[b, c]$ if, for each $\epsilon > 0$, there exists a number $N \in \mathbf{R}$ such that $\left| \int_a^t f(x, y)\, dx - F(y) \right| < \epsilon$ for all $t > N$ and all $y \in [b, c]$ (so that for each $y \in [b, c]$ the improper integral $\int_a^{+\infty} f(x, y)dx$ exists and equals $F(y)$ (cf. Prob. 28, Chap. VI)). Prove that if $\int_a^{+\infty} f(x, y)dx$ converges uniformly to $F(y)$ on $[b, c]$, then F is continuous.

38. Compute

$$\lim_{n\to\infty}\left(\frac{a^{1/n}+b^{1/n}+c^{1/n}}{3}\right)^n$$

if $a, b, c > 0$.

39. For $n = 0, 1, 2, \ldots$ let $I_n = \int_0^{\pi/2} \sin^n x\,dx$. Show that

(a) $\frac{d}{dx}(\cos x \sin^{n-1} x) = (n-1)\sin^{n-2} x - n \sin^n x$

(b) $I_n = \frac{n-1}{n} I_{n-2}$ if $n \geq 2$

(c)
$$\left.\begin{aligned} I_{2n} &= \frac{1\cdot 3\cdot 5\cdots(2n-1)}{2\cdot 4\cdot 6\cdots(2n)}\frac{\pi}{2} \\ I_{2n+1} &= \frac{2\cdot 4\cdot 6\cdots(2n)}{3\cdot 5\cdot 7\cdots(2n+1)}\end{aligned}\right\} \text{ for } n = 1, 2, 3, \ldots$$

(d) $I_0, I_1, I_2, \ldots$ is a decreasing sequence having the limit zero and

$$\lim_{n\to\infty}\frac{I_{2n+1}}{I_{2n}} = 1$$

(e) $\lim_{n\to\infty}\frac{2\cdot 2\cdot 4\cdot 4\cdot 6\cdot 6\cdots(2n)\cdot(2n)}{1\cdot 3\cdot 3\cdot 5\cdot 5\cdot 7\cdots(2n-1)\cdot(2n+1)} = \frac{\pi}{2}$ (Wallis' product).

40. (a) Show that if $f: \{x \in \mathbf{R} : x \geq 1\} \to \mathbf{R}$ is continuous, then

$$\sum_{i=1}^{n} f(i) = \int_1^{n+1} f(x)dx + \sum_{i=1}^{n}\left(f(i) - \int_i^{i+1} f(x)dx\right).$$

(b) Show that if $i > 1$ then $\log i - \int_i^{i+1} \log x\,dx$ differs from $-1/2i$ by less than $1/6i^2$. (*Hint:* Work out the integral using the Taylor series for $\log(1 + x)$ at the point 0.)

(c) Use part (a) with $f = \log$, part (b), and Prob. 22, Chap. VI to prove that

$$\lim_{n\to\infty}\left(\log n! - \left(n + \frac{1}{2}\right)\log n + n\right)$$

exists.

(d) Use part (e) of the preceding problem to compute the above limit, thus obtaining

$$\lim_{n\to\infty}\frac{n!}{n^n e^{-n}\sqrt{2\pi n}} = 1 \quad \text{(Stirling's formula).}$$

41. For $x \in \mathbf{R}$, let $\varphi(x) = \min\{|x - i| : i = 0, \pm 1, \pm 2, \ldots\}$ (which is the distance from x to the nearest integer). Show that

(a) there is a continuous function $f: \mathbf{R} \to \mathbf{R}$ given by

$$f(x) = \sum_{n=0}^{\infty} \varphi(10^n x)/10^n$$

(b) if x and y are real numbers which have decimal expansions which are equal except in their i^{th} decimal places, for some $i > 0$, their i^{th} digits differing by 1 and being distinct from the pair $\{4, 5\}$, then

$$\varphi(y) - \varphi(x) = \pm 10^{-i}$$

(c) if x and y are as in (b) and n is one of the integers $0, 1, \ldots, i-1$ then

$$\varphi(10^n y) - \varphi(10^n x) = \pm 10^{n-i},$$

while if $n \geq i$ then $\varphi(10^n y) = \varphi(10^n x)$

(d) if x and y are as in (b), then

$$f(y) - f(x) = \sum_{n=0}^{i-1} \pm 10^{-i},$$

so that $(f(y) - f(x))/(y - x)$ is an integer, odd or even according as i is odd or even

(e) the continuous function $f\colon \mathbf{R} \to \mathbf{R}$ is nowhere differentiable.

CHAPTER VIII

The Method of Successive Approximations

In this chapter a number of important existence theorems are proved by a successive approximation method. By way of introduction to successive approximations, consider Newton's method of solving an equation $f(x) = 0$: Suppose that f is a continuous real-valued function on an open interval U in $\mathbf{R}$ and that f attains the value zero at some point of U. We wish to find this point. Assume that f is differentiable on U and that f' is continuous and nowhere zero on U (so that f has the value zero at only one point of U). Let $x_0 \in U$ be some first approximation to the root of $f(x) = 0$. Then $f(x_0 + h) = 0$, for some small h. Since $f(x_0 + h)$ is "approximately" $f(x_0) + hf'(x_0)$, by setting the latter expression equal to zero we have h "approximately" $-f(x_0)/f'(x_0)$. Hence we get the next approximation to the root of $f(x) = 0$ to be $x_1 = x_0 - f(x_0)/f'(x_0)$. (Geometrically, x_1 is the point of intersection of the x-axis with the tangent to the curve $y = f(x)$ at the point $(x_0, f(x_0))$.) If $x_1 \in U$ we can try to get a better approximation to the root by setting $x_2 = x_1 - f(x_1)/f'(x_1)$. If $x_2 \in U$ we can similarly define x_3. Thus, provided we never leave the interval U, we get a sequence of points $x_0, x_1, x_2, \ldots$ of U such that

$$x_{n+1} = x_n - \frac{f(x_n)}{f'(x_n)}, \quad n = 0, 1, 2, \ldots.$$

If this sequence converges to a point $\xi \in U$, then by continuity we have

$$\xi = \xi - \frac{f(\xi)}{f'(\xi)},$$

so that $f(\xi) = 0$ and ξ is our desired root. It goes without saying that this procedure does not always work. Several possibilities are illustrated in the figure on the next page and only in case (a) do we arrive at a root.

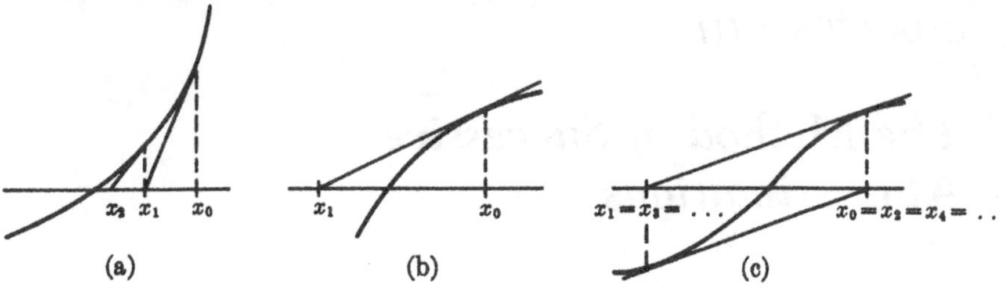

FIGURE 31. Various cases of Newton's method.

§1. THE FIXED POINT THEOREM.

The following theorem says that a certain rather general problem can always be solved by means of the most simple-minded kind of successive approximation. In the remainder of this chapter and in the next chapter we shall see how this easy result can be applied to a variety of special problems of considerable moment.

Theorem. *Let E be a nonempty complete metric space, $F: E \to E$ a function. Suppose there exists a real number k less than one such that for all $p, q \in E$ we have*

$$d(F(p), F(q)) \leq kd(p, q).$$

Then there exists a unique point $P \in E$ such that $F(P) = P$. Furthermore if p_0 is any point of E and $p_1 = F(p_0)$, $p_2 = F(p_1)$, $p_3 = F(p_2)$, etc., then

$$\lim_{n\to\infty} p_n = P.$$

If we apply the given inequality to distinct points p, q of E we get $k \geq 0$. If E does not contain distinct points, that is if E consists of a single point, the inequality holds for $k = 0$. Thus we may assume it given that $0 \leq k < 1$.

Let us start with the last part of the theorem, letting p_0 be an arbitrary point of E and letting $p_1, p_2, p_3, \ldots$ be given by

$$p_{n+1} = F(p_n), \quad n = 0, 1, 2, \ldots.$$

For any integer $n > 0$ we have

$$d(p_n, p_{n+1}) = d(F(p_{n-1}), F(p_n)) \leq kd(p_{n-1}, p_n).$$

Repeated application of this gives

$$d(p_n, p_{n+1}) \leq k\, d(p_{n-1}, p_n) \leq k^2\, d(p_{n-2}, p_{n-1}) \leq k^3\, d(p_{n-3}, p_{n-2}) \leq \cdots$$

so that

$$d(p_n, p_{n+1}) \leq k^n d(p_0, p_1).$$

It follows that for any integers $n > m > 0$ we have

$$\begin{aligned} d(p_m, p_n) &\leq d(p_m, p_{m+1}) + d(p_{m+1}, p_{m+2}) + \cdots + d(p_{n-1}, p_n) \\ &\leq k^m\, d(p_0, p_1) + k^{m+1}\, d(p_0, p_1) + \cdots + k^{n-1}\, d(p_0, p_1) \\ &\leq d(p_0, p_1)(k^m + k^{m+1} + k^{m+2} + \cdots) \\ &= \frac{d(p_0, p_1)k^m}{1 - k}, \end{aligned}$$

the last step using the equation for the sum of a geometric series. Since $\lim_{m\to\infty} k^m = 0$, the sequence $p_0, p_1, p_2, \ldots$ is a Cauchy sequence. E is complete, so this sequence converges to a limit, say P. That is

$$P = \lim_{n\to\infty} p_n.$$

The inequality $d(F(p), F(q)) \leq d(p, q)$ shows that F is uniformly continuous, hence continuous. Thus

$$F(P) = \lim_{n\to\infty} F(p_n) = \lim_{n\to\infty} p_{n+1} = P.$$

To show that P is the only point with the property that $F(P) = P$, suppose that $Q \in E$, $F(Q) = Q$. Then

$$d(P, Q) = d(F(P), F(Q)) \leq kd(P, Q).$$

Since $k < 1$ this implies that $d(P, Q) = 0$, so that $P = Q$.

A map F of a metric space E into itself is called a *contraction map* if there exists a real number $k < 1$ such that $d(F(p), F(q)) \leq kd(p, q)$ for all $p, q \in E$. A *fixed point* for F is a point $P \in E$ such that $F(P) = P$. The theorem says, in brief, that a contraction map of a nonempty complete metric space has a fixed point. The theorem also asserts that this fixed point is unique and it gives a simple successive approximation procedure for finding the fixed point. If we check the details of the proof we see that we can even estimate the accuracy of any approximation of the fixed point, for one direct consequence of the inequality

$$d(p_m, p_n) \leq \frac{d(p_0, p_1)k^m}{1 - k}$$

is

$$d(p_m, P) \leq \frac{d(p_0, p_1)k^m}{1 - k}.$$

Thus for any $p_0 \in E$ we have

$$d(p_0, P) \le \frac{d(p_0, F(p_0))}{1-k}.$$

Proposition. *Let $a, b \in \mathbf{R}$, $a < b$, and let $F\colon [a, b] \to [a, b]$ be a continuous function. Suppose that F is differentiable on (a, b) and that there exists a real number $k < 1$ such that $|F'(x)| \le k$ for all $x \in (a, b)$. Then F is a contraction map, so that the fixed point theorem is applicable to $[a, b]$ and F.*

The proof of this consists in showing that for all $p, q \in [a, b]$ we have $|F(p) - F(q)| \le k|p - q|$. This is clear if $p = q$, whereas if $p \ne q$ the mean value theorem gives us the existence of some ξ between p and q such that

$$F(p) - F(q) = F'(\xi)(p - q),$$

so that

$$|F(p) - F(q)| = |F'(\xi)||p - q| \le k|p - q|.$$

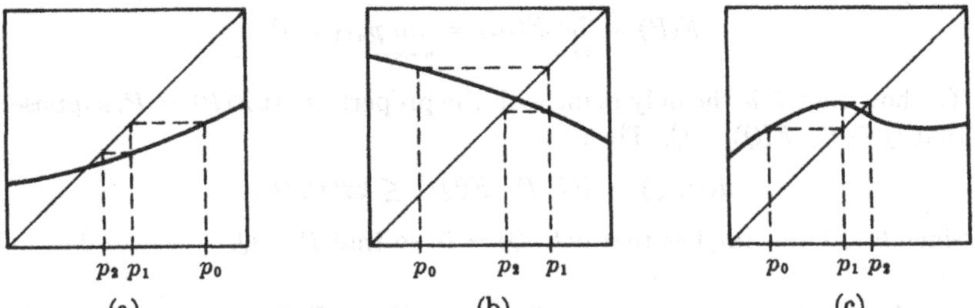

FIGURE 32. The fixed point theorem for contraction maps of a closed interval in **R**. In each diagram the curved line is the graph of the function.

The important point about the proposition is that it provides a specific procedure for finding the point $P \in [a, b]$ such that $F(P) = P$, not that it tells us that such a P exists. The mere existence of a fixed point P for *any* continuous map $F\colon [a, b] \to [a, b]$ can be deduced from the intermediate value theorem by noting that the real-valued function on $[a, b]$ whose value at any x is $F(x) - x$ is continuous, nonnegative at a, and nonpositive at b, so that it equals zero somewhere on $[a, b]$.

The proposition can be used to solve equations of the form $f(x) = 0$.

Suppose in fact that $a, b \in \mathbf{R}$, $a < b$, and that $f: [a, b] \to \mathbf{R}$ is a continuous function that changes sign on $[a, b]$. Suppose further that f is differentiable on (a, b) and that there exist $K_1, K_2 \in \mathbf{R}$ such that $0 < K_1 \leq f'(x) \leq K_2$ for all $x \in (a, b)$. (Thus f is an increasing function on $[a, b]$; changing f to $-f$ would enable us to handle decreasing functions that happen to satisfy analogous conditions.) Then we can show that the proposition, and hence the fixed point theorem, are applicable to $[a,b]$ and the function $F(x) = x - cf(x)$, where c is any constant such that $0 < c \leq 1/K_2$. To do this note first that F is continuous and that for any $x \in (a, b)$ the number $F'(x) = 1 - cf'(x)$ is at least $1 - cK_2$ and at most $1 - cK_1$, hence is nonnegative and less than some number less than one. In particular F is an increasing function. Since f also increases on $[a, b]$, $f(a) < 0 < f(b)$ and for any $x \in [a, b]$

$$a < a - cf(a) = F(a) \leq F(x) \leq F(b) = b - cf(b) < b,$$

so that F actually maps $[a, b]$ into itself. Thus the conditions of the proposition are indeed satisfied. The fixed point $\xi \in [a, b]$ is such that $\xi = F(\xi) = \xi - cf(\xi)$, that is $f(\xi) = 0$, as desired. We remark finally that if we choose $c = 1/f'(x_0)$ for some fixed $x_0 \in [a, b]$, set $F(x) = x - cf(x)$, and then try to define $x_1, x_2, x_3, \ldots$ by the recursion relation $x_{n+1} = F(x_n)$ for $n = 0, 1, 2, \ldots$, we obtain a well-known simplification of Newton's method which, however, does not always work.

§ 2. THE SIMPLEST CASE OF THE IMPLICIT FUNCTION THEOREM.

It often happens that for a given function f of two real variables we want to solve the equation $f(x, y) = 0$ for y in terms of x. That is, for a given real-valued function f on a subset of E^2 we want to find a real-valued function φ on a subset of $\mathbf{R}$ such that for all x in the latter subset we have $f(x, \varphi(x)) = 0$. The problem as thus posed is unwieldy. Among other difficulties, for a given real value of x there may not exist any $y \in \mathbf{R}$ such that $f(x, y) = 0$, or there may exist many such numbers y, the number of them possibly depending on x. Even if there exists such a function $\varphi(x)$ there is no reason to expect that it can be given "explicitly," that is by means of some sort of formula, so that actually "solving" for y in terms of x, or "finding" φ, is literally out of the question. The most that we can hope to do, and this would be of some moment, is to show that under certain general conditions there *exists* some function φ satisfying the equation $f(x, \varphi(x)) = 0$ and possessing other desirable qualities, such as being defined on a fairly large subset of $\mathbf{R}$, being continuous, or being unique. To be somewhat more specific as well as more practical, we reformulate our problem as follows: We assume that a real-valued function f is defined and continuous on a given open subset of E^2 and we ask whether there exists a continuous real-valued function φ on some nonempty open subset of $\mathbf{R}$ such that for all x in the latter subset the point $(x, \varphi(x))$ lies in the given open subset of E^2 and $f(x, \varphi(x)) = 0$. These conditions are not enough to assure the existence of a solution; for example if f is always positive then $\varphi(x)$ cannot be defined for any $x \in \mathbf{R}$. We therefore suppose we are given a point (a, b) in the given open subset of E^2 such that $f(a, b) = 0$ and we insist that the function φ be defined on some open interval containing a and that $\varphi(a) = b$. But even this is not enough. For example, if $f(x, y) = x^2 + y^2$ and $(a, b) = (0, 0)$ then φ cannot be defined for any real number $x \neq 0$. The trouble in this last example seems to be that for any given x near a the function f has an extreme value at $y = b$. Thus we need some condition guaranteeing that for any given x near a the function f actually goes both up and down as y varies near b, and the obvious way to do this is to suppose that $\partial f/\partial y$ exists near (a, b) and is continuous and different from zero. As a matter of fact this condition is sufficient for the solvability of our problem, as the following implicit function theorem shows.

Theorem. *Let f be a continuous real-valued function on an open subset of E^2 that contains the point (a, b), with $f(a, b) = 0$. Suppose that $\partial f/\partial y$ exists and is continuous on the given open subset and that $\dfrac{\partial f}{\partial y}(a, b) \neq 0$. Then there exist open intervals $U, V \subset \mathbf{R}$, with $a \in U$ and $b \in V$, such that there exists a unique function $\varphi\colon U \to V$ such that $f(x, \varphi(x)) = 0$ for all $x \in U$, and such that this function φ is continuous.*

We begin by defining another real-valued function F on the same open subset of E^2 on which f is defined by

$$F(x, y) = y - \frac{f(x, y)}{\dfrac{\partial f}{\partial y}(a, b)}.$$

This F has as basic properties that F and $\partial F/\partial y$ are continuous, $F(a, b) = b$, $\dfrac{\partial F}{\partial y}(a, b) = 0$, and for any (x, y) the equation $f(x, y) = 0$ holds if and only if $F(x, y) = y$. The last property indicates the main idea of the proof, which is a judicious application of the fixed point theorem. For this, we choose some $r > 0$ such that the open ball in E^2 of center (a, b) and radius r is entirely contained in the open set on which f is defined. Since $\partial F/\partial y$ is continuous and $\dfrac{\partial F}{\partial y}(a, b) = 0$ we may assume r taken so small that $|\partial F/\partial y| < 1/2$ at each point of the ball. Choose k such that $0 < k < r$, then choose h such that $0 < h < \sqrt{r^2 - k^2}$ and such that $|F(x, b) - b| < k/2$ whenever $|x - a| < h$, this last demand being justifiable by the continuity of F. We shall prove the theorem with $U = (a - h, a + h)$ and

$V = (b - k, b + k)$. Consider any fixed $x \in U$. For any $y \in \mathbf{R}$ such that $|y - b| \leq k$ we have

$$d((x, y), (a, b)) = \sqrt{(x - a)^2 + (y - b)^2} < \sqrt{h^2 + k^2} < r,$$

so that (x, y) is in our open ball of radius r. If also $y' \in \mathbf{R}$, $|y' - b| \leq k$, then by the mean value theorem we have

$$F(x, y) - F(x, y') = \frac{\partial F}{\partial y}(x, y'')(y - y')$$

for some y'' between y and y' (or, if $y = y'$, for $y'' = y = y'$). The point (x, y'') is also in our ball of radius r, so we deduce

$$|F(x, y) - F(x, y')| \leq \frac{1}{2}|y - y'|.$$

Also

$$\begin{aligned} |F(x, y) - b| &\leq |F(x, y) - F(x, b)| + |F(x, b) - b| \\ &< \frac{1}{2}|y - b| + \frac{k}{2} \leq \frac{k}{2} + \frac{k}{2} = k. \end{aligned}$$

Thus the fixed point theorem is applicable to the closed interval $[b - k, b + k]$ and the function that sends each y into $F(x, y)$, a function that maps this interval into itself. (Recall that x is fixed.) This gives us the existence of a unique $\bar{y}$ such that $|\bar{y} - b| \leq k$ and $F(x, \bar{y}) = \bar{y}$, that is $f(x, \bar{y}) = 0$. Notice that in fact $|\bar{y} - b| < k$ by the last displayed inequality; that is, $\bar{y} \in V$. Since this is valid for each $x \in U$ our desired function φ is defined by $\varphi(x) = \bar{y}$ and to complete the proof of the theorem it remains only to prove that φ is continuous. But the continuity of φ can easily be deduced from what has been proved already. Note first that $\partial f/\partial y$ is not zero at any point of the open ball with center (a, b) and radius r, since $|\partial F/\partial y| < 1/2$ there. To prove φ continuous at some $a' \in U$, for any $\epsilon > 0$ consider the same problem as in the statement of the theorem, with (a, b) replaced by (a', b'), where $b' = \varphi(a')$, and f replaced by its restriction to the open subset of E^2 given by

$$\{(x, y) \in E^2 : x \in U, y \in V, |y - b'| < \epsilon\}.$$

The procedure used to obtain U, V, φ gives us, analogously, U', V', φ', the latter being a function $\varphi' : U' \to V'$ such that $f(x, \varphi'(x)) = 0$ for all $x \in U'$. (In the present context the prime $'$ does *not* indicate differentiation.) But we are dealing here with the restriction of f to a smaller open subset of E^2 than given originally, so that $U' \subset U$, $V' \subset V$, and so that $|y - b'| < \epsilon$ for all $y \in V'$. The uniqueness property of φ implies that $\varphi'(x) = \varphi(x)$ for all $x \in U'$, so that $|\varphi(x) - \varphi(a')| < \epsilon$ for all $x \in U'$. Thus $|\varphi(x) - \varphi(a')| < \epsilon$ whenever x is in some open ball in $\mathbf{R}$ of center a'. Hence φ is continuous at a'. Since a' was an arbitrary point of U, the function φ is continuous.

***Corollary* (*Inverse function theorem*).** *Let g be a real-valued function on an open subset of* $\mathbf{R}$ *that contains the point b and suppose that g' exists and is continuous on this open subset, with $g'(b) \neq 0$. Then there exist open intervals U, V in* $\mathbf{R}$*, with $b \in V$, such that g is defined at each point of V and the restriction of g to V is a one-one map of V onto U whose inverse function $g^{-1}: U \to V$ is differentiable.*

On the open subset of E^2 consisting of all $(x, y) \in E^2$ such that y is in the open subset of $\mathbf{R}$ on which g is defined we define a function f by $f(x, y) = x - g(y)$. Set $a = g(b)$. We may apply the theorem to this f and the point (a, b) to obtain open intervals $U_1, V_1 \subset \mathbf{R}$, with $a \in U_1$ and $b \in V_1$, and a unique and continuous function $\varphi: U_1 \to V_1$ such that $x = g(\varphi(x))$ for all $x \in U_1$. The map φ is one-one from U_1 onto $\varphi(U_1) = g^{-1}(U_1) \cap V_1$. By the first proposition of Chapter IV the set $g^{-1}(U_1)$ is an open subset of the set on which g is defined, hence an open subset of $\mathbf{R}$. Therefore $\varphi(U_1) = g^{-1}(U_1) \cap V_1$ is an open subset of $\mathbf{R}$. $\varphi(U_1)$ is also connected, since it is a continuous image of a connected set. As a nonempty connected open subset of an open interval in $\mathbf{R}$, $\varphi(U_1)$ is itself an open interval (in fact it is the open interval $(\text{g.l.b. } \varphi(U_1), \text{l.u.b. } \varphi(U_1))$). If we set $U = U_1$, $V = \varphi(U_1)$, then the restriction of g to V is a one-one map onto U whose inverse map is φ and the whole of the corollary is proved except for the differentiability of φ. [It is only fair to remark that there are much more elementary proofs. For example, g' maintains the same sign on some open interval in $\mathbf{R}$ that contains b, so that we can assume that g is either strictly increasing or strictly decreasing. We can also assume that g is bounded. Using the intermediate value theorem we deduce that g is one-one from any open subinterval of the open set on which it is defined onto an open interval in $\mathbf{R}$. This enables us to define the inverse function g^{-1} and to prove that g^{-1} is continuous.] To prove φ differentiable we may suppose V chosen so that $g'(y) \neq 0$ if $y \in V$; indeed this is true for the V we have constructed, and we could in any case guarantee this by replacing U and V by suitable open subintervals. Then for $x, x_1 \in U, x \neq x_1$, we have

$$x - x_1 = g(\varphi(x)) - g(\varphi(x_1)) = \big(\varphi(x) - \varphi(x_1)\big)g'(\theta),$$

for some θ between $\varphi(x)$ and $\varphi(x_1)$. Since $\theta \in V$ we have $g'(\theta) \neq 0$ and we may write

$$\frac{\varphi(x) - \varphi(x_1)}{x - x_1} = \frac{1}{g'(\theta)}.$$

Since φ is continuous we have $\lim_{x \to x_1} \theta = \varphi(x_1)$, so since g' is continuous we have $\lim_{x \to x_1} g'(\theta) = g'\big(\varphi(x_1)\big)$. Hence

$$\lim_{x \to x_1} \frac{\varphi(x) - \varphi(x_1)}{x - x_1} = \frac{1}{g'(\varphi(x_1))}.$$

Thus φ is differentiable at each $x_1 \in U$, as was to be shown.

In the course of the above proof the equation

$$\varphi'(x) = \frac{1}{g'(\varphi(x))}$$

was obtained. This equation can itself be considered an immediate consequence of the corollary, since once it is known that φ is differentiable, the application of the chain rule to the equation $x = g(\varphi(x))$ gives $1 = g'(\varphi(x))\varphi'(x)$.

The above implicit function theorem and inverse function theorem generalize to functions of more than one variable. Their generalizations will be proved in the next chapter, after the necessary preliminaries on partial differentiation.

§ 3. EXISTENCE AND UNIQUENESS THEOREMS FOR ORDINARY DIFFERENTIAL EQUATIONS.

Suppose that f is a continuous real-valued function on a certain open subset of E^2 and let (a, b) be a point of this open subset. To solve the differential equation

$$\frac{dy}{dx} = f(x, y)$$

with the initial condition $y(a) = b$ means to find a differentiable real-valued function φ on some open interval in $\mathbf{R}$ containing a such that for all x in this interval we have $\varphi'(x) = f(x, \varphi(x))$ (this implies that the point $(x, \varphi(x))$ must lie in the given open subset of E^2) and in addition $\varphi(a) = b$. We note first that the interval in $\mathbf{R}$ on which a solution φ can be defined may be rather small, even if the function f is defined on the whole of E^2 and is very nicely behaved. For example, for any solution φ of the differential equation

$$\frac{dy}{dx} = 1 + y^2$$

we have

$$\frac{d \tan^{-1} \varphi(x)}{dx} = \frac{1}{1 + (\varphi(x))^2}\, \varphi'(x) = 1,$$

so that $\tan^{-1} \varphi(x) - x$ is constant on any open interval on which φ is defined; if we impose the initial condition $\varphi(0) = 0$, then the only solution on an open interval in $\mathbf{R}$ containing 0 is given by $\varphi(x) = \tan x$, thus restricting us to $|x| < \pi/2$. Therefore if we are interested in solving the above differential equation with initial condition, all we can hope for in general is that a solution exist on *some* open interval containing a. This is indeed the case, with no further conditions, although we shall not prove this fact in this text. However if we want the solution to be unique, which is highly desirable in many cases, some further conditions are necessary. For example, if $f(x, y) = 3|y|^{2/3}$ and $(a, b) = (0, 0)$, we have the two solutions $\varphi_1(x) = 0$ and $\varphi_2(x) = x^3$. Hence some condition must be imposed on f if we wish to guarantee a unique solution.

The condition we shall impose on f is the following so-called *Lipschitz condition*: there exists $M \in \mathbf{R}$ such that whenever (x, y) and (x, z) are in the open subset of E^2 on which f is defined we have

$$|f(x, y) - f(x, z)| \leq M|y - z|.$$

This condition is automatically satisfied if $\partial f/\partial y$ exists and is bounded in the given open subset of E^2 and if a vertical line segment lies entirely within this open set whenever its extremities do, for in this case the mean value theorem enables us to write

$$f(x, y) - f(x, z) = (y - z)\frac{\partial f}{\partial y}(x, \eta),$$

for some η between y and z (if $y = z$, we take $\eta = y = z$), so M may be taken to be any upper bound for $|\partial f/\partial y|$.

Theorem. *Let f be a continuous real-valued function on an open subset of E^2 that contains the point (a, b). Suppose there exists $M \in \mathbf{R}$ such that*

$$|f(x, y) - f(x, z)| \leq M|y - z|$$

whenever (x, y) and (x, z) are in the given open set. Then there exists $h \in \mathbf{R}$, $h > 0$, such that there exists one and only one real-valued function φ on $(a - h, a + h)$ such that $\varphi'(x) = f(x, \varphi(x))$ on this interval and $\varphi(a) = b$.

For a continuous real-valued function φ on an open interval in $\mathbf{R}$ that contains a, the equations $\varphi'(x) = f(x, \varphi(x))$ and $\varphi(a) = b$ hold if and only if $\varphi(x) = \int_a^x f(t, \varphi(t))dt + b$, as follows from the fundamental theorem of calculus. Thus solving the given differential equation with initial condition is equivalent to solving the "integral equation"

$$\varphi(x) = \int_a^x f(t, \varphi(t))dt + b.$$

If to a function ψ we associate another function $F(\psi)$ whose value at any x is $(F(\psi))(x) = \int_a^x f(t, \psi(t))dt + b$, we see that solving the integral equation is the same as finding a function φ such that $F(\varphi) = \varphi$, that is, solving a kind of fixed point problem; this is the basic idea of the proof, which we now proceed to work out in detail. We begin by choosing some $N \in \mathbf{R}$ such that $N > |f(a, b)|$, then some $r \in \mathbf{R}$, $r > 0$, such that the open ball in E^2 of center (a, b) and radius r is entirely contained in the open set on which f is defined and such that $|f(x, y)| < N$ whenever (x, y) is in the ball. Then choose $h \in \mathbf{R}$, $h > 0$, so that $h < r/2$, $h < r/2N$, and $hM < 1$. The rectangle

$$\{(x, y) \in E^2 : |x - a| \leq h, |y - b| \leq Nh\}$$

is then entirely contained in the open set on which f is defined and for each (x, y) in this rectangle we have $|f(x, y)| < N$. We are going to prove that there exists one and only one continuous function φ on the closed interval $[a - h, a + h]$ such that

$$\varphi(x) = \int_a^x f(t, \varphi(t))dt + b$$

for all $x \in [a - h, a + h]$. To do this, consider the complete metric space $C([a - h, a + h])$ of all continuous real-valued functions on the compact metric space $[a - h, a + h]$, as at the end of Chapter IV. Let B be the closed ball in $C([a - h, a + h])$ of radius Nh whose center is the constant function b, that is B is the set of all continuous functions

$$\psi: [a - h, a + h] \rightarrow [b - Nh, b + Nh].$$

Since B is a closed subset of a complete metric space, B is itself a complete metric space. We claim that any solution of the above integral equation must lie in B, that in fact if φ is as above then $|\varphi(x) - b| < Nh$ for all $x \in [a - h, a + h]$. For if there exist points $x \in [a - h, a + h]$ such that $|\varphi(x) - b| \geq Nh$, let γ be the greatest lower bound of $|x - a|$ for all such points. Since φ is continuous and $\varphi(a) = b$, it follows that $\gamma > 0$ and $|\varphi(a \pm \gamma) - b| = Nh$ for at least one choice of the sign $\pm$. Thus $Nh = |\varphi(a \pm \gamma) - \varphi(a)| = |\gamma\varphi'(\alpha)|$, for some α between a and $a \pm \gamma$, by the mean value theorem, and the latter expression equals $|\gamma f(\alpha, \varphi(\alpha))| < \gamma N \leq hN$, which is a contradiction. Thus any solution φ of the integral equation is in B. Now for any $\psi \in B$ define a new function

$$F(\psi): [a - h, a + h] \rightarrow \mathbf{R}$$

by

$$(F(\psi))(x) = \int_a^x f(t, \psi(t))dt + b.$$

Since $\psi \in B$, for any $t \in [a - h, a + h]$ we have $|\psi(t) - b| \leq Nh$, so that $f(t, \psi(t))$ is defined, is continuous as a function of t, and $|f(t, \psi(t))| < N$.

Hence for $x \in [a-h, a+h]$, $(F(\psi))(x)$ is defined and $|(F(\psi))(x) - b| = \left|\int_a^x f(t, \psi(t))dt\right| \leq N|x-a| \leq hN$. Since $F(\psi)$ is clearly continuous we have $F(\psi) \in B$. Thus $F: B \to B$. If $\psi, \omega \in B$ then for any element $x \in [a-h, a+h]$ we have

$$\begin{aligned}
|(F(\psi))(x) - (F(\omega))(x)| & \\
&= \left| \int_a^x (f(t, \psi(t)) - f(t, \omega(t)))dt \right| \\
&\leq |x-a| \max\{|f(t, \psi(t)) - f(t, \omega(t))| : t \in [a-h, a+h]\} \\
&\leq |x-a| M \max\{|\psi(t) - \omega(t)| : t \in [a-h, a+h]\} \\
&\leq h M d(\psi, \omega),
\end{aligned}$$

where d denotes the metric in B. Thus

$$d(F(\psi), F(\omega)) \leq h M d(\psi, \omega).$$

But $hM < 1$, so that F is a contraction map. The fixed point theorem thus assures us of the existence of a unique $\varphi \in B$ such that $\varphi = F(\varphi)$, that is such that

$$\varphi(x) = \int_a^x f(t, \varphi(t))dt + b$$

for all $x \in [a-h, a+h]$. For $x \in (a-h, a+h)$ we clearly have $\varphi'(x) = f(x, \varphi(x))$ and $\varphi(a) = b$. Thus the existence part of the proof is complete. However it is not immediately obvious that the restriction of φ to $(a-h, a+h)$ is the only solution on $(a-h, a+h)$ of our differential equation with initial condition. To see this, note that the above proof would have gone through with h replaced by any $h_1 \in \mathbf{R}$ such that $0 < h_1 < h$. Any solution on $(a-h, a+h)$ of the differential equation with initial condition gives a solution of the integral equation on $[a-h_1, a+h_1]$. But we know that the integral equation has a unique solution on $[a-h_1, a+h_1]$. Thus any two solutions on $(a-h, a+h)$ have equal restrictions to $[a-h_1, a+h_1]$. Since this is true for all h_1 such that $0 < h_1 < h$, there is at most one solution on $(a-h, a+h)$ and our proof is now complete.

The preceding theorem can be generalized to systems of first order differential equations of the form

$$\begin{aligned}
\frac{dy_1}{dx} &= f_1(x, y_1, \ldots, y_n) \\
\frac{dy_2}{dx} &= f_2(x, y_1, \ldots, y_n) \\
&\cdots \\
\frac{dy_n}{dx} &= f_n(x, y_1, \ldots, y_n)
\end{aligned}$$

with initial conditions $y_i(a) = b_i$, $i = 1, 2, \ldots, n$. Here functions $f_1, f_2, \ldots, f_n$ of $n+1$ variables are given, together with real numbers $a, b_1, \ldots, b_n$, and the problem is to find functions $y_1, \ldots, y_n$ of x satisfying the given equations. Except for notational complications, the generalization of the preceding theorem is straightforward. However we are also interested in getting sharper results than have so far been obtained for $n = 1$, so we begin with a rather specific lemma that isolates the technical details relating to the fixed point theorem.

Lemma. *Let $f_1, \ldots, f_n$ be continuous real-valued functions on an open subset U of E^{n+1} that contains the point $(a, b_1, \ldots, b_n)$. Suppose there exist $N, M \in \mathbf{R}$ such that for each $i = 1, \ldots, n$*

$$|f_i(x, y_1, \ldots, y_n)| < N$$

whenever $(x, y_1, \ldots, y_n) \in U$ and

$$|f_i(x, y_1, \ldots, y_n) - f_i(x, z_1, \ldots, z_n)| \leq M((y_1 - z_1)^2 + \cdots + (y_n - z_n)^2)^{1/2}$$

whenever $(x, y_1, \ldots, y_n), (x, z_1, \ldots, z_n) \in U$. Let $h \in \mathbf{R}$, $h > 0$, be such that

$$\{(x, y_1, \ldots, y_n) \in E^{n+1} : |x - a| \leq h, \\ |y_1 - b_1| \leq Nh, \ldots, |y_n - b_n| \leq Nh\} \subset U.$$

Then if $hM\sqrt{n} < 1$ there exists one and only one n-tuple $(\varphi_1, \ldots, \varphi_n)$ of real-valued functions on the interval $(a - h, a + h)$ such that for each $i = 1, \ldots, n$, $\varphi_i'(x) = f_i(x, \varphi_1(x), \ldots, \varphi_n(x))$ on this interval and $\varphi_i(a) = b_i$.

We want to find functions $\varphi_1, \ldots, \varphi_n$ satisfying the system of integral equations

$$\varphi_i(x) = \int_a^x f_i(t, \varphi_1(t), \ldots, \varphi_n(t))dt + b_i, \quad i = 1, \ldots, n.$$

Analogously to what was done in the proof of the preceding theorem, we consider the compact metric space $[a - h, a + h]$ and the complete metric space $\mathfrak{F}$ of all continuous functions from $[a - h, a + h]$ into E^n, as at the end of Chapter IV. We indicate a function into E^n by its n-tuple of component functions, so that an element ψ of $\mathfrak{F}$ is an n-tuple $(\psi_1, \ldots, \psi_n)$, where each ψ_i is a continuous real-valued function on $[a - h, a + h]$ and for any $x \in [a - h, a + h]$ we have $\psi(x) = (\psi_1(x), \ldots, \psi_n(x))$. Consider the subset B of $\mathfrak{F}$ consisting of all $\psi = (\psi_1, \ldots, \psi_n)$ such that $|\psi_i(x) - b_i| \leq Nh$ for all $x \in [a - h, a + h]$ and all $i = 1, \ldots, n$. In the present case B is not a ball in $\mathfrak{F}$, as it was in the preceding proof, but it is still a closed subset of $\mathfrak{F}$: for if $\psi^1, \psi^2, \psi^3, \ldots$ is a sequence of elements of B that converges to the element ψ of $\mathfrak{F}$ then for each $x \in [a - h, a + h]$ we have

$$\lim_{j\to\infty} \psi^j(x) = \psi(x),$$

so that

$$\lim_{j\to\infty} \psi^j{}_i(x) = \psi_i(x), \quad i = 1, \ldots, n,$$

and from the inequalities $|\psi^j_i(x) - b_i| \leq Nh$ for all j we can therefore deduce $|\psi_i(x) - b_i| \leq Nh$, so that $\psi \in B$, thus verifying the criterion for closure of the theorem of § 3 of Chapter III. Since B is a closed subset of a complete metric space, B is itself a complete metric space. We claim that if $\varphi = (\varphi_1, \ldots, \varphi_n) \in \mathfrak{F}$ satisfies the above system of integral equations then $\varphi \in B$, and in fact $|\varphi_i(x) - b_i| < Nh$ for all $x \in [a - h, a + h]$ and all $i = 1, \ldots, n$. For if there exist points $x \in [a - h, a + h]$ such that $|\varphi_i(x) - b_i| \geq Nh$ for some i, let γ be the greatest lower bound of $|x - a|$ for all such points x. Since each φ_i is continuous and $\varphi_i(a) = b_i$ we have $\gamma > 0$ and $|\varphi_i(a \pm \gamma) - b_i| = Nh$ for some $i = 1, \ldots, n$ and at least one of the two choices of sign $\pm$. Thus $Nh = |\varphi_i(a \pm \gamma) - \varphi_i(a)| = |\gamma\varphi_i'(\alpha)|$, for some α between a and $a \pm \gamma$, by the mean value theorem, and the latter expression equals $|\gamma f_i(\alpha, \varphi_1(\alpha), \ldots, \varphi_n(\alpha))| < \gamma N \leq hN$, a contradiction. Thus any solution φ of the system of integral equations on $[a - h, a + h]$ is in B. Now for any $\psi \in B$ define another n-tuple of functions $F(\psi) = (F_1(\psi), \ldots, F_n(\psi))$ from $[a - h, a + h]$ into $\mathbf{R}$ by

$$(F_i(\psi))(x) = \int_a^x f_i(t, \psi_1(t), \ldots, \psi_n(t))dt + b_i.$$

Since $\psi \in B$, for any $t \in [a - h, a + h]$ and any $i = 1, \ldots, n$ we have $|\psi_i(t) - b_i| \leq Nh$, so that $f(t, \psi_1(t), \ldots, \psi_n(t))$ is defined, is continuous as a function of t, and $|f_i(t, \psi_1(t), \ldots, \psi_n(t))| < N$. Hence for $x \in [a - h, a + h]$, $(F_i(\psi))(x)$ is defined and

$$|(F_i(\psi))(x) - b_i| = \left| \int_a^x f_i(t, \psi_1(t), \ldots, \psi_n(t))dt \right| \leq Nh.$$

Since each $F_i(\psi)$ is clearly continuous, we have $F(\psi) \in B$. That is, $F: B \to B$. We now show that F is a contraction map. Let $\psi, \omega \in B$. For any $x \in [a - h, a + h]$ and $i = 1, \ldots, n$,

$$\begin{aligned}
&|(F_i(\psi))(x) - (F_i(\omega))(x)| \\
&\quad = \left| \int_a^x (f_i(t, \psi_1(t), \ldots, \psi_n(t)) - f_i(t, \omega_1(t), \ldots, \omega_n(t)))dt \right| \\
&\quad \leq |x - a| \max \{ |f_i(t, \psi_1(t), \ldots, \psi_n(t)) \\
&\qquad\qquad - f_i(t, \omega_1(t), \ldots, \omega_n(t))| : t \in [a - h, a + h]\} \\
&\quad \leq |x - a| M \max \{d(\psi(t), \omega(t)) : t \in [a - h, a + h]\} \\
&\quad \leq h\, M\, d(\psi, \omega).
\end{aligned}$$

(We have here used the same letter d to denote the metrics in E^n and in B.) Thus

$$\begin{aligned}
d((F(\psi))(x), (F(\omega))(x)) &= \left(\sum_{i=1}^n ((F_i(\psi))(x) - (F_i(\omega))(x))^2 \right)^{1/2} \\
&\leq h\, M \sqrt{n}\, d(\psi, \omega),
\end{aligned}$$

so

$$d(F(\psi), F(\omega)) \leq h\, M \sqrt{n}\, d(\psi, \omega).$$

Since we have assumed that $hM\sqrt{n} < 1$, F is indeed a contraction map. Thus by the fixed point theorem there is a unique $\varphi = (\varphi_1, \ldots, \varphi_n) \in B$ such that $\varphi = F(\varphi)$, that is, such that

$$\varphi_i(x) = \int_a^x f_i(t, \varphi_1(t), \ldots, \varphi_n(t))dt + b_i$$

for all $x \in [a - h, a + h]$ and all $i = 1, \ldots, n$. For $x \in (a - h, a + h)$ we clearly have $\varphi_i'(x) = f_i(x, \varphi_1(x), \ldots, \varphi_n(x))$ and $\varphi_i(a) = b_i$, so the existence part of the proof is finished. To prove that the restrictions of $\varphi_1, \ldots, \varphi_n$ to $(a - h, a + h)$ are the only functions with the desired properties, note that the above proof would have gone through with h replaced by any $h_1 \in \mathbf{R}$ such that $0 < h_1 < h$. Any solution on $(a - h, a + h)$ of the system of differential equations with initial conditions gives a solution of the system of integral equations on $[a - h_1, a + h_1]$. But we know that the system of integral equations has a unique solution on $[a - h_1, a + h_1]$. Thus any two solutions on $(a - h, a + h)$ of the system of differential equations with initial conditions have equal restrictions to $[a - h_1, a + h_1]$. Since this is true for all h_1 such that $0 < h_1 < h$ there is at most one solution on $(a - h, a + h)$ and our proof is complete.

Generalizing our previous definition, we say that a real-valued function f on an open subset of E^{n+1} satisfies a *Lipschitz condition* if there exists a number $M \in \mathbf{R}$ such that whenever $(x, y_1, \ldots, y_n)$ and $(x, z_1, \ldots, z_n)$ are in the open set on which f is defined we have

$$|f(x, y_1, \ldots, y_n) - f(x, z_1, \ldots, z_n)| \leq M\big((y_1 - z_1)^2 + \cdots + (y_n - z_n)^2\big)^{1/2}.$$

This condition can be given in another way, for since

$$\begin{aligned} |y_1 - z_1| + \cdots + |y_n - z_n| &\geq ((y_1 - z_1)^2 + \cdots + (y_n - z_n)^2)^{1/2} \\ &\geq \max\{|y_1 - z_1|, \ldots, |y_n - z_n|\} \\ &\geq \frac{1}{n}(|y_1 - z_1| + \cdots + |y_n - z_n|), \end{aligned}$$

we see that f satisfies a Lipschitz condition if and only if there exists $M' \in \mathbf{R}$ such that whenever f is defined at $(x, y_1, \ldots, y_n)$ and $(x, z_1, \ldots, z_n)$ we have

$$|f(x, y_1, \ldots, y_n) - f(x, z_1, \ldots, z_n)| \leq M'(|y_1 - z_1| + \cdots + |y_n - z_n|).$$

As a consequence it is possible to state that a rather large class of functions satisfy Lipschitz conditions: a real-valued function f on an open subset of E^{n+1} satisfies a Lipschitz condition if $\partial f/\partial y_1, \ldots, \partial f/\partial y_n$ exist and are bounded on the open set and if whenever $(x, y_1, \ldots, y_n)$ and $(x, z_1, \ldots, z_n)$ are in the open set so is the entire line segment between these two points.

For

$$\begin{aligned}|f(x, y_1, \ldots, y_n) - f(x, z_1, \ldots, z_n)|\\ &\leq |f(x, y_1, \ldots, y_n) - f(x, z_1, y_2, \ldots, y_n)|\\ &\quad + |f(x, z_1, y_2, \ldots, y_n) - f(x, z_1, z_2, y_3, \ldots, y_n)| + \cdots\\ &\quad + |f(x, z_1, \ldots, z_{n-1}, y_n) - f(x, z_1, \ldots, z_n)|\\ &= \left|(y_1 - z_1)\frac{\partial f}{\partial y_1}(x, \eta_1, y_2, \ldots, y_n)\right| + \cdots\\ &\quad + \left|(y_n - z_n)\frac{\partial f}{\partial y_n}(x, z_1, \ldots, z_{n-1}, \eta_n)\right|,\end{aligned}$$

where for $i = 1, \ldots, n$, η_i is between y_i and z_i (or equal to them if these latter are equal), so that if M'' is an upper bound for $|\partial f/\partial y_1|, \ldots, |\partial f/\partial y_n|$ we have

$$|f(x, y_1, \ldots, y_n) - f(x, z_1, \ldots, z_n)| \leq M''(|y_1 - z_1| + \cdots + |y_n - z_n|).$$

Theorem. *Let $f_1, \ldots, f_n$ be continuous real-valued functions on an open subset of E^{n+1} that contains the point $(a, b_1, \ldots, b_n)$. Suppose that $f_1, \ldots, f_n$ satisfy Lipschitz conditions, that is there exists $M \in \mathbf{R}$ such that*

$$|f_i(x, y_1, \ldots, y_n) - f_i(x, z_1, \ldots, z_n)| \leq M\left((y_1 - z_1)^2 + \cdots + (y_n - z_n)^2\right)^{1/2}$$

for $i = 1, \ldots, n$ whenever $(x, y_1, \ldots, y_n)$ and $(x, z_1, \ldots, z_n)$ are in the given open set. Then there exists $h \in \mathbf{R}$, $h > 0$, such that there exists one and only one n-tuple of real-valued functions $(\varphi_1, \ldots, \varphi_n)$ on $(a - h, a + h)$ such that for $i = 1, \ldots, n$ we have $\varphi_i'(x) = f_i(x, \varphi_1(x), \ldots, \varphi_n(x))$ on this interval and $\varphi_i(a) = b_i$.

To prove this choose some $N \in \mathbf{R}$ such that

$$N > \max\{|f_1(a, b_1, \ldots, b_n)|, \ldots, |f_n(a, b_1, \ldots, b_n)|\},$$

then some $r \in \mathbf{R}$, $r > 0$, such that the open ball in E^{n+1} of center $(a, b_1, \ldots, b_n)$ and radius r is entirely contained in the open set on which $f_1, \ldots, f_n$ are defined and such that $|f_i(x, y_1, \ldots, y_n)| < N$ for each $i = 1, \ldots, n$ whenever $(x, y_1, \ldots, y_n)$ is in the ball. If $h \in \mathbf{R}$, $h > 0$, any point $(x, y_1, \ldots, y_n) \in E^{n+1}$ such that $|x - a| \leq h$, $|y_1 - b_1| \leq Nh, \ldots, |y_n - b_n| \leq Nh$ will automatically be in our open ball if $h^2 + nN^2h^2 < r^2$. Hence the theorem results directly from the lemma if we take the U of the lemma to be our open ball, take the same $f_1, \ldots, f_n$, a, $b_1, \ldots, b_n$, N, M, and choose $h \in \mathbf{R}$, $h > 0$, such that $h < r/(1 + nN^2)^{1/2}$ and $hM\sqrt{n} < 1$.

Corollary 1. *Let $f_1, \ldots, f_n$ be continuous real-valued functions on an open subset of E^{n+1} that contains the point $(a, b_1, \ldots, b_n)$. Suppose that $f_1, \ldots, f_n$ satisfy Lipschitz conditions. Then if S is any open interval in $\mathbf{R}$ that contains the point a there is at most one n-tuple of real-valued functions $(\varphi_1, \ldots, \varphi_n)$ on S such that for each $i = 1, \ldots, n$ we have $\varphi_i'(x) = f_i(x, \varphi_1(x), \ldots, \varphi_n(x))$ on S and $\varphi_i(a) = b_i$.*

For suppose that $(\varphi_1, \ldots, \varphi_n)$ and $(\psi_1, \ldots, \psi_n)$ are two n-tuples of functions each of which satisfies the given conditions. We must show that $\varphi_1 = \psi_1, \ldots, \varphi_n = \psi_n$. This can be accomplished by a very simple argument as follows: We begin by noting that by the uniqueness part of the theorem the subset of S given by

$$\{\alpha \in S : \varphi_i(\alpha) = \psi_i(\alpha),\ i = 1, \ldots, n\}$$

is open. By the continuity of $\varphi_1, \ldots, \varphi_n, \psi_1, \ldots, \psi_n$, this subset is closed. Since S is connected this subset must be S itself or the empty set. Since the subset includes the point a, we are forced to the conclusion that it must be S itself.

Corollary 2. *Let $f_1, \ldots, f_n$ be continuous real-valued functions on an open subset U of E^{n+1} that contains the point $(a, b_1, \ldots, b_n)$. Suppose that $f_1, \ldots, f_n$ satisfy Lipschitz conditions. Let $N_1, \ldots, N_n \in \mathbf{R}$ be such that*

$$|f_i(x, y_1, \ldots, y_n)| \leq N_i$$

for all $i = 1, \ldots, n$ and all $(x, y_1, \ldots, y_n) \in U$. Let $S \subset \mathbf{R}$ be an open interval containing the point a such that

$$\{(x, y_1, \ldots, y_n) \in E^{n+1} : x \in S,\ |y_i - b_i| \leq N_i|x - a|,\ i = 1, \ldots, n\} \subset U.$$

Then there exist unique functions $\varphi_1 : S \to \mathbf{R}, \ldots, \varphi_n : S \to \mathbf{R}$ such that for each $i = 1, \ldots, n$ we have $\varphi_i'(x) = f_i(x, \varphi_1(x), \ldots, \varphi_n(x))$ on S and $\varphi_i(a) = b_i$.

If there exist functions $\varphi_1, \ldots, \varphi_n$ with the stated properties then they must be unique, by Corollary 1. Also the equations

$$\varphi_i(x) = \int_a^x f_i(x, \varphi_1(x), \ldots, \varphi_n(x))dx + b_i, \quad x \in S, \quad i = 1, \ldots, n$$

imply $|\varphi_i(x) - b_i| \leq N_i|x - a|$, so that for all $x \in S$ we will have $(x, \varphi_1(x), \ldots, \varphi_n(x)) \in Q$, where Q is the set defined by

$$Q = \{(x, y_1, \ldots, y_n) \in E^{n+1} : x \in S,\ |y_i - b_i| \leq N_i|x - a|,\ i = 1, \ldots, n\}.$$

(See Figure 33, which illustrates the case $n = 1$). One consequence of this is that $\varphi_1, \ldots, \varphi_n$ do not depend at all on the values of $f_1, \ldots, f_n$ outside the set Q. That is, if we consider a similar problem, with all the same data as at present except that the values of $f_1, \ldots, f_n$ are altered on $U - Q$, then the same functions $\varphi_1, \ldots, \varphi_n$ will solve both problems. But to go through with the proof we must take into account the behavior of $f_1, \ldots, f_n$ outside Q, so we go to the trouble of modifying $f_1, \ldots, f_n$ outside Q in such a way

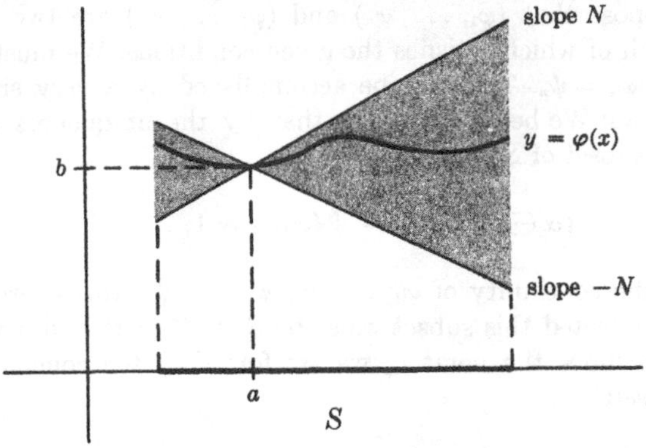

FIGURE 33. The set $Q = \{(x, y) \in E^2 : x \in S, |y - b| \leq N|x - a|\}$ for the case $n = 1$ of Corollary 2 is indicated by shading.

that our proof will work. As auxiliaries, we define functions $\mu_i\colon E^2 \to \mathbf{R}$, $i = 1, \ldots, n$, by

$$\mu_i(x, y) = \begin{cases} N_i|x - a| + b_i & \text{if } \ y - b_i > \ \ N_i|x - a| \\ y & \text{if } |y - b_i| \leq \ \ N_i|x - a| \\ -N_i|x - a| + b_i & \text{if } \ y - b_i < -N_i|x - a|. \end{cases}$$

These functions $\mu_1, \ldots, \mu_n$ are continuous and for any $x \in S$ and $y_1, \ldots, y_n \in \mathbf{R}$ we have $(x, \mu_1(x, y_1), \ldots, \mu_n(x, y_n)) \in Q$. Setting

$$g_i(x, y_1, \ldots, y_n) = f_i(x, \mu_1(x, y_1), \ldots, \mu_n(x, y_n))$$

for $x \in S$, $y_1, \ldots, y_n \in \mathbf{R}$ and $i = 1, \ldots, n$, we get each g_i continuous, $|g_i(x, y_1, \ldots, y_n)| \leq N_i$, and $g_i(x, y_1, \ldots, y_n) = f_i(x, y_1, \ldots, y_n)$ whenever $(x, y_1, \ldots, y_n) \in Q$. Now let $M \in \mathbf{R}$ be such that

$$|f_i(x, y_1, \ldots, y_n) - f_i(x, z_1, \ldots, z_n)| \leq M((y_1 - z_1)^2 + \cdots + (y_n - z_n)^2)^{1/2}$$

for all $(x, y_1, \ldots, y_n)$, $(x, z_1, \ldots, z_n) \in U$ and $i = 1, \ldots, n$. It is a fact that for each $(x, y), (x, z) \in E^2$ and each $i = 1, \ldots, n$ we have

$$|\mu_i(x, y) - \mu_i(x, z)| \leq |y - z|;$$

to prove this it suffices to suppose $y \geq z$, $\mu_i(x, y) \neq \mu_i(x, z)$, so that

$$y \geq \mu_i(x, y) \geq \mu_i(x, z) \geq z.$$

Thus for any $x \in S$, $y_1, \ldots, y_n, z_1, \ldots, z_n \in \mathbf{R}$ and $i = 1, \ldots, n$, we have

$$\begin{aligned} &|g_i(x, y_1, \ldots, y_n) - g_i(x, z_1, \ldots, z_n)| \\ &\qquad = |f_i(x, \mu_1(x, y_1), \ldots, \mu_n(x, y_n)) - f_i(x, \mu_1(x, z_1), \ldots, \mu_n(x, z_n))| \\ &\qquad \leq M((\mu_1(x, y_1) - \mu_1(x, z_1))^2 + \cdots + (\mu_n(x, y_n) - \mu_n(x, z_n))^2)^{1/2} \\ &\qquad \leq M((y_1 - z_1)^2 + \cdots + (y_n - z_n)^2)^{1/2}, \end{aligned}$$

so that $g_1, \ldots, g_n$ also satisfy Lipschitz conditions, with the same M. Since f_i and g_i have the same restrictions to Q we may, if necessary, replace each f_i by g_i so as to be able to assume that the open set U in the statement of the corollary is $\{(x, y_1, \ldots, y_n) \in E^{n+1} : x \in S\}$. Making this assumption, choose $h \in \mathbf{R}$, $h > 0$, such that S contains the closed interval $[a - h, a + h]$ and such that $hM\sqrt{n} < 1$, M being the constant of the Lipschitz conditions. Then the lemma tells us that given any open subinterval of S of length $2h$ whose extremities are in S there exists a unique solution of the system of differential equations $y_i' = f_i(x, y_1, \ldots, y_n)$, $i = 1, \ldots, n$ on the subinterval with arbitrarily prescribed values of $y_1, \ldots, y_n$ at the center of the subinterval. Suppose now that $(\varphi_1, \ldots, \varphi_n)$ is a solution of the system of differential equations on an open subinterval S' of S such that $a \in S'$ and $\varphi_i(a) = b_i$, $i = 1, \ldots, n$. (For example, one possible such S' is $(a - h, a + h)$.) For any $\alpha \in S'$ such that $[\alpha - h, \alpha + h] \subset S$ we can find a solution $(\psi_1, \ldots, \psi_n)$ of the system of differential equations on $(\alpha - h, \alpha + h)$ such that $\psi_i(\alpha) = \varphi_i(\alpha)$, $i = 1, \ldots, n$. By uniqueness (Corollary 1), $\psi_i(x) = \varphi_i(x)$ for all $x \in S' \cap (\alpha - h, \alpha + h)$, so that we can put the φ_i's and ψ_i's together to get a solution on the open interval $S' \cup (\alpha - h, \alpha + h)$. Choosing α close to the extremities of S', we see that we can extend $(\varphi_1, \ldots, \varphi_n)$ to a solution of the system of differential equations on the open interval got by lengthening S' by a distance h at either extremity, provided we still remain in the given interval S; otherwise we can lengthen S' up to an extremity of S. Repeating this procedure will give us a unique solution on all of S.

Corollary 3. *Let $S \subset \mathbf{R}$ be an open interval containing a and let $f_1, \ldots, f_n$ be continuous real-valued functions on $\{(x, y_1, \ldots, y_n) \in E^{n+1} : x \in S\}$ that satisfy Lipschitz conditions. Then for any $b_1, \ldots, b_n \in \mathbf{R}$ there exist unique functions $\varphi_1 : S \to \mathbf{R}, \ldots, \varphi_n : S \to \mathbf{R}$ such that for each $i = 1, \ldots, n$ we have $\varphi_i'(x) = f_i(x, \varphi_1(x), \ldots, \varphi_n(x))$ on S and $\varphi_i(a) = b_i$.*

First suppose that this has been proved in the special case that $f_1, \ldots, f_n$ are bounded on the subset of E^{n+1} given by $\{(x, b_1, \ldots, b_n) : x \in S\}$. Then for any $a_1, a_2 \in S$ such that $a_1 < a < a_2$, the functions $f_1, \ldots, f_n$ are bounded on the compact subset of E^{n+1} given by $\{(x, b_1, \ldots, b_n) : x \in [a_1, a_2]\}$, so that there is a solution of the given system of differential equations with initial conditions on the subinterval (a_1, a_2) of S. By Corollary 1, if we choose different a_1, a_2 the solutions we get will be the same on the intersection of the two intervals (a_1, a_2). Since any point of S is contained in some subinterval (a_1, a_2) of S, we thus get a unique solution on all of S. Hence we may suppose to begin with that $f_1, \ldots, f_n$ are bounded on $\{(x, b_1, \ldots, b_n) : x \in S\}$. Let $M \in \mathbf{R}$ be such that if $x \in S$, $y_1, \ldots, y_n$, $z_1, \ldots, z_n \in \mathbf{R}$ and $i = 1, \ldots, n$ we have

$$|f_i(x, y_1, \ldots, y_n) - f_i(x, z_1, \ldots, z_n)| \leq M((y_1 - z_1)^2 + \cdots + (y_n - z_n)^2)^{1/2}.$$

Let $h \in \mathbf{R}$, $h > 0$, be such that S contains the open interval $(a - h, a + h)$ and $hM\sqrt{n} < 1$. We shall show that given any open subinterval of S of length $2h$ there exists a unique solution of the system of differential equations $y_i' = f_i(x, y_1, \ldots, y_n)$, $i = 1, \ldots, n$, on the subinterval having arbitrarily prescribed values of $y_1, \ldots, y_n$ at the center of the subinterval. Granting this and reasoning as at the end of the proof of the preceding corollary, given any solution of the system of differential equations with initial conditions on an open subinterval S' of S such that $a \in S'$ and given any $\alpha \in S'$ we can get a solution on the interval $S' \cup (\alpha - h, \alpha + h)$, provided this latter interval is contained in S. We can repeat this procedure to get a unique solution on all of S. Thus we are reduced to proving the corollary in the special case where $S = (a - h, a + h)$, M is as above, $hM\sqrt{n} < 1$, and there is a number $A \in \mathbf{R}$ such that $|f_i(x, b_1, \ldots, b_n)| < A$ for all $x \in (a - h, a + h)$ and $i = 1, \ldots, n$. For some positive real number N to be determined later define

$$U = \{(x, y_1, \ldots, y_n) \in E^{n+1} : |x - a| < h,\ |y_i - b_i| < Nh,\ i = 1, \ldots, n\}.$$

Then if $(x, y_1, \ldots, y_n) \in U$ and $i = 1, \ldots, n$ we have

$$|f_i(x, y_1, \ldots, y_n) - f_i(x, b_1, \ldots, b_n)| \\ \leq M\Big(\sum_{i=1}^{n} (y_i - b_i)^2\Big)^{1/2} < M(nN^2h^2)^{1/2} = hMN\sqrt{n},$$

implying

$$|f_i(x, y_1, \ldots, y_n)| < hMN\sqrt{n} + A.$$

We can now try to apply Corollary 2 to the present $S, f_1, \ldots, f_n, a, b_1, \ldots, b_n$ and U, taking $N_1 = \cdots = N_n = N$. All that is wanting for Corollary 2 to go through, thereby completing the proof of Corollary 3, is that the inequalities

$$|f_i(x, y_1, \ldots, y_n)| \leq N$$

hold for all $(x, y_1, \ldots, y_n) \in U$ and all $i = 1, \ldots, n$. But these are valid if

$$hMN\sqrt{n} + A \leq N.$$

Since $hM\sqrt{n} < 1$ the last inequality will be guaranteed by taking $N \geq A/(1 - hM\sqrt{n})$, so we are done.

An almost immediate consequence of the last result is the following main theorem on systems of first order ordinary linear differential equations

$$\frac{dy_i}{dx} = u_{i1}(x)y_1 + \cdots + u_{in}(x)y_n + v_i(x), \quad i = 1, \ldots, n.$$

Corollary 4. *Let $S \subset \mathbf{R}$ be an open interval containing the point a and for each $i, j = 1, \ldots, n$ let u_{ij} and v_i be continuous real-valued functions on S. Then for any $b_1, \ldots, b_n \in \mathbf{R}$ there exist unique functions $\varphi_1: S \to \mathbf{R}, \ldots, \varphi_n: S \to \mathbf{R}$ such that for each $i = 1, \ldots, n$ we have*

$$\varphi_i'(x) = \sum_{j=1}^{n} u_{ij}(x)\varphi_j(x) + v_i(x)$$

on S and $\varphi_i(a) = b_i$.

If all of the functions u_{ij} are bounded on S, say $|u_{ij}(x)| \leq M$ for all $x \in S$ and all $i, j = 1, \ldots, n$, then if $x \in S$ and $y_1, \ldots, y_n, z_1, \ldots, z_n \in \mathbf{R}$ we have

$$\left|\left(\sum_{j=1}^{n} u_{ij}(x)y_j + v_i(x)\right) - \left(\sum_{j=1}^{n} u_{ij}(x)z_j + v_i(x)\right)\right| \leq M(|y_1 - z_1| + \cdots + |y_n - z_n|),$$

our desired Lipschitz condition, so this corollary comes directly from the preceding one. If the u_{ij}'s are not bounded on S they are nevertheless bounded on the interval (a_1, a_2) whenever $a_1, a_2 \in S$ and $a_1 < a < a_2$ (for each u_{ij} is bounded on $[a_1, a_2]$), so we have a unique solution on (a_1, a_2). We therefore have a unique solution on the union of all such open intervals (a_1, a_2), which is S.

Higher order differential equations are equivalent to systems of first order differential equations. For example, setting

$$y_1 = y,\ y_2 = \frac{dy}{dx},\ y_3 = \frac{d^2y}{dx^2},\ \ldots,\ y_n = \frac{d^{n-1}y}{dx^{n-1}},$$

the nth order differential equation

$$\frac{d^ny}{dx^n} = F\left(x, y, \frac{dy}{dx}, \ldots, \frac{d^{n-1}y}{dx^{n-1}}\right)$$

is equivalent to the system of first order differential equations

$$\frac{dy_1}{dx} = y_2$$

$$\frac{dy_2}{dx} = y_3$$

$$\cdots$$

$$\frac{dy_{n-1}}{dx} = y_n$$

$$\frac{dy_n}{dx} = F(x, y_1, y_2, \ldots, y_n).$$

Thus the next two corollaries are immediate consequences of the theorem and the last corollary respectively.

Corollary 5. *Let f be a continuous real-valued function on an open subset of E^{n+1} that contains the point $(a, c_0, \ldots, c_{n-1})$. Suppose that f satisfies a Lipschitz condition, that is there exists $M \in \mathbf{R}$ such that*

$$|f(x, y_1, \ldots, y_n) - f(x, z_1, \ldots, z_n)| \leq M((y_1 - z_1)^2 + \cdots + (y_n - z_n)^2)^{1/2}$$

whenever $(x, y_1, \ldots, y_n)$ and $(x, z_1, \ldots, z_n)$ are in the given open set. Then there exists $h \in \mathbf{R}$, $h > 0$, such that there exists one and only one function $\varphi: (a - h, a + h) \to \mathbf{R}$ such that

$$\frac{d^n\varphi(x)}{dx^n} = f\left(x, \varphi(x), \frac{d\varphi(x)}{dx}, \ldots, \frac{d^{n-1}\varphi(x)}{dx^{n-1}}\right)$$

for all $x \in (a - h, a + h)$ and $\varphi(a) = c_0$, $\varphi'(a) = c_1, \ldots, \varphi^{(n-1)}(a) = c_{n-1}$.

Corollary 6. *Let $S \subset \mathbf{R}$ be an open interval containing the point a and let $u_1, u_2, \ldots, u_n, v$ be continuous real-valued functions on S. Then for any $c_0, \ldots, c_{n-1} \in \mathbf{R}$ there exists a unique function $\varphi: S \to \mathbf{R}$ such that*

$$\varphi^{(n)} + u_1\varphi^{(n-1)} + \cdots + u_{n-1}\varphi' + u_n\varphi = v$$

and $\varphi(a) = c_0$, $\varphi'(a) = c_1, \ldots, \varphi^{(n-1)}(a) = c_{n-1}$.

This last result is of course the main theorem on ordinary linear differential equations. We had previously considered two notable special cases, namely the differential equations

$$y' = f(x) \quad \text{(in Chapter VI, § 4)}$$

and

$$y'' + y = 0 \qquad \text{(in Chapter VII, § 4).}$$

PROBLEMS

1. (a) Draw diagrams to verify that Newton's method of solving an equation $f(x) = 0$ works if f is a twice differentiable real-valued function on an open interval U in $\mathbf{R}$ that changes sign, whose derivative is nowhere zero, and whose second derivative does not change sign, provided that the point $x_0 \in U$ is so chosen that also $x_1 \in U$; indeed under these circumstances the sequence $x_1, x_2, x_3, \ldots$ is monotonic.
   (b) Use Problem 12, Chapter V to prove these facts.
2. Let $a \in \mathbf{R}$, $a > 0$. Show that applying Newton's method to the function $x^2 - a$ gives the formula $x_{n+1} = \frac{1}{2}\left(x_n + \frac{a}{x_n}\right)$. Prove that Newton's method works for any $x_0 > 0$ by showing that then $x_1 \geq \sqrt{a}$ and the map sending x into $\frac{1}{2}\left(x + \frac{a}{x}\right)$ is a contraction map of $\{x \in \mathbf{R} : x \geq \sqrt{a}\}$. (This method of finding square roots occurs in ancient Babylonian manuscripts.)

3. Prove that the equation $\cos x - x - \frac{1}{2} = 0$ has a unique real solution. Show that the fixed point theorem is applicable to the function $F(x) = \cos x - \frac{1}{2}$ and the interval $[0, \pi/4]$ and thereby find this solution to three decimal places.

4. Find the "maximal" U, φ of the implicit function theorem if $f(x, y) = x^2 + y^2 - 1$ and $(a, b) = (0, 1)$.

5. Generalize the proof of the implicit function theorem to get the following result: Let f be a continuous real-valued function on an open subset of E^{n+1} that contains the point $(a_1, \ldots, a_n, b)$, with $f(a_1, \ldots, a_n, b) = 0$. Suppose that $\partial f/\partial y$ exists and is continuous on the given open subset and that
$$\frac{\partial f}{\partial y}(a_1, \ldots, a_n, b) \neq 0.$$
Then there exist positive real numbers h and k such that there exists a unique function
$$\varphi: \{(x_1, \ldots, x_n) \in E^n : (x_1 - a_1)^2 + \cdots + (x_n - a_n)^2 < h^2\} \to \{y \in \mathbf{R} : |y - b| < k\}$$
such that $f(x_1, \ldots, x_n, \varphi(x_1, \ldots, x_n)) = 0$ for all $(x_1, .., x_n)$ in question.

6. Expand the following argument into a proof of the implicit function theorem that avoids the use of the fixed point theorem: Take $r > 0$ such that f is defined on the entire open ball in E^2 of center (a, b) and radius r and such that $\partial f/\partial y$ is never zero on this ball. Choose k such that $0 < k < r$, then choose h such that $0 < h < \sqrt{r^2 - k^2}$ and f is nowhere zero on the set
$$\{(x, y) \in E^2 : |x - a| < h, |y - b| = k\}.$$
Then f is zero at precisely one point of each vertical section of the rectangle
$$\{(x, y) \in E^2 : |x - a| < h, |y - b| < k\}.$$

7. Find all solutions on $\mathbf{R}$ of the differential equation $y' = 3|y|^{2/3}$.

8. Solve the system $y' = 2\sqrt{|y|}$, $y(0) = 0$. (There is an infinity of answers, with essentially four different ones near $x = 0$.)

9. Apply the method of proof of the first theorem of § 3 to solve the system $y' = y$, $y(0) = 1$, starting the successive approximations with $\psi_0 = 0$, obtaining thereby a power series expansion of the solution.

10. Modify the proof of the first theorem of § 3 to show that we may take for h any positive real number less than $1/M$ such that the open subset of E^2 on which f is defined contains $\{(x, y) \in E^2 : |x - a| \leq h, y \in \mathbf{R}\}$ by showing that for such an h the given formula for F defines a contraction map on all of $C([a - h, a + h])$, not just on a ball B.

11. Suppose that the conditions of the first theorem of § 3 obtain and that ψ is a real-valued function on some open interval of $\mathbf{R}$ one of whose extremities is a, ψ having the properties that $\psi'(x) = f(x, \psi(x))$ and $\lim_{x \to a} \psi(x) = b$. Prove that $\psi(x) = \varphi(x)$ whenever both expressions are defined.

12. Show that Corollaries 3, 4 and 6 of the last theorem remain valid if, instead of being an open interval, S is $\mathbf{R}$ itself.

13. Prove that if $u_1, u_2, \ldots, u_n, v$ are real-valued functions on **R** that are m times differentiable, then any solution of the differential equation

$$y^{(n)} + u_1(x)y^{(n-1)} + \cdots + u_n(x)y = v(x)$$

is $(n + m)$ times differentiable.

14. Let $[a, b]$ be a closed interval in **R** and let A and K be continuous real-valued functions on $[a, b]$ and $\{(x, y) \in E^2 : x, y \in [a, b]\}$ respectively. If $\psi \in C([a, b])$, define $F(\psi) \in C([a, b])$ by

$$(F(\psi))(x) = A(x) + \int_a^b K(x, y)\, \psi(y)\, dy$$

(the continuity of $F(\psi)$ following from Prob. 15, Chap. VI). Show that if $|(b - a)\, K(x, y)| < 1$ for all $x, y \in [a, b]$ then F is a contraction map, and therefore there is a unique $\varphi \in C([a, b])$ such that

$$\varphi(x) = A(x) + \int_a^b K(x, y)\, \varphi(y)\, dy$$

for all $x \in [a, b]$.

15. Let $[a, b]$ be a closed interval in **R** and let A and K be continuous real-valued functions on $[a, b]$ and $\{(x, y) \in E^2 : a \leq y \leq x \leq b\}$ respectively. Prove that there is a unique $\varphi \in C([a, b])$ such that

$$\varphi(x) = A(x) + \int_a^x K(x, y)\, \varphi(y)\, dy$$

for all $x \in [a, b]$. (*Hint:* Imitate the procedure of the preceding problem if $|(b - a)\, K(x, y)| < 1$ whenever $a \leq y \leq x \leq b$. To do the general case, note that for any $a_1 \in (a, b)$, the problem reduces to proving the existence of a unique $\varphi_1 \in C([a, a_1])$ such that

$$\varphi_1(x) = A(x) + \int_a^x K(x, y)\, \varphi_1(y)\, dy$$

for all $x \in [a, a_1]$ and the existence of a unique $\varphi_2 \in C([a_1, b])$ such that

$$\varphi_2(x) = A(x) + \int_a^{a_1} K(x, y)\, \varphi_1(y)\, dy + \int_{a_1}^x K(x, y)\, \varphi_2(y)\, dy$$

for all $x \in [a_1, b]$.)

CHAPTER IX

Partial Differentiation

This chapter is concerned with extending the methods of one-variable differential calculus to functions of more than one variable. There are few difficulties, once one has the correct definition of differentiability for functions of several variables.

§ 1. DEFINITIONS AND BASIC PROPERTIES.

Partial derivatives are themselves a matter of one-variable differential calculus. As such they have already made their appearance in this text in our discussions of differentiation under the integral sign and the implicit function theorem. They were also alluded to in our discussion of differential equations, in connection with Lipschitz conditions. Let us recall their definition, restricting ourselves for convenience to functions on *open* subsets of E^n.

For any positive integer n, any open subset U of E^n, any real-valued function f on U, any point $a = (a_1, \ldots, a_n) \in U$ and any $i = 1, \ldots, n$, the *i^{th} partial derivative of f at a* is defined to be the derivative at a_i of the real-valued function which sends x_i into $f(a_1, \ldots, a_{i-1}, x_i, a_{i+1}, \ldots, a_n)$, if this derivative exists. (The expression $f(a_1, \ldots, a_{i-1}, x_i, a_{i+1}, \ldots, a_n)$ is of course to be understood as $f(x_1, a_2, \ldots, a_n)$ if $i = 1$ and in like manner as $f(a_1, \ldots, a_{n-1}, x_n)$ if $i = n$. Note that the function sending x_i into $f(a_1, \ldots, a_{i-1}, x_i, a_{i+1}, \ldots, a_n)$ is defined on an open subset of $\mathbf{R}$ that contains a_i, so that it makes sense to speak of the derivative, if it exists.) The ith partial derivative of f at a is often denoted $f_i'(a)$ or $\frac{\partial f}{\partial x_i}(a)$. Thus we can write

$$f_i'(a) = \frac{\partial f}{\partial x_i}(a) = \lim_{x_i \to a_i} \frac{f(a_1, \ldots, a_{i-1}, x_i, a_{i+1}, \ldots, a_n) - f(a_1, \ldots, a_n)}{x_i - a_i}.$$

If $f_i'(a)$ exists for each $a \in U$ we get a real-valued function f_i' on U (also denoted $\partial f/\partial x_i$) whose value at any $a \in U$ is $f_i'(a)$; this is the *i^{th} partial derivative of f*.

We remark that there are many other notations for partial derivatives, none of which shall be used in this text. Alternate notations for $f_i' = \partial f/\partial x_i$ include

$$f'_{x_i},\ f_{x_i},\ \text{and}\ D_i f,$$

the analogous notations for $f_i'(a) = \frac{\partial f}{\partial x_i}(a)$ being

$$f'_{x_i}(a),\ f_{x_i}(a),\ \text{and}\ (D_i f)(a).$$

These are often expanded to $f_i'(a_1, \ldots, a_n)$, $\frac{\partial f}{\partial x_i}(a_1, \ldots, a_n)$, etc., and one even finds

$$\frac{\partial f(x_1, \ldots, x_n)}{\partial x_i}(a_1, \ldots, a_n), \quad \frac{\partial f(u_1, \ldots, u_n)}{\partial u_i}(a_1, \ldots, a_n), \text{ etc.}$$

There are clearly many possibilities for confusion and more will appear later. No systematic notation is perfect, although some are better than others. The only essential is that we know exactly what is meant in any given instance.

How should the notion of differentiability be defined for functions of several variables? The original definition by means of difference quotients $(f(x) - f(x_0))/(x - x_0)$ does not generalize immediately. A definition of differentiability for functions of more than one variable must be given which does more than refer to partial derivatives, for all the partial derivatives of a function may exist at a point without the function being well-behaved there. For example, the function on E^2 which has value zero at the origin $(0, 0)$ and the value $xy/(x^2 + y^2)$ at any other point (x, y) is such that $\frac{\partial f}{\partial x}(0, 0)$ and $\frac{\partial f}{\partial y}(0, 0)$ exist (and equal zero), but f is not even continuous at $(0, 0)$. It turns out that the property of being closely approximable by linear functions can be generalized and this will give us the desired definition, as follows.

Definition. Let f be a real-valued function on an open subset U of E^n. Let $a = (a_1, \ldots, a_n) \in U$. Then f is *differentiable at* a if there exist $c_1, \ldots, c_n \in \mathbf{R}$ such that

$$\lim_{x \to a} \frac{|f(x) - (f(a) + c_1(x_1 - a_1) + \cdots + c_n(x_n - a_n))|}{d(x, a)} = 0.$$

The $x_1, \ldots, x_n$ in this definition are the coordinates of x, so that $x = (x_1, \ldots, x_n)$. The d denotes the metric in E^n, that is $d(x, a) = ((x_1 - a_1)^2 + \cdots + (x_n - a_n)^2)^{1/2}$.

The limit condition in the above definition is sometimes more conveniently stated as follows: given any $\epsilon > 0$, there exists a $\delta > 0$ such that if $x \in U$ and $d(x, a) < \delta$ then

$$|f(x) - (f(a) + c_1(x_1 - a_1) + \cdots + c_n(x_n - a_n))| \leq \epsilon d(x, a).$$

The symbol $\leq$ is used here rather than $<$ in order to include the case $x = a$.

If for any $i = 1, \ldots, n$ we set $x = (a_1, \ldots, a_{i-1}, x_i, a_{i+1}, \ldots, a_n)$ in the above definition, we get

$$\lim_{x_i \to a_i} \frac{|f(a_1, \ldots, a_{i-1}, x_i, a_{i+1}, \ldots, a_n) - f(a) - c_i(x_i - a_i)|}{|x_i - a_i|} = 0,$$

or

$$\lim_{x_i \to a_i} \left| \frac{f(a_1, \ldots, a_{i-1}, x_i, a_{i+1}, \ldots, a_n) - f(a)}{x_i - a_i} - c_i \right| = 0,$$

so that $f_i'(a)$ exists, and indeed $f_i'(a) = c_i$. Thus if f is differentiable at a then the coefficients $c_1, \ldots, c_n$ are unique and equal to $f_1'(a), \ldots, f_n'(a)$ respectively.

The following technical lemma will prove useful on several occasions.

Lemma. *Let f be a real-valued function on an open subset U of E^n and let $a \in U$. Then f is differentiable at a if and only if there exist real-valued functions $A_1, \ldots, A_n$ on U, continuous at a, such that*

$$f(x) - f(a) = A_1(x)(x_1 - a_1) + A_2(x)(x_2 - a_2) + \cdots + A_n(x)(x_n - a_n)$$

for all $x \in U$. In this case, for each $i = 1, \ldots, n$ we have $f_i'(a) = A_i(a)$.

If f is differentiable at a we have

$$\lim_{x \to a} \frac{|f(x) - (f(a) + f_1'(a)(x_1 - a_1) + \cdots + f_n'(a)(x_n - a_n))|}{d(x, a)} = 0.$$

Since

$$d(x, a) = ((x_1 - a_1)^2 + \cdots + (x_n - a_n)^2)^{1/2} \leq |x_1 - a_1| + \cdots + |x_n - a_n|,$$

if we define the function $\epsilon \colon U \to \mathbf{R}$ by

$$\epsilon(x) = \frac{f(x) - (f(a) + f_1'(a)(x_1 - a_1) + \cdots + f_n'(a)(x_n - a_n))}{|x_1 - a_1| + \cdots + |x_n - a_n|}$$

for $x \neq a$ and $\epsilon(a) = 0$ we have $\lim_{x \to a} \epsilon(x) = 0$ and

$$f(x) = f(a) + f_1'(a)(x_1 - a_1) + \cdots + f_n'(a)(x_n - a_n) + \epsilon(x)(|x_1 - a_1| + \cdots + |x_n - a_n|)$$

for all $x \in U$. Setting

$$A_i(x) = f_i'(a) \pm \epsilon(x)$$

for $i = 1, \ldots, n$, with the plus sign being chosen if $x_i - a_i \geq 0$, otherwise the minus sign, we get

$$f(x) - f(a) = A_1(x)(x_1 - a_1) + \cdots + A_n(x)(x_n - a_n)$$

and $\lim_{x \to a} A_i(x) = f_i'(a) = A_i(a)$, which proves half of the lemma. For the converse, if $A_1, \ldots, A_n \colon U \to \mathbf{R}$ are functions continuous at a such that

$$f(x) - f(a) = A_1(x)(x_1 - a_1) + \cdots + A_n(x)(x_n - a_n)$$

then for $x \in U$, $x \neq a$, we have

$$\begin{aligned}
&\frac{|f(x) - (f(a) + A_1(a)(x_1 - a_1) + \cdots + A_n(a)(x_n - a_n))|}{d(x, a)} \\
&\quad = \frac{|(A_1(x) - A_1(a))(x_1 - a_1) + \cdots + (A_n(x) - A_n(a))(x_n - a_n)|}{d(x, a)} \\
&\quad \leq |A_1(x) - A_1(a)| \frac{|x_1 - a_1|}{d(x, a)} + \cdots + |A_n(x) - A_n(a)| \frac{|x_n - a_n|}{d(x, a)} \\
&\quad \leq |A_1(x) - A_1(a)| + \cdots + |A_n(x) - A_n(a)|.
\end{aligned}$$

By the continuity of $A_1, \ldots, A_n$ at a, the last expression approaches the limit zero as x approaches a, proving that f is differentiable at a and also that $f_i'(a) = A_i(a)$ for $i = 1, \ldots, n$.

Note that if $n > 1$, the functions $A_1, \ldots, A_n$ appearing in the lemma are certainly not unique.

Proposition. *Let U be an open subset of E^n, $f: U \to \mathbf{R}$. If f is differentiable at $a \in U$, then f is continuous at a.*

For if $A_1, \ldots, A_n$ are as in the lemma, then

$$\begin{aligned}\lim_{x\to a} f(x) &= \lim_{x\to a} (f(a) + A_1(x)(x_1 - a_1) + \cdots + A_n(x)(x_n - a_n)) \\ &= f(a) + A_1(a) \cdot 0 + \cdots + A_n(a) \cdot 0 = f(a),\end{aligned}$$

so f is continuous at a.

It would obviously be of great value to have some practical criterion for the differentiability of a function at a point. Such a criterion is afforded by the following result.

Theorem. *Let U be an open subset of E^n, $f: U \to \mathbf{R}$ a function whose partial derivatives $f_1', \ldots, f_n'$ exist on U and are continuous at the point $a \in U$. Then f is differentiable at a.*

Without loss of generality we may assume that U is an open ball in E^n of center a. Then for any $x = (x_1, \ldots, x_n) \in U$, all of the points

$$(x_1, \ldots, x_n),\ (a_1, x_2, \ldots, x_n),\ (a_1, a_2, x_3, \ldots, x_n),\ \ldots,\ (a_1, \ldots, a_{n-1}, x_n),\ (a_1, \ldots, a_n)$$

are in U and so are all points of all line segments between any consecutive two of these points. Writing

$$\begin{aligned}f(x) - f(a) = (f(x_1, \ldots, x_n) - f(a_1, x_2, \ldots, x_n)) \\ + (f(a_1, x_2, \ldots, x_n) - f(a_1, a_2, x_3, \ldots, x_n)) \\ + \cdots + (f(a_1, \ldots, a_{n-1}, x_n) - f(a_1, \ldots, a_n))\end{aligned}$$

and applying the mean value theorem to write

$$\begin{aligned}f(x_1, \ldots, x_n) - f(a_1, x_2, \ldots, x_n) &= f_1'(\xi_1, x_2, \ldots, x_n)(x_1 - a_1) \\ f(a_1, x_2, \ldots, x_n) - f(a_1, a_2, x_3, \ldots, x_n) &= f_2'(a_1, \xi_2, x_3, \ldots, x_n)(x_2 - a_2) \\ &\cdots \\ f(a_1, \ldots, a_{n-1}, x_n) - f(a_1, \ldots, a_n) &= f_n'(a_1, \ldots, a_{n-1}, \xi_n)(x_n - a_n),\end{aligned}$$

where each ξ_i is between a_i and x_i (or $\xi_i = a_i = x_i$ if $a_i = x_i$), we obtain

$$\begin{aligned}f(x) - f(a) = f_1'(\xi_1, x_2, \ldots, x_n)(x_1 - a_1) + f_2'(a_1, \xi_2, x_3, \ldots, x_n)(x_2 - a_2) \\ + \cdots + f_n'(a_1, \ldots, a_{n-1}, \xi_n)(x_n - a_n).\end{aligned}$$

Suppose that for each $x \in U$ a specific choice of $\xi_1, \ldots, \xi_n$ is made. Since $f_1', \ldots, f_n'$ are continuous at a we have

$$\lim_{x\to a} f_1'(\xi_1, x_2, \ldots, x_n) = f_1'(a)$$
$$\lim_{x\to a} f_2'(a_1, \xi_2, x_3, \ldots, x_n) = f_2'(a)$$
$$\ldots$$
$$\lim_{x\to a} f_n'(a_1, \ldots, a_{n-1}, \xi_n) = f_n'(a).$$

Hence the differentiability of f at a follows from the lemma to the preceding proposition, taking

$$A_1(x) = f_1'(\xi_1, x_2, \ldots, x_n),$$
$$A_2(x) = f_2'(a_1, \xi_2, x_3, \ldots, x_n),$$
$$\ldots$$
$$A_n(x) = f_n'(a_1, \ldots, a_{n-1}, \xi_n).$$

A real-valued function f on an open subset U of E^n is called *differentiable on* U (or just *differentiable*) if it is differentiable at each point of U. A necessary condition for this is that $f_1', \ldots, f_n'$ exist on U. f is called *continuously differentiable on* U (or just *continuously differentiable*) if $f_1', \ldots, f_n'$ exist and are continuous on U; this terminology is reasonable because the theorem implies that such a function is necessarily differentiable. It is easy to give many examples of continuously differentiable functions. For instance, any polynomial in the coordinate functions $x_1, \ldots, x_n$ on E^n is continuously differentiable on E^n.

Recall that a map f of any set U into E^m is determined by its m component functions $f_1, \ldots, f_m \colon U \to \mathbf{R}$ and we often write $f = (f_1, \ldots, f_m)$, this meaning that $f(p) = (f_1(p), \ldots, f_m(p))$ for any $p \in U$. If U is an open subset of E^n and $a \in U$ we know that f is continuous at a if and only if $f_1, \ldots, f_m$ are continuous at a. It is therefore reasonable to define f to be *differentiable at* a if $f_1, \ldots, f_m$ are differentiable at a. Similarly, f is said to be *differentiable* if $f_1, \ldots, f_m$ are differentiable. f is called *continuously differentiable* if $f_1, \ldots, f_m$ are continuously differentiable.

Exactly as in one-variable calculus, a differentiable function of a differentiable function is differentiable. More precisely, we have the following "chain rule".

Theorem. *Let U, V be open subsets of E^n, E^m respectively and let $f \colon U \to V$, $g \colon V \to \mathbf{R}$ be functions. Let $a \in U$ be such that f is differentiable at a and g is differentiable at $f(a)$. Then $g \circ f$ is differentiable at a and for $j = 1, \ldots, n$*

$$(g \circ f)_j'(a) = \sum_{i=1}^{m} g_i'(f(a))(f_i)_j'(a).$$

Since g is differentiable at $f(a) = (f_1(a), \ldots, f_m(a))$, the lemma proved earlier implies the existence of functions $A_1, \ldots, A_m: V \to \mathbf{R}$, each continuous at $f(a)$, such that

$$g(y) - g(f(a)) = A_1(y)(y_1 - f_1(a)) + \cdots + A_m(y)(y_m - f_m(a))$$

for all $y \in V$. Similarly for each $i = 1, \ldots, m$ the function f_i is differentiable at a, so there exist functions $B_{i1}, \ldots, B_{in}: U \to \mathbf{R}$, each continuous at a, such that

$$f_i(x) - f_i(a) = B_{i1}(x)(x_1 - a_1) + \cdots + B_{in}(x)(x_n - a_n)$$

for all $x \in U$. Therefore

$$\begin{aligned} g(f(x)) - g(f(a)) &= \sum_{i=1}^{m} A_i(f(x))(f_i(x) - f_i(a)) \\ &= \sum_{i=1}^{m} A_i(f(x)) \sum_{j=1}^{n} B_{ij}(x)(x_j - a_j) \\ &= \sum_{j=1}^{n} \left(\sum_{i=1}^{m} A_i(f(x))B_{ij}(x) \right)(x_j - a_j) \end{aligned}$$

for all $x \in U$. Since each B_{ij} is continuous at a, since f is continuous at a, and since each A_i is continuous at $f(a)$, we deduce that each function

$$\sum_{i=1}^{m} (A_i \circ f) \cdot B_{ij}: U \to \mathbf{R}$$

is continuous at a. By the lemma, $g \circ f$ is differentiable at a. Moreover, again by the lemma, each $g_i'(f(a)) = A_i(f(a))$ and each $(f_i)_j'(a) = B_{ij}(a)$, so that for each $j = 1, \ldots, n$

$$(g \circ f)_j'(a) = \sum_{i=1}^{m} A_i(f(a))B_{ij}(a) = \sum_{i=1}^{m} g_i'(f(a))(f_i)_j'(a).$$

If the f and g of the theorem are differentiable functions the conclusion can be written in the slightly simpler form

$$(g \circ f)_j' = \sum_{i=1}^{m} (g_i' \circ f)(f_i)_j'.$$

The chain rule is the occasion for much imprecision in notation. For example, the equation of the theorem is often written in the form

$$\frac{\partial g}{\partial x_j} = \sum_{i=1}^{m} \frac{\partial g}{\partial f_i} \frac{\partial f_i}{\partial x_j}.$$

This is not literally correct since, among other faults, g is not a function of $x = (x_1, \ldots, x_n)$. A correct version of this formula using the ∂ notation is the cumbersome

$$\frac{\partial g(f_1(x_1, \ldots, x_n), \ldots, f_m(x_1, \ldots, x_n))}{\partial x_j}$$

$$= \sum_{i=1}^{m} \frac{\partial g(y_1, \ldots, y_m)}{\partial y_i}(f_1(x_1, \ldots, x_n), \ldots, f_m(x_1, \ldots, x_n))\frac{\partial f_i(x_1, \ldots, x_n)}{\partial x_j}.$$

The kind of oversimplification appearing in the first formula often results in confusion in practice. Here is a typical example: Suppose that $z = 2x + u$ and that $u = x + y$, so that $z = 3x + y$. Regarding z as a function of x and u we have $\partial z/\partial x = 2$, but regarding z as a function of x and y we have $\partial z/\partial x = 3$. The explanation of this anomaly is that the single symbol z is used here to represent two distinct functions. In fact let $f\colon E^2 \to E^2$ be given by $f(x, y) = (x, x + y)$ and let $g\colon E^2 \to \mathbf{R}$ be given by $g(x, y) = 2x + y$. Then the symbol z stands for both $g(x, u)$ (i.e., the function g) and $g(f(x, y))$ (i.e., the function $g \circ f$), so that the two uses of $\partial z/\partial x$ above really represent the distinct functions g_1' and $(g \circ f)_1'$.

The following paragraph, which will not be required in what follows, is intended for students adept in linear algebra. We consider E^n a vector space over $\mathbf{R}$, defining addition and scalar multiplication by the formulas

$$(x_1, \ldots, x_n) + (y_1, \ldots, y_n) = (x_1 + y_1, \ldots, x_n + y_n)$$
$$c(x_1, \ldots, x_n) = (cx_1, \ldots, cx_n).$$

If U is an open subset of E^n and $f\colon U \to E^m$ is differentiable at the point $a \in U$, define a linear transformation $f'(a)\colon E^n \to E^m$ by means of the $m \times n$ matrix

$$((f_i)_j'(a))_{i=1,\ldots,m;j=1,\ldots,n} = \left(\frac{\partial f_i}{\partial x_j}(a)\right)$$

(called the *jacobian matrix* of f). That is, if $(x_1, \ldots, x_n) \in E^n$ we set $f'(a)(x_1, \ldots, x_n) = (y_1, \ldots, y_m) \in E^m$, with $y_1, \ldots, y_m$ defined by the matrix product

$$((f_i)_j'(a)) \begin{pmatrix} x_1 \\ \cdot \\ \cdot \\ \cdot \\ x_n \end{pmatrix} = \begin{pmatrix} y_1 \\ \cdot \\ \cdot \\ \cdot \\ y_m \end{pmatrix},$$

so that for each $i = 1, \ldots, m$ we have $y_i = \sum_{j=1}^{n} (f_i)_j'(a)x_j$. Then the differentiability of f at a implies that

$$\lim_{x \to a} \frac{d(f(x), f(a) + f'(a)(x - a))}{d(x, a)} = 0.$$

(Note that in the top line of the formula d refers to distance in E^m, while in the bottom line it refers to distance in E^n.) Conversely if U is an open subset of E^n, $f: U \to E^m$ a map, and if we have a point $a \in U$ and a linear transformation $T: E^n \to E^m$ such that

$$\lim_{x\to a} \frac{d(f(x), f(a) + T(x - a))}{d(x, a)} = 0,$$

then we verify immediately by looking at the coordinates of $f(x)$ that f is differentiable at a and that $T = f'(a)$. (Thus the differentiability of f at a and the linear transformation $f'(a)$ could have been defined without resort to component functions.) Our statement of the chain rule translates to the statement that the $1 \times n$ matrix for $(g \circ f)'(a)$ is the product of the $1 \times m$ matrix for $g'(f(a))$ by the $m \times n$ matrix for $f'(a)$; in other words we have the equality of linear transformations

$$(g \circ f)'(a) = g'(f(a)) \cdot f'(a).$$

This last statement has an immediate generalization to the following neat version of the chain rule, which is exactly analogous to the corresponding one-variable result (cf. last proposition of § 2, Chapter V): If U, V are open subsets of E^n, E^m respectively and $f: U \to V$, $g: V \to E^p$ are functions, with f differentiable at the point $a \in U$ and g differentiable at $f(a)$, then $g \circ f$ is differentiable at a and

$$(g \circ f)'(a) = g'(f(a)) \cdot f'(a).$$

§ 2. HIGHER DERIVATIVES.

Higher order partial derivatives are defined similarly to higher order ordinary derivatives. Let U be an open subset of E^n, $f: U \to \mathbf{R}$ a function, and i, j integers among $1, 2, \ldots, n$. If f_i' exists on U we may be able to define the ij^{th} *second order partial derivative of f at a* $(f_i')_j'(a)$ for some point $a \in U$. If $(f_i')_j'(a)$ exists for all $a \in U$ we get a new function $(f_i')_j'$ on U. If $(f_i')_j'$ exists and $k = 1, \ldots, n$, we may be able to define the ijk^{th} *third order partial derivative of f at a* $((f_i')_j')_k'(a)$, and if this exists for all $a \in U$ we have a function $((f_i')_j')_k'$. And so on for still higher order derivatives.

$(f_i')_j'$ can also be written $\dfrac{\partial}{\partial x_j}\left(\dfrac{\partial f}{\partial x_i}\right)$, which is usually abbreviated to $\dfrac{\partial^2 f}{\partial x_j \partial x_i}$. Other notations for this, which we shall refrain from using, include D_jD_if, $D_{i,j}f$, $f''_{x_ix_j}$, and $f_{x_ix_j}$. When still higher order derivatives are in question certain obvious abbreviations are used. For example

$$\frac{\partial^4 f}{\partial x \partial y^2 \partial z} \quad \text{means} \quad \frac{\partial}{\partial x}\left(\frac{\partial}{\partial y}\left(\frac{\partial}{\partial y}\left(\frac{\partial f}{\partial z}\right)\right)\right).$$

The large number of possible higher order partial derivatives of a function of several variables is much reduced by the circumstance that the order of performing the partial differentiations is usually irrelevant. The simplest case of this is the equation

$$\frac{\partial^2 f}{\partial x \partial y} = \frac{\partial^2 f}{\partial y \partial x},$$

repeated application of which yields

$$\frac{\partial^3 f}{\partial x \partial y \partial z} = \frac{\partial^3 f}{\partial z \partial y \partial x}$$

and all similar results. Of course some mild conditions must be satisfied to guarantee this irrelevance of order. The conditions in the following theorem are not the weakest known but are sufficient for all practical purposes. We note that slightly weaker conditions have already appeared in one of the exercises (Chapter VII, Problem 36).

Theorem. *Let f be a real-valued function on an open subset of E^n that contains the point a and let i, j be among $1, \ldots, n$. If $(f_i')_j'$ and $(f_j')_i'$ exist on our open subset and are continuous at a then $(f_i')_j'(a) = (f_j')_i'(a)$.*

There is nothing to prove if $i = j$, so we may suppose $i \neq j$. Also all variables but x_i and x_j are held fixed in the various limit processes by which we arrive at $(f_i')_j'(a)$ and $(f_j')_i'(a)$, so we are reduced to the case $n = 2$. Therefore we may suppose that f is defined on a certain open ball in E^2 of center $a = (a_1, a_2)$ and that $(f_1')_2'$ and $(f_2')_1'$ exist on this ball and are continuous at a and we must prove they are equal at a. We introduce the function Δ, given by

$$\Delta(x) = \frac{f(x_1, x_2) - f(x_1, a_2) - f(a_1, x_2) + f(a_1, a_2)}{(x_1 - a_1)(x_2 - a_2)},$$

defined at all points $x = (x_1, x_2)$ of our ball of center a for which $x_1 \neq a_1$, $x_2 \neq a_2$. In the rest of the proof we restrict ourselves to such points x. If we set

$$\varphi(x_1, x_2) = f(x_1, x_2) - f(x_1, a_2)$$

we have

$$\Delta(x) = \frac{\varphi(x_1, x_2) - \varphi(a_1, x_2)}{(x_1 - a_1)(x_2 - a_2)}.$$

Now the entire line segment in E^2 between (x_1, x_2) and (a_1, x_2) is in our open ball, so the mean value theorem enables us to write

$$\varphi(x_1, x_2) - \varphi(a_1, x_2) = (x_1 - a_1)\varphi_1'(\xi_1, x_2)$$

for some ξ_1 between a_1 and x_1. Therefore

$$\Delta(x) = \frac{\varphi_1'(\xi_1, x_2)}{x_2 - a_2} = \frac{f_1'(\xi_1, x_2) - f_1'(\xi_1, a_2)}{x_2 - a_2}.$$

Since the entire line segment in E^2 between (ξ_1, x_2) and (ξ_1, a_2) is in our open ball and since by assumption $(f_1')_2'$ exists in our ball, another application of the mean value theorem gives

$$\Delta(x) = (f_1')_2'(\xi_1, \xi_2)$$

for some ξ_2 between a_2 and x_2. That is

$$\Delta(x) = (f_1')_2'(\xi_1, \xi_2)$$

for some ξ_1 between a_1 and x_1 and some ξ_2 between a_2 and x_2. Since $(f_1')_2'$ is assumed continuous at a we deduce

$$\lim_{x \to a} \Delta(x) = (f_1')_2'(a).$$

But by the symmetry of Δ this limit is independent of the order of the two differentiations. (This can also be proved explicitly by going through the same argument as above with $\varphi(x_1, x_2)$ replaced by $\psi(x_1, x_2) = f(x_1, x_2) - f(a_1, x_2)$.) That is, we also have

$$\lim_{x \to a} \Delta(x) = (f_2')_1'(a).$$

Thus $(f_1')_2'(a) = (f_2')_1'(a)$.

In the following theorem, which is a version of Taylor's theorem for functions of several variables, we shall find it convenient to use a "differential operator" of the type

$$c_1\frac{\partial}{\partial x_1} + c_2\frac{\partial}{\partial x_2} + \cdots + c_m\frac{\partial}{\partial x_m}.$$

Here $c_1, \ldots, c_m \in \mathbf{R}$ and for any real-valued function f on an open subset of E^m on which all the first partial derivatives $f_1', \ldots, f_m'$ of f exist, we set

$$\left(c_1\frac{\partial}{\partial x_1} + \cdots + c_m\frac{\partial}{\partial x_m}\right)f = c_1\frac{\partial f}{\partial x_1} + \cdots + c_m\frac{\partial f}{\partial x_m} = c_1f_1' + \cdots + c_mf_m',$$

another function on the same open subset of E^m. Similarly if all the second partial derivatives $(f_i')_j'$ exist on this open set we can define

$$\left(c_1\frac{\partial}{\partial x_1} + \cdots + c_m\frac{\partial}{\partial x_m}\right)^2 f$$
$$= \left(c_1\frac{\partial}{\partial x_1} + \cdots + c_m\frac{\partial}{\partial x_m}\right)\left(\left(c_1\frac{\partial}{\partial x_1} + \cdots + c_m\frac{\partial}{\partial x_m}\right)f\right).$$

Similarly for higher iterates of $c_1\frac{\partial}{\partial x_1} + \cdots + c_m\frac{\partial}{\partial x_m}$. One verifies immediately the explicit relation

$$\left(c_1\frac{\partial}{\partial x_1} + \cdots + c_m\frac{\partial}{\partial x_m}\right)^\nu f = \sum_{i_1, \ldots, i_\nu = 1, \ldots, m} c_{i_1}c_{i_2}\cdots c_{i_\nu}\frac{\partial^\nu f}{\partial x_{i_1}\partial x_{i_2}\cdots\partial x_{i_\nu}}.$$

Theorem. *Let U be an open subset of E^m and let $f: U \to \mathbf{R}$ be a function all of whose partial derivatives of order $n + 1$ exist and are continuous on U. Then if $a = (a_1, \ldots, a_m)$, $b = (b_1, \ldots, b_m)$, and the entire line segment between a and b are in U there exists a point c on this line segment such that*

$$f(b) = f(a) + \frac{1}{1!}\left(\left((b_1 - a_1)\frac{\partial}{\partial x_1} + \cdots + (b_m - a_m)\frac{\partial}{\partial x_m}\right)f\right)(a)$$
$$+ \frac{1}{2!}\left(\left((b_1 - a_1)\frac{\partial}{\partial x_1} + \cdots + (b_m - a_m)\frac{\partial}{\partial x_m}\right)^2 f\right)(a) + \cdots$$
$$+ \frac{1}{n!}\left(\left((b_1 - a_1)\frac{\partial}{\partial x_1} + \cdots + (b_m - a_m)\frac{\partial}{\partial x_m}\right)^n f\right)(a)$$
$$+ \frac{1}{(n+1)!}\left(\left((b_1 - a_1)\frac{\partial}{\partial x_1} + \cdots + (b_m - a_m)\frac{\partial}{\partial x_m}\right)^{n+1} f\right)(c).$$

Define a map $h: \mathbf{R} \to E^m$ by

$$h(t) = (a_1 + (b_1 - a_1)t, \ldots, a_m + (b_m - a_m)t),$$

so that h is differentiable, $h(0) = a$, $h(1) = b$, and h maps $[0, 1]$ onto the line segment between a and b. Since U is open and contains this line segment, the composite function $f \circ h$ is defined on some open interval in $\mathbf{R}$ containing $[0, 1]$. The function f is differentiable since it has continuous first partial derivatives. By the chain rule, $f \circ h$ is differentiable and we have

$$(f \circ h)'(t) = \sum_{i=1}^{m} f_i'(h(t))(b_i - a_i)$$
$$= \left(\left((b_1 - a_1)\frac{\partial}{\partial x_1} + \cdots + (b_m - a_m)\frac{\partial}{\partial x_m}\right)f\right)(h(t))$$

or

$$(f \circ h)' = \left(\left((b_1 - a_1)\frac{\partial}{\partial x_1} + \cdots + (b_m - a_m)\frac{\partial}{\partial x_m}\right)f\right) \circ h.$$

Repeating this, for $\nu = 1, \ldots, n + 1$ we get

$$(f \circ h)^{(\nu)} = \left(\left((b_1 - a_1)\frac{\partial}{\partial x_1} + \cdots + (b_m - a_m)\frac{\partial}{\partial x_m}\right)^{\nu} f\right) \circ h.$$

By Taylor's theorem

$$(f \circ h)(1) = (f \circ h)(0) + \frac{(f \circ h)'(0)}{1!} + \frac{(f \circ h)''(0)}{2!} + \cdots$$
$$+ \frac{(f \circ h)^{(n)}(0)}{n!} + \frac{(f \circ h)^{(n+1)}(\zeta)}{(n+1)!}$$

for some ζ between 0 and 1. Hence the present theorem, with $c = h(\xi)$.

The special case $n = 0$ is of particular importance and is often called the "mean value theorem for functions of several variables". It states that there exists a point c on the line segment between a and b such that

$$f(b) - f(a) = (b_1 - a_1)\frac{\partial f}{\partial x_1}(c) + \cdots + (b_m - a_m)\frac{\partial f}{\partial x_m}(c).$$

If we wish to prove only this special case the proof is especially easy: we just apply the ordinary mean value theorem to the function $f(h(t))$. This proof shows that in the present special case the hypotheses may be weakened somewhat, for all we need is that f be continuously differentiable on some open subset of E^m that contains all points of the line segment between a and b, except possibly a and b themselves, and continuous on a larger set containing a and b.

REMARK. In applications of Taylor's theorem the set U is often a ball, so that it is useful to know that *the entire line segment between two points of E^m is contained in a given ball (open or closed) if its extremities are in the ball.* For if the points are the distinct points $(a_1, \ldots, a_m)$ and $(b_1, \ldots, b_m)$ and the center of the ball is $(c_1, \ldots, c_m)$ then for any $t \in \mathbf{R}$

$$(d((a_1 + (b_1 - a_1)t, \ldots, a_m + (b_m - a_m)t), (c_1, \ldots, c_m)))^2$$
$$= \sum_{i=1}^{m} (a_i + (b_i - a_i)t - c_i)^2,$$

which can be written $\alpha(t - \beta)^2 + \gamma$ for certain $\alpha, \beta, \gamma \in \mathbf{R}$, $\alpha > 0$, and this clearly attains its maximum on any closed interval of $\mathbf{R}$ at one of the extremities.

§ 3. THE IMPLICIT FUNCTION THEOREM.

To simplify the following exposition, if $x = (x_1, \ldots, x_m) \in E^m$ and $y = (y_1, \ldots, y_n) \in E^n$ we denote by (x, y) the point $(x_1, \ldots, x_m, y_1, \ldots, y_n) \in E^{m+n}$.

Theorem. *Let m, n be positive integers, $a \in E^m$, $b \in E^n$, and let $f_1, \ldots, f_n$ be continuous real-valued functions on an open subset of E^{m+n} that contains the point (a, b), with $f_1(a, b) = \cdots = f_n(a, b) = 0$. Suppose that for each $i, j = 1, \ldots, n$*

$$\frac{\partial f_i(x_1, \ldots, x_m, y_1, \ldots, y_n)}{\partial y_j} = (f_i)'_{m+j}$$

exists and is continuous on the given open subset and that the $n \times n$ determinant

$$\det\left(\frac{\partial f_i}{\partial y_j}(a, b)\right)$$

is not zero. Then there exist open subsets $U \subset E^m$, $V \subset E^n$, with $a \in U$ and $b \in V$, such that there exists a unique function $\varphi: U \to V$ such that $f_i(x, \varphi(x)) = 0$ for each $i = 1, \ldots, n$ and each $x \in U$, and such that this function φ is continuous.

The proof is a straightforward generalization of that previously given for the special case $m = n = 1$. We begin by defining real-valued functions $F_1, \ldots, F_n$ on the same open subset of E^{m+n} on which $f_1, \ldots, f_n$ are defined by

$$F_i(x, y) = y_i - \sum_{j=1}^{n} c_{ij} f_j(x, y),$$

where $\{c_{ij}\}_{i,j=1,\ldots,n}$ are certain real numbers to be determined in such a manner that $F_1, \ldots, F_n$ satisfy the following basic properties:

(1) each F_i and each $\dfrac{\partial F_i}{\partial y_j}$ is continuous
(2) for each $i = 1, \ldots, n$, $F_i(a, b) = b_i$
(3) for each $i, j = 1, \ldots, n$ we have $\dfrac{\partial F_i}{\partial y_j}(a, b) = 0$
(4) for any x, y we have $f_i(x, y) = 0$ for $i = 1, \ldots, n$ if and only if $F_i(x, y) = y_i$ for $i = 1, \ldots, n$.

Note that properties (1) and (2) hold for any choice of the c_{ij}'s. For property (3) to hold we need

$$\sum_{j=1}^{n} c_{ij} \frac{\partial f_j}{\partial y_k}(a, b) = \delta_{ik}, \quad i, k = 1, \ldots, n$$

where δ_{ik} is the Kronecker delta, equal to 1 if $i = k$, 0 if $i \neq k$. Those who know linear algebra will see in the last equations the statement that the $n \times n$ matrix (c_{ij}) times the $n \times n$ matrix $\left(\dfrac{\partial f_i}{\partial y_j}(a, b)\right)$ is the $n \times n$ identity matrix, so that (c_{ij}) is to be taken to be the inverse of the second matrix, which indeed has an inverse since its determinant is not zero. But the c_{ij}'s may also be found in a more "elementary" way by noting that for any fixed i we get n linear equations in the unknowns $c_{i1}, c_{i2}, \ldots, c_{in}$ and we can solve these equations for $c_{i1}, c_{i2}, \ldots, c_{in}$ provided the determinant of the coefficients is not zero. But this determinant is that obtained from the square array $\left(\dfrac{\partial f_i}{\partial y_j}(a, b)\right)$ by first interchanging rows and columns, and we know that this new determinant has the same value as the original one, which was given to be nonzero. Thus c_{ij}'s may be found such that (3) holds. As for (4), it is clear that if $f_i(x, y) = 0$ for $i = 1, \ldots, n$ then $F_i(x, y) = y_i$ for $i = 1, \ldots, n$. To prove the converse we must show that if $\sum_{j=1}^{n} c_{ij} f_j(x, y) = 0$ for $i = 1, \ldots, n$, then $f_1(x, y) = \cdots = f_n(x, y) = 0$. For those familiar with linear algebra, this is an immediate consequence of the nonsingularity of the matrix (c_{ij}). Those who prefer to reason otherwise may note that we can find $u_1, \ldots, u_n \in \mathbf{R}$ such that

$$\sum_{k=1}^{n} \frac{\partial f_j}{\partial y_k}(a, b)u_k = f_j(x, y), \quad j = 1, \ldots, n,$$

for this involves solving a system of n linear equations in n unknowns, which is possible since the system of equations has nonzero determinant $\det\left(\frac{\partial f_j}{\partial y_k}(a, b)\right)$; this enables us to compute

$$0 = \sum_{j=1}^{n} c_{ij} f_j(x, y) = \sum_{j,k=1}^{n} c_{ij} \frac{\partial f_j}{\partial y_k}(a, b)u_k = \sum_{k=1}^{n} \delta_{ik}u_k = u_i,$$

which in turn implies that each $f_j(x, y) = 0$. This completes the argument that c_{ij}'s can be found as desired, so that we may take for granted in the rest of the proof that $F_1, \ldots, F_n$ have properties (1)–(4).

Choose some $r > 0$ such that the open ball in E^{m+n} of center (a, b) and radius r is entirely contained in the open set on which $f_1, \ldots, f_n$ are defined. Since each $\partial F_i/\partial y_j$ is continuous and $\frac{\partial F_i}{\partial y_j}(a, b) = 0$ we may assume r taken so small that for each $i, j = 1, \ldots, n$ we have $\left|\frac{\partial F_i}{\partial y_j}\right| < \frac{1}{2n^2}$ at each point of the ball. We further assume r is so small that the continuous function $\det\left(\frac{\partial f_i}{\partial y_j}\right)$ is nowhere zero on the ball, this being possible since this determinant is not zero at (a, b). Choose k such that $0 < k < r$ is true and then choose h such that $0 < h < \sqrt{r^2 - k^2}$ and such that $d((F_1(x, b), \ldots, F_n(x, b)), b) < k/2$ whenever $x \in E^m$ and $d(x, a) < h$, this last demand being justifiable by the continuity of $F_1, \ldots, F_n$. We shall prove the theorem with U the open ball in E^m of center a and radius h, and V the open ball in E^n of center b and radius k.

Consider any fixed $x \in U$. If $y \in E^n$ and $d(y, b) \leq k$ we have

$$\begin{aligned} d((x, y), (a, b)) &= ((x_1 - a_1)^2 + \cdots + (x_m - a_m)^2 + (y_1 - b_1)^2 \\ &\qquad + \cdots + (y_n - b_n)^2)^{1/2} \\ &= ((d(x, a))^2 + (d(y, b))^2)^{1/2} < (h^2 + k^2)^{1/2} < r, \end{aligned}$$

so that (x, y) is in our ball of radius r. If also $y' \in E^n$ and $d(y', b) \leq k$, then by the remark at the end of the last section the entire line segment in E^n between y and y' is in the closed ball of center b and radius k. For our fixed x, each F_i is a differentiable function of the last n variables on an open subset of E^n containing the latter closed ball. Thus for each $i = 1, \ldots, n$, the several-variable version of the mean value theorem (which immediately follows the preceding theorem) implies the existence of a point y'' on the line segment between y and y' such that

$$F_i(x, y) - F_i(x, y') = (y_1 - y'_1)\frac{\partial F_i}{\partial y_1}(x, y'') + \cdots + (y_n - y'_n)\frac{\partial F_i}{\partial y_n}(x, y'').$$

Thus

$$
\begin{aligned}
|F_i(x, y) - F_i(x, y')| &\le |y_1 - y'_1| \cdot \left| \frac{\partial F_i}{\partial y_1}(x, y'') \right| + \cdots + |y_n - y'_n| \cdot \left| \frac{\partial F_i}{\partial y_n}(x, y'') \right| \\
&\le \frac{1}{2n^2}(|y_1 - y'_1| + \cdots + |y_n - y'_n|) \\
&\le \frac{1}{2n} \max \{ |y_1 - y'_1|, \ldots, |y_n - y'_n| \} \\
&\le \frac{1}{2n} d(y, y').
\end{aligned}
$$

Therefore

$$
\begin{aligned}
d((F_1(x, y), &\ldots, F_n(x, y)), (F_1(x, y'), \ldots, F_n(x, y'))) \\
&= ((F_1(x, y) - F_1(x, y'))^2 + \cdots + (F_n(x, y) - F_n(x, y'))^2)^{1/2} \\
&\le \left(n \cdot \left(\frac{d(y, y')}{2n} \right)^2 \right)^{1/2} \le \frac{1}{2} d(y, y').
\end{aligned}
$$

Also

$$
\begin{aligned}
d((F_1(x, y), &\ldots, F_n(x, y)), b) \\
&\le d((F_1(x, y), \ldots, F_n(x, y)), (F_1(x, b), \ldots, F_n(x, b))) \\
&\qquad + d((F_1(x, b), \ldots, F_n(x, b)), b) \\
&< \frac{1}{2} d(y, b) + \frac{k}{2} \le \frac{k}{2} + \frac{k}{2} = k.
\end{aligned}
$$

Thus the fixed point theorem is applicable to the closed ball in E^n of center b and radius k and the map which sends any y in this ball into $(F_1(x, y), \ldots, F_n(x, y))$. (Recall that x is fixed.) This gives us the existence of a unique $\bar{y} \in E^n$ such that $d(\bar{y}, b) \le k$ and $F_i(x, \bar{y}) = \bar{y}_i$ for $i = 1, \ldots, n$, that is $f_i(x, \bar{y}) = 0$ for $i = 1, \ldots, n$. (Notice in fact that $d(\bar{y}, b) < k$ by the last inequality. That is, $\bar{y} \in V$.) Since this is valid for each $x \in U$ we can define our function $\varphi \colon U \to V$ by $\varphi(x) = \bar{y}$, and to complete the proof it remains only to prove that φ is continuous.

The continuity of φ can be deduced from what has already been proved. To prove φ continuous at some $a' \in U$, for any $\epsilon > 0$ consider the same problem as in the statement of the theorem of this section, with (a, b) replaced by (a', b'), where $b' = \varphi(a')$, and each f_i replaced by its restriction to the open subset of E^{m+n} given by

$$\{(x, y) \in E^{m+n} : x \in U, y \in V, d(y, b') < \epsilon\}.$$

Note that our choice of r above guarantees that $\det\left(\frac{\partial f_i}{\partial y_j}(a', b')\right) \neq 0$, so all the analogs of the original hypotheses hold. Analogous to U, V, φ we obtain U', V', φ', where U' and V' are open subsets of U and V respectively and φ' is a function $\varphi'\colon U' \to V'$ such that for each $x \in U'$ we have $f_1(x, \varphi'(x)) = \cdots = f_n(x, \varphi'(x)) = 0$ and $d(\varphi'(x), b') < \epsilon$. By the uniqueness property of φ we get that $\varphi(x) = \varphi'(x)$ for all $x \in U'$. Therefore $d(\varphi(x), \varphi(a')) < \epsilon$ if $x \in U'$. Hence φ is continuous at a'. Since a' was an arbitrary point of U, the function φ is continuous.

Corollary 1. *If the hypotheses of the theorem are strengthened by the assumption that $f_1, \ldots, f_n$ are continuously differentiable on the given open subset of E^{m+n}, then U, V may be chosen so that φ is continuously differentiable.*

We first show that if U and V are chosen suitably then φ is differentiable. First choose U and V as in the conclusion of the theorem. Then the continuous function on U whose value at x is $\det\left(\dfrac{\partial f_i}{\partial y_j}(x, \varphi(x))\right)$ is not zero at a, therefore nowhere zero in some open ball in E^m of center a. It therefore suffices to prove that φ is differentiable at any point $x \in U$ at which this determinant is not zero. Hence it suffices to prove φ differentiable at a, under no further conditions than those given in the theorem and corollary. Now each point of $E^{n(m+n)}$ may be considered an n-tuple of points of E^{m+n}, the coordinates of the first point of E^{m+n} being the first $(m+n)$ coordinates of the point of $E^{n(m+n)}$, the coordinates of the second point of E^{m+n} being the second $(m+n)$ coordinates of the point of $E^{n(m+n)}$, etc. Consider the subset of $E^{n(m+n)}$ consisting of all points $(z^1, \ldots, z^n)$ such that each z^i is a point of the open subset of E^{m+n} on which $f_1, \ldots, f_n$ are defined. We have a continuous function on this subset of $E^{n(m+n)}$ whose value at $(z^1, \ldots, z^n)$ is $\det\left(\dfrac{\partial f_i}{\partial y_j}(z^i)\right)$ and this function is not zero at $((a, b), (a, b), \ldots, (a, b))$, hence not zero if $z^1, \ldots, z^n$ are all sufficiently near (a, b). Thus by restricting the set on which $f_1, \ldots, f_n$ are defined we may assume that for any $z^1, \ldots, z^n$ in this set we have $\det\left(\dfrac{\partial f_i}{\partial y_j}(z^i)\right) \neq 0$. Let U, V be as in the conclusion of the theorem. Without loss of generality we may assume that U, V are open balls in E^m, E^n respectively, with centers a, b respectively, for otherwise V may be replaced by an open ball of center b that is entirely contained in V and U replaced by any open ball of center a of sufficiently small radius. The set of points $\{(x, y) \in E^{m+n} : x \in U, y \in V\}$ has the property that it contains the entire line segment between any of its points and (a, b). Write $\varphi = (\varphi_1, \ldots, \varphi_n)$ where each φ_i is a real-valued function on U, so that for any $x \in U$ we have $\varphi(x) = (\varphi_1(x), \ldots, \varphi_n(x))$. For each $i = 1, \ldots, n$ and any $x \in U$ we have $f_i(x, \varphi(x)) = 0$, so by the several-variable version of the mean value theorem

$$\begin{aligned}0 &= f_i(x, \varphi(x)) - f_i(a, \varphi(a)) \\ &= \frac{\partial f_i}{\partial x_1}(z^i)(x_1 - a_1) + \cdots + \frac{\partial f_i}{\partial x_m}(z^i)(x_m - a_m) \\ &\quad + \frac{\partial f_i}{\partial y_1}(z^i)\big(\varphi_1(x) - b_1\big) + \cdots + \frac{\partial f_i}{\partial y_n}(z^i)\big(\varphi_n(x) - b_n\big)\end{aligned}$$

for some z^i on the line segment between $(x, \varphi(x))$ and (a, b). For each $x \in U$ we choose specific $z^1, \ldots, z^n$. Since $\det\left(\frac{\partial f_i}{\partial y_j}(z^i)\right) \neq 0$, the system of n equations

$$\begin{aligned}&\frac{\partial f_i}{\partial y_1}(z^i)\big(\varphi_1(x) - b_1\big) + \cdots + \frac{\partial f_i}{\partial y_n}(z^i)\big(\varphi_n(x) - b_n\big) \\ &\qquad = -\frac{\partial f_i}{\partial x_1}(z^i)(x_1 - a_1) - \cdots - \frac{\partial f_i}{\partial x_m}(z^i)(x_m - a_m)\end{aligned}$$

can be solved for $\varphi_1(x) - b_1, \ldots, \varphi_n(x) - b_n$. We get, for each $i = 1, \ldots, n$,

$$\varphi_i(x) - b_i = A_{i1}(x)(x_1 - a_1) + A_{i2}(x)(x_2 - a_2) + \cdots + A_{im}(x)(x_m - a_m),$$

where each $A_{ij}(x)$ is the quotient by $\det\left(\frac{\partial f_i}{\partial y_j}(z^i)\right)$ of a specific polynomial expression in the various partial derivatives of $f_1, \ldots, f_n$ evaluated at various points $z^1, \ldots, z^n$. Since $\lim_{x \to a} z^i = (a, b)$ for $i = 1, \ldots, n$ and the partial derivatives of $f_1, \ldots, f_n$ were assumed continuous, the various A_{ij}'s are continuous at a. The lemma of § 1 implies that $\varphi_1, \ldots, \varphi_n$ are differentiable at a. Thus φ is differentiable at a, and we have completed the proof that φ is differentiable for suitably chosen U, V.

Having proved φ differentiable, it is easy to see how to compute the various partial derivatives $\frac{\partial \varphi_i}{\partial x_j}$. We apply the chain rule to the equations

$$f_i(x_1, \ldots, x_m, \varphi_1(x), \ldots, \varphi_n(x)) = 0$$

to get

$$\frac{\partial f_i}{\partial x_j}(x, \varphi(x)) + \frac{\partial f_i}{\partial y_1}(x, \varphi(x))\frac{\partial \varphi_1}{\partial x_j} + \cdots + \frac{\partial f_i}{\partial y_n}(x, \varphi(x))\frac{\partial \varphi_n}{\partial x_j} = 0$$

(equivalently, for anyone likely to be confused by the ∂ notation,

$$(f_i)'_j(x, \varphi(x)) + (f_i)'_{m+1}(x, \varphi(x))(\varphi_1)'_j(x) + \cdots + (f_i)'_{m+n}(x, \varphi(x))(\varphi_n)'_j(x) = 0).$$

For any fixed j and varying i we get a system of n linear equations in the n unknowns $\frac{\partial \varphi_1}{\partial x_j}, \ldots, \frac{\partial \varphi_n}{\partial x_j}$, the determinant of this system being $\det\left(\frac{\partial f_i}{\partial y_j}(x, \varphi(x))\right) \neq 0$. Thus we can solve explicitly for the various $\frac{\partial \varphi_i}{\partial x_j}$. In doing this we get the desired information that under the conditions of the corollary each $\partial\varphi_i/\partial x_j$ is continuous on U, and this completes the proof.

Corollary 2 (Inverse function theorem). *Let* $g = (g_1, \ldots, g_n)$ *be a continuously differentiable function from an open subset of* E^n *which contains the point* b *into* E^n, *each* g_i *being a real-valued function on this open subset, and suppose*

$$\det\left((g_i)'_j(b)\right) \neq 0$$

(that is, $\det\left(\frac{\partial g_i}{\partial y_j}(b)\right) \neq 0$*). Then there exist open subsets* U, V *of* E^n, *with* $b \in V$, *such that* g *is defined at each point of* V *and the restriction of* g *to* V *is a one-one map of* V *onto* U *whose inverse function* $g^{-1}\colon U \to V$ *is continuously differentiable.*

On the open subset of E^{2n} consisting of all points (x, y) such that $x \in E^n$ and y is in the open subset of E^n on which g is defined, we define functions $f_1, \ldots, f_n$ by

$$f_i(x, y) = x_i - g_i(y).$$

Set $a = g(b)$. Applying Corollary 1 to $f_1, \ldots, f_n$ and the point (a, b), we get open subsets U_1, V_1 of E^n, with $a \in U_1$ and $b \in V_1$, such that there exists a unique function $\varphi\colon U_1 \to V_1$ such that $x = g(\varphi(x))$ for all $x \in U_1$ and φ is continuously differentiable. The map φ is one-one from U_1 onto $\varphi(U_1) = g^{-1}(U_1) \cap V_1$. By the first proposition of Chapter IV, $g^{-1}(U_1)$ is an open subset of the set on which g is defined, hence an open subset of E^n. Therefore $\varphi(U_1)$ is an open subset of E^n. If we set $U = U_1$, $V = \varphi(U_1)$, then the restriction of g to V is one-one onto U; furthermore the inverse function $g^{-1}\colon U \to V$ is just φ, which is continuously differentiable.

The determinant $\det\left((g_i)'_j\right) = \det\left(\frac{\partial g_i}{\partial y_j}\right)$ is called the *jacobian determinant* (or *jacobian*) of g. It is frequently denoted

$$\frac{\partial(g_1, \ldots, g_n)}{\partial(y_1, \ldots, y_n)} \quad \text{or} \quad J_g.$$

The inverse function theorem implies that, if g is a continuously differentiable function from a connected open subset W of E^n into E^n and the jacobian of g is nowhere zero, then $g(W)$ is a connected open subset of E^n and g is one-one on some open ball centered at any given point of W. However, g need not be one-one on all of W. If $n = 1$, g is indeed one-one on all of W, for then the jacobian g' is either positive or negative on all of W, so that g is either strictly increasing or strictly decreasing. But if $n = 2$ the "polar coordinate map," which sends any $(r, \theta) \in E^2$ such that $r > 0$ into the point $(r \cos \theta, r \sin \theta)$, is *not* one-one, although it is one-one if θ is restricted to any open interval in $\mathbf{R}$ of length 2π.

PROBLEMS

1. Show that the function $f\colon E^2 \to \mathbf{R}$ given by
$$f(x,y) = \begin{cases} \dfrac{xy}{|x|+|y|} & \text{if } (x,y) \neq (0,0) \\ 0 & \text{if } (x,y) = (0,0) \end{cases}$$
is continuous. Where is it differentiable?

2. For which real numbers $\alpha > 0$ is the function $f\colon E^2 \to \mathbf{R}$ that is given by $f(x,y) = (x^2 + y^2)^\alpha$ differentiable?

3. Show that if f is a real-valued function on a connected open subset of E^n and $f_1' = f_2' = \cdots = f_n' = 0$ then f is constant.

4. Let f be a differentiable real-valued function on the open ball in E^n of center $(a_1, \ldots, a_n)$ and radius r and suppose that $f_n' = \partial f/\partial x_n = 0$. Prove that there is a unique real-valued function g on the open ball in E^{n-1} of center $(a_1, \ldots, a_{n-1})$ and radius r such that $f(x_1, \ldots, x_n) = g(x_1, \ldots, x_{n-1})$, and this g is differentiable.

5. Let f be a real-valued function on an open subset U of E^n. Prove that f is continuously differentiable if and only if there exist continuous real-valued functions $A_1, \ldots, A_n$ on the set
$$\{(x_1, \ldots, x_n, y_1, \ldots, y_n) \in E^{2n} : (x_1, \ldots, x_n), (y_1, \ldots, y_n) \in U\}$$
such that
$$f(x) - f(y) = A_1(x,y)(x_1 - y_1) + \cdots + A_n(x,y)(x_n - y_n)$$
for all $x, y \in U$.

6. Let U be an open subset of $\mathbf{R}$, let $\alpha, \beta\colon U \to \mathbf{R}$ be differentiable functions, let V be an open subset of E^2 that for each $y \in U$ contains the entire line segment between the points $(\alpha(y), y)$ and $(\beta(y), y)$, and let $f\colon V \to \mathbf{R}$ be a continuous function such that $\partial f/\partial y$ exists and is continuous. Prove that if $F\colon U \to \mathbf{R}$ is defined by
$$F(y) = \int_{\alpha(y)}^{\beta(y)} f(x,y)\,dx,$$
then
$$F'(y) = \int_{\alpha(y)}^{\beta(y)} \frac{\partial f}{\partial y}(x,y)\,dx + f(\beta(y), y)\,\beta'(y) - f(\alpha(y), y)\,\alpha'(y).$$

7. Let V, W be normed vector spaces (Prob. 22, Chap. III), let U be an open subset of V, and let $a \in U$. Call a function $f\colon U \to W$ *differentiable at* a if there exists a continuous linear transformation (Prob. 22, Chap. IV) $T\colon V \to W$ such that
$$\lim_{x \to a} \frac{\|f(x) - f(a) - T(x-a)\|}{\|x - a\|} = 0.$$

(a) Prove that if f is differentiable at a, then T is unique (so that we may write $T = f'(a)$, generalizing what was done in the last paragraph of § 1).

(b) Prove that if f is differentiable at a then f is continuous at a.

(c) Prove that if $W = E^n$ then f is differentiable at a if and only if the component functions of f are differentiable at a.

(d) Prove the following generalization of the chain rule: If V, W, Z are normed vector spaces, if U and U' are open subsets of V and W respectively, and if $f\colon U \to U'$ is differentiable at the point $a \in U$ and $g\colon U' \to Z$ is differentiable at $f(a)$, then $g \circ f$ is differentiable at a and $(g \circ f)' = g'(f(a))\,f'(a)$.

8. Verify that if $\varphi, \psi\colon \mathbf{R} \to \mathbf{R}$ are twice differentiable functions, if $a \in \mathbf{R}$, and if $f(x, y) = \varphi(x - ay) + \psi(x + ay)$ for all $(x, y) \in E^2$, then

$$\frac{\partial^2 f}{\partial y^2} = a^2 \frac{\partial^2 f}{\partial x^2}.$$

9. Verify that the function $u(x, y) = e^{-x^2/4y}/\sqrt{y}$ satisfies the differential equation

$$\frac{\partial^2 u}{\partial x^2} = \frac{\partial u}{\partial y}.$$

Do the same for the function $\int_a^b f(t)e^{-(x-t)^2/4y}y^{-1/2}\,dt$, where $[a, b]$ is a closed interval in $\mathbf{R}$ and $f\colon [a, b] \to \mathbf{R}$ is continuous.

10. Show that if f is a continuously differentiable real-valued function on an open interval in E^2 and $\partial^2 f/\partial x \partial y = 0$, then there are continuously differentiable real-valued functions f_1, f_2 on open intervals in $\mathbf{R}$ such that

$$f(x, y) = f_1(x) + f_2(y).$$

11. Prove that if U is an open ball in E^n and $f_1, \ldots f_n\colon U \to \mathbf{R}$ are continuously differentiable functions such that

$$\frac{\partial f_i}{\partial x_j} = \frac{\partial f_j}{\partial x_i}$$

for all $i, j = 1, \ldots, n$, then there exists a function $F\colon U \to \mathbf{R}$ such that $f_i = \partial F/\partial x_i$ for $i = 1, \ldots, n$. (*Hint:* If $(a_1, \ldots, a_n)$ is the center of the ball, try defining F by

$$F(x_1, \ldots, x_n) = \int_{a_1}^{x_1} f_1(t, a_2, \ldots, a_n)\,dt + \int_{a_2}^{x_2} f_2(x_1, t, a_3, \ldots, a_n)\,dt$$
$$+ \int_{a_3}^{x_3} f_3(x_1, x_2, t, a_4, \ldots, a_n)\,dt + \cdots + \int_{a_n}^{x_n} f_n(x_1, \ldots, x_{n-1}, t)\,dt.)$$

12. Show that the function $f\colon E^2 \to \mathbf{R}$ given by

$$f(x, y) = \begin{cases} \dfrac{x^3 y}{x^2 + y^2} & \text{if } (x, y) \neq (0, 0) \\ 0 & \text{if } (x, y) = (0, 0) \end{cases}$$

is continuously differentiable and has all its second order partial derivatives, but that

$$\frac{\partial^2 f}{\partial x \partial y}(0, 0) \neq \frac{\partial^2 f}{\partial y \partial x}(0, 0).$$

13. Let f be a real-valued function on an open subset of E^3 containing the point (a_1, a_2, a_3). Suppose that f possesses all its third order partial derivatives and that these are continuous. Compute

$$\lim_{(x_1,x_2,x_3)\to(a_1,a_2,a_3)} (x_1-a_1)^{-1}(x_2-a_2)^{-1}(x_3-a_3)^{-1}\big(f(x_1,x_2,x_3) - f(a_1,x_2,x_3) - f(x_1,a_2,x_3) - f(x_1,x_2,a_3) + f(x_1,a_2,a_3) + f(a_1,x_2,a_3) + f(a_1,a_2,x_3) - f(a_1,a_2,a_3)\big).$$

14. (a) Using the notation of the theorem giving the chain rule and assuming that $f_1, \ldots, f_m$ and g have continuous second order partial derivatives, work out the expression for $((g \circ f)'_j)'_k$.

(b) Use part (a) to express the laplacian $\dfrac{\partial^2 u}{\partial x^2} + \dfrac{\partial^2 u}{\partial y^2}$ in polar coordinates.

(Here $u = g \circ f$, where $f(x, y) = (r, \theta) = \left(\sqrt{x^2+y^2}, \tan^{-1} y/x\right)$, and the Laplacian is to be expressed in terms of the partial derivatives of g with respect to r and θ.)

15. Write down explicitly all terms of the multivariable Taylor formula if $m = 3$ and $n = 2$, collecting terms together wherever possible.

16. Let U be an open subset of E^n, $f: U \to \mathbf{R}$ a differentiable function. Prove that if f attains a maximum or a minimum at the point $a \in U$, then $f'_1(a) = \cdots = f'_n(a) = 0$. Prove conversely that if f has all its second order partial derivatives continuous at a and $f'_1(a) = \cdots = f'_n(a) = 0$, then the restriction of f to some open ball of center a attains a maximum at a if the $n \times n$ symmetric matrix $\left((f'_i)'_j(a)\right)$ is negative definite and attains a minimum at a if this matrix is positive definite, while f does not attain either a maximum or a minimum at a if this matrix is neither positive nor negative semidefinite.

17. Prove that if the functions $f_1, \ldots, f_n$ in the statement of the implicit function theorem are assumed to be k times continuously differentiable (i.e., all partial derivatives of order k exist and are continuous), then the same is true of the component functions $\varphi_1, \ldots, \varphi_n$ of φ.

18. (a) Compute $\partial(x, y)/\partial(r, \theta)$ if $x = r\cos\theta$, $y = r\sin\theta$.

(b) Compute $\partial(x, y, z)/\partial(r, \theta, \varphi)$ if $x = r\cos\theta\sin\varphi$, $y = r\sin\theta\sin\varphi$, and $z = r\cos\varphi$.

19. "If $f(x, y, z) = 0$, then

$$\frac{\partial z}{\partial y}\cdot\frac{\partial y}{\partial x}\cdot\frac{\partial x}{\partial z} = -1."$$

Make sense out of this nonsense and prove.

20. Let S be a closed subset of E^n which contains the entire line segment between any two of its points and let f be a continuously differentiable map from an open subset of E^n containing S into E^n. Suppose that $f(S) \subset S$ and that there is a real number $k < 1$ such that

$$\sum_{i,j=1}^{n} \left((f_i)'_j(x)\right)^2 \le k$$

for all $x \in S$. Prove that the restriction of f to S is a contraction map, so that the fixed point theorem is applicable. (*Hint:* You may want to use Prob. 20, Chap. VI.)

CHAPTER X

Multiple Integrals

In our treatment of one-variable integration we were primarily concerned with continuous functions. Step functions appeared in the proofs, but their use was an easily avoidable technical device. In multivariable integration we are of course still primarily interested in continuous functions, but the necessity for integrating over figures with curved boundaries forces us to consider fairly general noncontinuous functions. In this chapter we start with a straightforward mimicking of what was previously done for one variable. At a certain point the need for generality entails some special consideration of sets of volume zero, but this hurdle is quickly passed. We end up with stronger results than before for the one-variable case, in addition to all the essential specifically multidimensional statements.

§ 1. RIEMANN INTEGRATION ON A CLOSED INTERVAL IN E^n. EXAMPLES AND BASIC PROPERTIES.

Recall that a closed interval in E^n is a subset of E^n of the form

$$\{(x_1, \ldots, x_n) \in E^n : a_i \leq x_i \leq b_i \text{ for each } i = 1, \ldots, n\},$$

where $a_1, \ldots, a_n, b_1, \ldots, b_n$ are fixed real numbers such that $a_1 < b_1, \ldots, a_n < b_n$; we call this closed interval the *closed interval determined by* $a_1, \ldots, a_n, b_1, \ldots, b_n$ and we note that the numbers $a_1, \ldots, a_n, b_1, \ldots, b_n$ are themselves determined by the closed interval. The notion of an open interval in E^n is obtained in a similar manner by replacing each symbol $\leq$ by the symbol $<$.

Definitions. Let $a_1, \ldots, a_n, b_1, \ldots, b_n \in \mathbf{R}$, with $a_1 < b_1, \ldots, a_n < b_n$. By a *partition of the closed interval* $I \subset E^n$ determined by $a_1, \ldots, a_n, b_1, \ldots, b_n$ we mean an n-tuple of partitions of the closed intervals $[a_1, b_1], \ldots, [a_n, b_n]$ in $\mathbf{R}$, that is, an ordered set of n finite sequences of real numbers

$$(x_1^0, x_1^1, x_1^2, \ldots, x_1^{N_1}),\ (x_2^0, x_2^1, x_2^2, \ldots, x_2^{N_2}),\ \ldots,\ (x_n^0, x_n^1, x_n^2, \ldots, x_n^{N_n})$$

(where the superscripts are indices, not exponents) such that for each $i = 1, \ldots, n$ we have

$$a_i = x_i^0 < x_i^1 < x_i^2 < \cdots < x_i^{N_i} = b_i.$$

The *width* of this partition is defined to be

$$\max \{x_i^j - x_i^{j-1} : i = 1, \ldots, n \text{ and } j = 1, \ldots, N_i\}.$$

If f is a real-valued function on I, by a *Riemann sum for f corresponding to the given partition* we mean a sum

$$\sum_{j_1=1,\ldots,N_1;\ldots;j_n=1,\ldots,N_n} f(y_1^{j_1\cdots j_n}, \ldots, y_n^{j_1\cdots j_n})\,(x_1^{j_1} - x_1^{j_1-1}) \cdots (x_n^{j_n} - x_n^{j_n-1}),$$

where each $y_i^{j_1\cdots j_n} \in [x_i^{j_i-1}, x_i^{j_i}]$. We say that f is *Riemann integrable on I* if there exists a number $A \in \mathbf{R}$ such that, given any $\epsilon > 0$, there exists a $\delta > 0$ such that $|S - A| < \epsilon$ whenever S is any Riemann sum for f corresponding to any partition of I of width less than δ; in this case A is called the *Riemann integral of f on I* (or *over I*), and is denoted $\int_I f$.

Note that if $n = 1$ we have exactly what we had in Chapter VI, except for a slight change of notation. For $n > 1$ we have an immediate generalization of what was done earlier. The above partition of the closed interval I in E^n subdivides I into $N_1 N_2 \cdots N_n$ closed subintervals no two of which overlap, except possibly at extremities, that is points $(x_1, \ldots, x_n)$

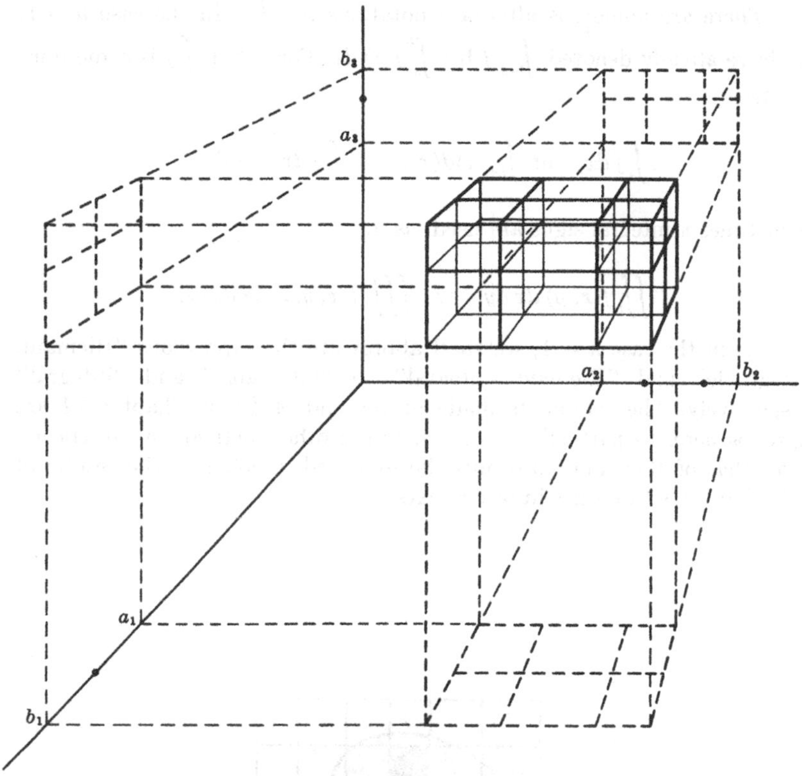

FIGURE 34. A partition of a closed interval in E^3 induces a subdivision into closed subintervals no two of which overlap, except possibly at extremities.

such that $x_i = x_i^j$ for some i, j, as is illustrated in Figure 34. For a Riemann sum corresponding to a specific partition, the various points $(y_1^{j_1\cdots j_n}, \ldots, y_n^{j_1\cdots j_n})$ are obtained by choosing one point from each of the closed subintervals. The expression $(x_1^{j_1} - x_1^{j_1-1}) \cdots (x_n^{j_n} - x_n^{j_n-1})$ represents the "volume" of a subinterval. (Cf. Figure 35.)

The Riemann integral of f on I, if it exists, is unique: If A, A' are Riemann integrals of f on I, then for each $\epsilon > 0$ there exists a number $\delta > 0$ such that if S is any Riemann sum for f corresponding to a partition of I of width less than δ (such partitions exist!) then we have $|S - A|$, $|S - A'| < \epsilon$, as a consequence of which $|A - A'| < 2\epsilon$, and since this is true for each $\epsilon > 0$ we must have $A = A'$.

There are numerous alternate notations for $\int_I f$. In the case $n = 1$, we have already denoted $\int_{[a,b]} f$ by $\int_a^b f(x)\,dx$. For $n > 1$, $\int_I f$ is sometimes written

$$\int_I f\,dx, \quad \text{or} \quad \int_I f(x)d(x), \quad \text{or} \quad \int_I f\,dx_1 \cdots dx_n.$$

Sometimes n integral signs are used, as in

$$\iint_I f(x, y)\,dx\,dy \quad \text{or} \quad \iiint_I f(x, y, z)\,dx\,dy\,dz.$$

As in the case $n = 1$, we shall abbreviate the expressions "Riemann integrable" and "Riemann integral" to "integrable" and "integral" respectively. The comments made at the end of § 1 of Chapter VI are apropos here. In particular, since there are other methods of integration than that of Riemann, care must be exercised in collating the results of this chapter with results in other texts.

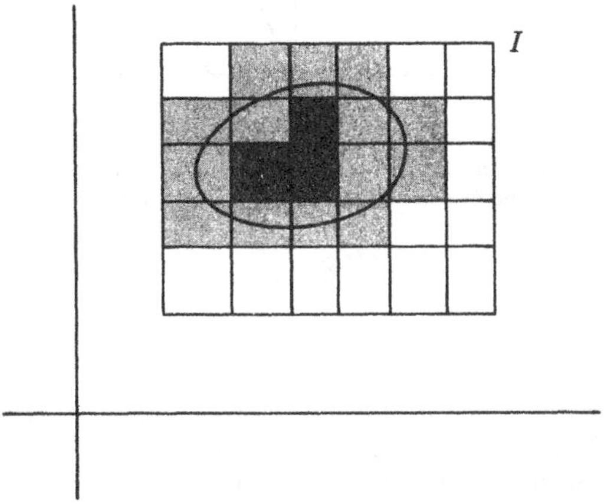

FIGURE 35. Examples of Riemann sums. If the function f on the given closed interval I in E^2 has the value 1 at each point on or within the indicated oval and the value zero at all other points of I then any Riemann sum for f corresponding to the indicated partition of I is the sum of the areas of certain of the rectangles into which I is subdivided. Which rectangles are to be included in the sum depends on how the points $(y_1^{i_1 i_2}, y_2^{i_1 i_2})$ are chosen. For different choices of these points the unshaded rectangles are never included, the lightly shaded rectangles may or may not be included, and the darkly shaded rectangles are always included.

EXAMPLE 1. If I is as above and $f(x_1, \ldots, x_n) = c$, a constant, for all $(x_1, \ldots, x_n) \in I$, then for any partition of I, say the partition in the above definition, we have

$$\sum_{j_1=1,\ldots,N_1;\ldots;j_n=1,\ldots,N_n} f(y_1^{j_1\cdots j_n}, \ldots, y_n^{j_1\cdots j_n})(x_1^{j_1} - x_1^{j_1-1}) \cdots (x_n^{j_n} - x_n^{j_n-1})$$

$$= \sum_{j_1=1,\ldots,N_1;\ldots;j_n=1,\ldots,N_n} c(x_1^{j_1} - x_1^{j_1-1}) \cdots (x_n^{j_n} - x_n^{j_n-1})$$

$$= c\left(\sum_{j_1=1}^{N_1} (x_1^{j_1} - x_1^{j_1-1})\right) \cdots \left(\sum_{j_n=1}^{N_n} (x_n^{j_n} - x_n^{j_n-1})\right)$$

$$= c(b_1 - a_1) \cdots (b_n - a_n).$$

Since all Riemann sums equal $c(b_1 - a_1) \cdots (b_n - a_n)$, we have a function that is integrable on I and

$$\int_I c = c(b_1 - a_1) \cdots (b_n - a_n).$$

EXAMPLE 2. Let I be as above, let $\xi_1 \in [a_1, b_1]$, and let $f: I \to \mathbf{R}$ be a bounded function such that $f(x_1, \ldots, x_n) = 0$ if $x_1 \neq \xi_1$. Suppose $|f(x_1, \ldots, x_n)| \leq M$ for all $(x_1, \ldots, x_n) \in I$. Consider the partition of I appearing in the definition, supposed to be of width less than δ, and the Riemann sum in the definition. We have $f(y_1^{j_1\cdots j_n}, \ldots, y_n^{j_1\cdots j_n}) = 0$ unless $y_1^{j_1\cdots j_n} = \xi_1$, which can be true for at most two distinct j_1's, so that

$$\left|\sum_{j_1=1,\ldots,N_1;\ldots;j_n=1,\ldots,N_n} f(y_1^{j_1\cdots j_n}, \ldots, y_n^{j_1\cdots j_n})(x_1^{j_1} - x_1^{j_1-1}) \cdots (x_n^{j_n} - x_n^{j_n-1})\right|$$

$$\leq \sum_{j_2=1,\ldots,N_2;\ldots;j_n=1,\ldots,N_n} 2M\,\delta(x_2^{j_2} - x_2^{j_2-1}) \cdots (x_n^{j_n} - x_n^{j_n-1})$$

$$= 2M\,\delta\left(\sum_{j_2=1}^{N_2} (x_2^{j_2} - x_2^{j_2-1})\right) \cdots \left(\sum_{j_n=1}^{N_n} (x_n^{j_n} - x_n^{j_n-1})\right)$$

$$= 2M\,\delta(b_2 - a_2) \cdots (b_n - a_n).$$

Clearly $\int_I f = 0$. More generally, $\int_I f = 0$ for any bounded function $f: I \to \mathbf{R}$ for which there exists some $i = 1, \ldots, n$ and some $\xi_i \in [a_i, b_i]$ such that $f(x_1, \ldots, x_n) = 0$ if $x_i \neq \xi_i$.

EXAMPLE 3. Let I be as above, and let $\alpha_1, \ldots, \alpha_n, \beta_1, \ldots, \beta_n \in \mathbf{R}$, with $a_i \leq \alpha_i < \beta_i \leq b_i$ for $i = 1, \ldots, n$. Define $f: I \to \mathbf{R}$ by

$$f(x_1, \ldots, x_n) = \begin{cases} 1 & \text{if } x_i \in (\alpha_i, \beta_i) \text{ for each } i = 1, \ldots, n \\ 0 & \text{if } (x_1, \ldots, x_n) \in I \text{ and } x_i \notin (\alpha_i, \beta_i) \\ & \quad \text{for some } i = 1, \ldots, n. \end{cases}$$

Consider the partition $(x_1^0, x_1^1, \ldots, x_1^{N_1}), \ldots, (x_n^0, x_n^1, \ldots, x_n^{N_n})$ of I, supposed of width less than δ, and a corresponding Riemann sum

$$S = \sum_{j_1=1,\ldots,N_1;\ldots;j_n=1,\ldots,N_n} f(y_1^{j_1\cdots j_n}, \ldots, y_n^{j_1\cdots j_n})\,(x_1^{j_1} - x_1^{j_1-1}) \cdots (x_n^{j_n} - x_n^{j_n-1}),$$

where $y_i^{j_1\cdots j_n} \in [x_i^{j_i-1}, x_i^{j_i}]$ for all $i, j_1, \ldots, j_n$. Since $f(y_1^{j_1\cdots j_n}, \ldots, y_n^{j_1\cdots j_n})$ is 1 or 0, according as the point $(y_1^{j_1\cdots j_n}, \ldots, y_n^{j_1\cdots j_n})$ is or is not in the open subinterval of I determined by $\alpha_1, \ldots, \alpha_n, \beta_1, \ldots, \beta_n$, we have

$$S = \sum_{j_1,\ldots,j_n}{}^{*} (x_1^{j_1} - x_1^{j_1-1}) \cdots (x_n^{j_n} - x_n^{j_n-1}),$$

the asterisk indicating that we include only those $j_1, \ldots, j_n$ for which $y_1^{j_1\cdots j_n} \in (\alpha_1, \beta_1), \ldots, y_n^{j_1\cdots j_n} \in (\alpha_n, \beta_n)$. For $i = 1, \ldots, n$, choose p_i, q_i from among $1, 2, \ldots, N_i$ so that

$$x_i^{p_i-1} \leq \alpha_i < x_i^{p_i}, \quad x_i^{q_i-1} < \beta_i \leq x_i^{q_i}.$$

Then $y_i^{j_1\cdots j_n} \in (\alpha_i, \beta_i)$ if $p_i + 1 \leq j_i \leq q_i - 1$ and $y_i^{j_1\cdots j_n} \notin (\alpha_i, \beta_i)$ if $j_i < p_i$ or $j_i > q_i$. Therefore

$$\sum_{p_1+1 \leq j_1 \leq q_1-1;\ldots;p_n+1 \leq j_n \leq q_n-1} (x_1^{j_1} - x_1^{j_1-1}) \cdots (x_n^{j_n} - x_n^{j_n-1}) \leq S$$
$$\leq \sum_{p_1 \leq j_1 \leq q_1;\ldots;p_n \leq j_n \leq q_n} (x_1^{j_1} - x_1^{j_1-1}) \cdots (x_n^{j_n} - x_n^{j_n-1})$$

or

$$\Big(\sum_{j_1=p_1+1}^{q_1-1} (x_1^{j_1} - x_1^{j_1-1})\Big) \cdots \Big(\sum_{j_n=p_n+1}^{q_n-1} (x_n^{j_n} - x_n^{j_n-1})\Big) \leq S$$
$$\leq \Big(\sum_{j_1=p_1}^{q_1} (x_1^{j_1} - x_1^{j_1-1})\Big) \cdots \Big(\sum_{j_n=p_n}^{q_n} (x_n^{j_n} - x_n^{j_n-1})\Big).$$

By the choice of p_i and q_i we have $\beta_i - \alpha_i \leq x_i^{q_i} - x_i^{p_i-1} < (q_i - p_i + 1)\delta$, so that we must have $q_i - 1 \geq p_i + 1$ for $i = 1, \ldots, n$ if δ is less than each $(\beta_i - \alpha_i)/2$, in which case the last inequalities become

$$(x_1^{q_1-1} - x_1^{p_1}) \cdots (x_n^{q_n-1} - x_n^{p_n}) \leq S \leq (x_1^{q_1} - x_1^{p_1-1}) \cdots (x_n^{q_n} - x_n^{p_n-1}).$$

Since our partition has width less than δ, for each $i = 1, \ldots, n$ we have

$$\alpha_i - \delta < x_i^{p_i-1} < x_i^{p_i} < \alpha_i + \delta \text{ and } \beta_i - \delta < x_i^{q_i-1} < x_i^{q_i} < \beta_i + \delta.$$

Therefore

$$(\beta_1 - \alpha_1 - 2\delta) \cdots (\beta_n - \alpha_n - 2\delta) \leq S \leq (\beta_1 - \alpha_1 + 2\delta) \cdots (\beta_n - \alpha_n + 2\delta).$$

Since the real-valued function on $\mathbf{R}$ which sends any point $t \in \mathbf{R}$ into $(\beta_1 - \alpha_1 + 2t) \cdots (\beta_n - \alpha_n + 2t)$ is continuous at 0, given any $\epsilon > 0$ we can find a $\delta > 0$ such that for the above Riemann sum S we have $|S - (\beta_1 - \alpha_1) \cdots (\beta_n - \alpha_n)| < \epsilon$. Thus f is integrable on I and

$$\int_I f = (\beta_1 - \alpha_1) \cdots (\beta_n - \alpha_n).$$

EXAMPLE 4. If I is as above and $f: I \to \mathbf{R}$ is defined by

$$f(x_1, \ldots, x_n) = \begin{cases} 1 & \text{if } x_1, \ldots, x_n \text{ are rational} \\ 0 & \text{otherwise} \end{cases}$$

then any open subinterval of I contains points at which f takes on the value 1 and also points at which f takes on the value 0 (this is a simple consequence of the corresponding fact for $n = 1$). Therefore both $(b_1 - a_1) \cdots (b_n - a_n)$ and zero are Riemann sums corresponding to *any* partition of I. It follows that f is not integrable.

Proposition. *Riemann integration has the following properties:*

(*1*) *If f and g are integrable real-valued functions on the closed interval I of E^n then $f + g$ is integrable on I and*

$$\int_I (f + g) = \int_I f + \int_I g.$$

(*2*) *If f is an integrable real-valued function on the closed interval I of E^n and $c \in \mathbf{R}$ then cf is integrable on I and*

$$\int_I cf = c \int_I f.$$

Given any partition of I, a Riemann sum for $f + g$ corresponding to this partition is the sum of a Riemann sum for f corresponding to this partition plus a Riemann sum for g corresponding to this partition, and similarly a Riemann sum for cf corresponding to this partition is c times a Riemann sum for f corresponding to this partition. Hence the proposition is quite trivial. (Those wishing to write down a proof in all detail may do so by effecting suitable minor changes in the proof of the corresponding result of Chapter VI.)

An immediate consequence of the proposition is that if f and g are integrable on I, then

$$\int_I (f - g) = \int_I f - \int_I g.$$

Proposition. *If f is an integrable real-valued function on the closed interval I of E^n and $f(x) \geq 0$ for all $x \in I$, then $\int_I f \geq 0$.*

For if S is any Riemann sum for f corresponding to any partition of I then $S \geq 0$.

As in Chapter VI, there are two immediate corollaries.

Corollary 1. *If f and g are integrable real-valued functions on the closed interval I of E^n and $f(x) \leq g(x)$ for all $x \in I$, then $\int_I f \leq \int_I g$.*

Corollary 2. *If f is an integrable real-valued function on the closed interval I of E^n that is determined by $a_1, \ldots, a_n, b_1, \ldots, b_n$ and $m \leq f(x) \leq M$ for all $x \in I$, then*

$$m(b_1 - a_1) \cdots (b_n - a_n) \leq \int_I f \leq M(b_1 - a_1) \cdots (b_n - a_n).$$

§ 2. EXISTENCE OF THE INTEGRAL. INTEGRATION ON ARBITRARY SUBSETS OF E^n. VOLUME.

Lemma 1. *A real-valued function f on a closed interval I of E^n is integrable on I if and only if, given any $\epsilon > 0$, there exists a number $\delta > 0$ such that $|S_1 - S_2| < \epsilon$ whenever S_1 and S_2 are Riemann sums for f corresponding to partitions of I of width less than δ.*

The proof of Lemma 1 of § 3, Chapter VI applies verbatim in the present case, if we change the symbol $[a, b]$, wherever it occurs, to I, and the symbol $\int_a^b f(x)dx$ to $\int_I f$.

Definition. A real-valued function f on a closed interval I of E^n is called a *step function* if there exists a partition $(x_1^0, x_1^1, \ldots, x_1^{N_1}), \ldots, (x_n^0, x_n^1, \ldots, x_n^{N_n})$ of I such that for any $(x_1, \ldots, x_n), (y_1, \ldots, y_n) \in I$ we have $f(x_1, \ldots, x_n) \neq f(y_1, \ldots, y_n)$ only if for some $i = 1, \ldots, n$ such that $x_i \neq y_i$, the closed interval in $\mathbf{R}$ whose extremities are x_i and y_i contains at least one of the points $x_i^0, x_i^1, \ldots, x_i^{N_i}$.

In other words, f is constant on any subset of I consisting of points $(x_1, \ldots, x_n)$ in which each x_i is restricted to one of the subsets (x_i^0, x_i^1), $(x_i^1, x_i^2), \ldots, (x_i^{N_i-1}, x_i^{N_i}), \{x_i^0\}, \{x_i^1\}, \ldots, \{x_i^{N_i}\}$. In particular, f takes on only a finite number of distinct values. The functions of Examples 1 and 3 of § 1 are step functions.

Lemma 2. *A step function on a closed interval I in E^n is integrable. In particular, if $(x_1^0, x_1^1, \ldots, x_1^{N_1}), \ldots, (x_n^0, x_n^1, \ldots, x_n^{N_n})$ is a partition of I, if $\{c_{j_1 \ldots j_n}\}_{j_1 = 1, \ldots, N_1; \ldots; j_n = 1, \ldots, N_n} \subset \mathbf{R}$, and if the step function $f: I \to \mathbf{R}$ is such that for any $j_1 = 1, \ldots, N_1; \ldots; j_n = 1, \ldots, N_n$ we have $f(x_1, \ldots, x_n) = c_{j_1 \ldots j_n}$ if $x_i^{j_i - 1} < x_i < x_i^{j_i}$ for each $i = 1, \ldots, n$, then*

$$\int_I f = \sum_{j_1 = 1, \ldots, N_1; \ldots; j_n = 1, \ldots, N_n} c_{j_1 \ldots j_n}(x_1^{j_1} - x_1^{j_1 - 1}) \cdots (x_n^{j_n} - x_n^{j_n - 1}).$$

For if we define $\varphi_{j_1 \ldots j_n}(x_1, \ldots, x_n)$ to be 1 if $x_i^{j_i - 1} < x_i < x_i^{j_i}$ for each $i = 1, \ldots, n$ and zero otherwise, then $f - \sum_{j_1, \ldots, j_n} c_{j_1 \ldots j_n} \varphi_{j_1 \ldots j_n}$ is a function on I that has the value zero at each point $(x_1, \ldots, x_n)$ of I for which all the inequalities $x_i \neq x_i^j$ hold, for $i = 1, \ldots, n$ and $j = 0, \ldots, N_i$. By the additivity of the integral and Example 2 of the previous section, we have $\int_I (f - \sum_{j_1,\ldots,j_n} c_{j_1\ldots j_n} \varphi_{j_1\ldots j_n}) = 0$. By Example 3 of the previous section, for all $j_1, \ldots, j_n$ we have $\int_I \varphi_{j_1\ldots j_n} = (x_1{}^{j_1} - x_1{}^{j_1-1}) \cdots (x_n{}^{j_n} - x_n{}^{j_n-1})$. Thus the given expression for $\int_I f$ results from the linearity of the integral, that is, the first proposition of the last section.

Proposition. *The real-valued function f on the closed interval I of E^n is integrable on I if and only if, for each $\epsilon > 0$, there exist step functions f_1, f_2 on I such that*

$$f_1(x) \leq f(x) \leq f_2(x) \quad \textit{for each } x \in I$$

and

$$\int_I (f_2 - f_1) < \epsilon.$$

The proof of this is exactly the same as that of the analogous proposition in § 3 of Chapter VI if we make a few appropriate changes in notation. Since we shall refrain from making the precise transcription, the reader should carefully check this statement.

Corollary 1. *If the real-valued function f on the closed interval I of E^n is integrable on I, then it is bounded on I.*

The following simple result could have been proved much earlier, but it is especially easy to prove at this point.

Corollary 2. *If $I \subset J$ are closed intervals in E^n and $f\colon J \to \mathbf{R}$ is such that $f(x) = 0$ for all $x \in J - I$, then the integral of f on J exists if and only if the integral of the restriction of f to I on I exists, in which case they are equal.*

Denoting the restriction of f to I by the same letter f when no confusion is possible, this corollary states simply that $\int_J f = \int_I f$ if $J \supset I$ and f is zero outside I. To prove this, first note that Lemma 2 implies the truth of Corollary 2 if f is a step function. Next suppose that $\int_I f$ exists. Then for any $\epsilon > 0$ there exist step functions f_1, f_2 on I such that $f_1(x) \leq f(x) \leq f_2(x)$ for each $x \in I$ and such that $\int_I (f_2 - f_1) < \epsilon$. Extend f_1, f_2 to functions on J by setting $f_1(x) = f_2(x) = 0$ if $x \in J - I$. Then f_1, f_2 are step functions on J, $f_1(x) \leq f(x) \leq f_2(x)$ for all $x \in J$, and $\int_J (f_2 - f_1) = \int_I (f_2 - f_1) < \epsilon$. Thus $\int_J f$ exists. Since $\int_I f_1 \leq \int_I f \leq \int_I f_2$ and $\int_I f_1 = \int_J f_1 \leq \int_J f \leq \int_J f_2 = \int_I f_2$, we have $\left| \int_J f - \int_I f \right| \leq \int_I f_2 - \int_I f_1 = \int_I (f_2 - f_1) < \epsilon$. Since the last inequality is true for each $\epsilon > 0$, we have $\int_J f = \int_I f$. Finally suppose that $\int_J f$ exists. Then for any $\epsilon > 0$ there exist step functions g_1, g_2 on J such that $g_1(x) \leq f(x) \leq g_2(x)$ for each $x \in J$ and $\int_J (g_2 - g_1) < \epsilon$. The restrictions of g_1, g_2 to I are step functions on I. By Lemma 2, $\int_I (g_2 - g_1) \leq \int_J (g_2 - g_1)$, so that $\int_I (g_2 - g_1) < \epsilon$. Hence $\int_I f$ exists, and this completes the proof.

It is now convenient to extend the notion of integral. First let $f: E^n \rightarrow \mathbf{R}$ be a function which is zero outside some bounded subset of E^n. We can then find a closed interval I of E^n such that $f(x) = 0$ for all $x \in \mathcal{C}I$. We call f *integrable on* E^n and define $\int_{E^n} f = \int_I f$ if the latter integral exists. This makes sense, for suppose $I' \subset E^n$ is another closed interval such that $f(x) = 0$ for all $x \in \mathcal{C}I'$. Let J be still another closed interval of E^n, such that $J \supset I \cup I'$. By the last corollary, $\int_I f$ exists if and only if $\int_J f$ exists, in which case these are equal, and similarly $\int_{I'} f$ exists if and only if $\int_J f$ exists, in which case they are equal. Therefore $\int_I f$ exists if and only if $\int_{I'} f$ exists, in which case they are equal.

Now consider an arbitrary subset $A \subset E^n$ and an arbitrary real-valued function f on some subset of E^n that contains A. Define $\bar{f}: E^n \rightarrow \mathbf{R}$ by setting $\bar{f}(x) = f(x)$ if $x \in A$, $\bar{f}(x) = 0$ if $x \notin A$. We say that f is *integrable on* A and define $\int_A f$ to be $\int_{E^n} \bar{f}$ if the latter integral exists. This agrees with the previous definition if A is a closed interval of E^n. For any A and f, $\int_A f$ can exist only if the set of points of A at which f is not zero is bounded and if f is bounded on A; this follows from the present definition.

For an arbitrary subset $A \subset E^n$, we say that A *has volume*, and define the *volume of* A to be

$$\text{vol}\,(A) = \int_A 1,$$

if this integral exists. A necessary condition that vol (A) exist is that A be bounded. If I is the open or closed interval in E^n determined by $a_1, \ldots, a_n, b_1, \ldots, b_n$, then $\text{vol}\,(I) = (b_1 - a_1) \cdots (b_n - a_n)$. An example of a bounded subset of E^n having no volume is the set of all points of a given closed interval I of E^n all of whose coordinates are rational numbers (cf. Example 4 of § 1). The word *volume*, as used here, is often replaced by *n-dimensional volume*, or *Jordan measure*. If $n = 1$, one often uses the word *length* instead, and if $n = 2$ the word *area*.

The two propositions of § 1 possess the following immediate generalizations.

Proposition. *Integration has the following linearity properties:*

(1) *If $A \subset E^n$ and f and g are integrable real-valued functions on A, then $f + g$ is integrable on A and*

$$\int_A (f+g) = \int_A f + \int_A g.$$

(2) *If $A \subset E^n$, f is an integrable real-valued function on A and $c \in \mathbf{R}$, then cf is integrable on A and*

$$\int_A cf = c \int_A f.$$

As usual, it follows that if f and g are integrable on A then

$$\int_A (f-g) = \int_A f - \int_A g.$$

Proposition. *If f is an integrable real-valued function on the subset A of E^n and $f(x) \geq 0$ for all $x \in A$, then $\int_A f \geq 0$.*

Corollary 1. *If f, g are integrable real-valued functions on the subset A of E^n and $f(x) \leq g(x)$ for all $x \in A$, then $\int_A f \leq \int_A g$.*

Corollary 2. *Let f be an integrable real-valued function on the subset A of E^n. Suppose that $m \leq f(x) \leq M$ for all $x \in A$ and that A has volume. Then*

$$m \operatorname{vol}(A) \leq \int_A f \leq M \operatorname{vol}(A).$$

Subsets of E^n of volume zero are especially important. We list together some of their properties.

Proposition. *The following statements hold:*

(1) *A subset A of E^n has volume zero if and only if, given any $\epsilon > 0$, there exists a finite number of closed (or open) intervals in E^n whose union contains A and the sum of whose volumes is less than ϵ.*

(2) *Any subset of a subset of E^n of volume zero is of volume zero.*

(3) *The union of a finite number of subsets of E^n of volume zero is of volume zero.*

(4) *If $A \subset E^n$ has volume zero and the set $B \subset E^n$ has volume, then* $\operatorname{vol}(B \cup A) = \operatorname{vol}(B - A) = \operatorname{vol}(B)$.

(5) *If $A \subset E^n$ has volume zero and $f\colon A \to \mathbf{R}$ is a bounded function, then* $\int_A f = 0$.

(6) *If $S \subset E^{n-1}$ is compact and $f\colon S \to \mathbf{R}$ is continuous, then the graph of f in E^n, i.e., the set*

$$\{(x_1, \ldots, x_n) \in E^n : (x_1, \ldots, x_{n-1}) \in S, f(x_1, \ldots, x_{n-1}) = x_n\},$$

has volume zero.

To prove (1) we may suppose $A \subset I$, for some closed interval I of E_n, since only bounded sets have volume. Let $f\colon I \to \mathbf{R}$ be defined by setting $f(x) = 1$ if $x \in A$, $f(x) = 0$ if $x \in I - A$, so that vol $(A) = \int_I f$. If vol $(A) = 0$ then for any $\epsilon > 0$ there is a partition of I such that any Riemann sum for f corresponding to this partition has absolute value less than ϵ. But one such Riemann sum for f corresponding to this partition is the sum of the volumes of those closed subintervals of I (for the subdivision of I corresponding to the given partition) which contain points of A. Hence A is contained in the union of a finite number of closed intervals the sum of whose volumes is less than ϵ. Conversely, suppose that for each $\epsilon > 0$ we can write $A \subset I_1 \cup \cdots \cup I_N$, where each I_j is a closed interval in E^n and $\sum_{j=1}^{N} \text{vol}\,(I_j) < \epsilon$. For $j = 1, \ldots, N$ we define $f_j\colon I \to \mathbf{R}$ by $f_j(x) = 1$ if $x \in I \cap I_j$ and $f_j(x) = 0$ if $x \in I - I_j$, so that $\sum_{j=1}^{N} f_j$ is a step function on I, $0 \leq f(x) \leq \sum_{j=1}^{N} f_j(x)$ for all $x \in I$, and $\int_I \left(\sum_{j=1}^{N} f_j - 0 \right) = \sum_{j=1}^{N} \text{vol}\,(I \cap I_j) \leq \sum_{j=1}^{N} \text{vol}\,(I_j) < \epsilon$. Since for each $\epsilon > 0$ f is sandwiched between the two step functions 0 and $\sum_{j=1}^{N} f_j$, we know that $\int_I f$ exists. Since $\int_I 0 \leq \int_I f \leq \int_I \sum_{j=1}^{N} f_j < \epsilon$ for each $\epsilon > 0$, we have $\int_I f = 0$, that is vol $(A) = 0$. This proves the part of (1) dealing with closed intervals. As for open intervals, the result for closed intervals plus the remarks that any open interval in E^n is contained in a closed interval of the same volume and that any closed interval in E^n is contained in an open interval of twice the volume (namely the open interval having the same center as the closed interval, with dimensions those of the closed interval times $2^{1/n}$) prove (1) for open intervals. Thus the proof of (1) is complete. Parts (2) and (3) follow immediately from (1). To prove (4), first note that we have vol $(A - B) =$ vol $(A \cap B) = 0$, by (2). Define functions $f_1, f_2, f_3\colon E^n \to \mathbf{R}$ by setting

$$f_1(x) = \begin{cases} 1 & \text{if } x \in B \\ 0 & \text{if } x \notin B \end{cases}$$

$$f_2(x) = \begin{cases} 1 & \text{if } x \in A - B \\ 0 & \text{if } x \notin A - B \end{cases}$$

$$f_3(x) = \begin{cases} 1 & \text{if } x \in A \cap B \\ 0 & \text{if } x \notin A \cap B. \end{cases}$$

Then $\int_{E^n} f_1 = \int_B 1 = \text{vol}(B)$, $\int_{E^n} f_2 = \int_{A-B} 1 = \text{vol}(A - B) = 0$, and $\int_{E^n} f_3 = \int_{A \cap B} 1 = \text{vol}(A \cap B) = 0$. We have $f_1(x) + f_2(x) = 1$ if $x \in B \cup A$ and $f_1(x) + f_2(x) = 0$ if $x \notin B \cup A$, so that $\text{vol}(B \cup A) = \int_{E^n}(f_1 + f_2) = \int_{E^n} f_1 + \int_{E^n} f_2 = \text{vol}(B)$. Also $f_1(x) - f_3(x) = 1$ if $x \in B - A$ and $f_1(x) - f_3(x) = 0$ if $x \notin B - A$, so that $\text{vol}(B - A) = \int_{E^n}(f_1 - f_3) = \int_{E^n} f_1 - \int_{E^n} f_3 = \text{vol}(B)$, which completes the proof of (4). For (5), suppose that A is contained in the closed interval I of E^n and suppose that $|f(x)| < M$ for each $x \in A$. Define $\bar{f}: E^n \to \mathbf{R}$ by setting $\bar{f}(x) = f(x)$ if $x \in A$, $\bar{f}(x) = 0$ if $x \notin A$. For any $\epsilon > 0$, part (1) tells us that there exists a step function $g: I \to \mathbf{R}$ such that $g(x) \geq 0$ for all $x \in I$, $g(x) \geq 1$ for $x \in A$, and $\int_I g < \epsilon$. Then $-Mg(x) \leq \bar{f}(x) \leq Mg(x)$ for all $x \in I$; $\int_I (Mg - (-Mg)) = 2M \int_I g < 2M\epsilon$ follows. This being true for each $\epsilon > 0$, $\int_I \bar{f}$ exists. Since $\int_I (-Mg) \leq \int_I \bar{f} \leq \int_I Mg$, we have $\left|\int_I \bar{f}\right| \leq M \int_I g < M\epsilon$, and since this is true for all $\epsilon > 0$ we have $\int_I \bar{f} = 0$. But $\int_A f$ is by definition $\int_I \bar{f}$, so $\int_A f = 0$, finishing (5). To prove (6), we may suppose that $S \subset I$, where I is some closed interval in E^{n-1}. Given any $\epsilon > 0$, by uniform continuity we can find a number $\delta > 0$ such that $|f(p) - f(q)| < \epsilon$ whenever $p, q \in S$ are such that $d(p, q) < \delta$. Choose a partition of I of width less than $\delta/\sqrt{n-1}$. Let this partition of I subdivide I into the closed subintervals $I_1, \ldots, I_N$, so that $I_1, \ldots, I_N$ are closed subintervals of I whose sides are all less than $\delta/\sqrt{n-1}$, $I = I_1 \cup \cdots \cup I_N$, and $\text{vol}(I) = \sum_{j=1}^{N} \text{vol}(I_j)$. If $p, q \in I_j \cap S$ then $d(p, q) < \delta$, so that $|f(p) - f(q)| < \epsilon$. Hence if $I_j \cap S$ is nonempty, the graph of the restriction of f to $I_j \cap S$, that is,

$$\{(x_1, \ldots, x_n) \in E^n : (x_1, \ldots, x_{n-1}) \in I_j \cap S,\ f(x_1, \ldots, x_{n-1}) = x_n\}$$

is contained in the set

$$\{(x_1, \ldots, x_n) \in E^n : (x_1, \ldots, x_{n-1}) \in I_j,\ m_j \leq x_n \leq M_j\},$$

where m_j and M_j are respectively the minimum and the maximum values attained by f on $I_j \cap S$, and the latter set is a closed interval in E^n of volume $(M_j - m_j)\text{vol}(I_j) < \epsilon\,\text{vol}(I_j)$. Hence the graph of f is contained in the union of a finite number of closed intervals in E^n the sum of whose volumes is at most $\sum_{j=1}^{N} \epsilon\,\text{vol}(I_j) = \epsilon\,\text{vol}(I)$. Since ϵ is an arbitrary positive number, (6) is implied by (1).

Proposition. *If A, B are subsets of E^n such that* vol $(A \cap B) = 0$ *and $f\colon A \cup B \to \mathbf{R}$ is integrable on A and on B, then*

$$\int_{A\cup B} f = \int_A f + \int_B f.$$

To prove this, define $f_1, f_2, f_3\colon E^n \to \mathbf{R}$ by

$$f_1(x) = \begin{cases} f(x) & \text{if } x \in A \\ 0 & \text{if } x \notin A \end{cases}$$

$$f_2(x) = \begin{cases} f(x) & \text{if } x \in B \\ 0 & \text{if } x \notin B \end{cases}$$

$$f_3(x) = \begin{cases} f(x) & \text{if } x \in A \cap B \\ 0 & \text{if } x \notin A \cap B. \end{cases}$$

Then $\int_{E^n} f_1 = \int_A f$ and $\int_{E^n} f_2 = \int_B f$. The existence of the last two integrals implies that f is bounded, so by (5) of the previous proposition we have $\int_{E^n} f_3 = \int_{A\cap B} f = 0$. Since $f_1(x) + f_2(x) - f_3(x) = f(x)$ if $x \in A \cup B$ and $f_1(x) + f_2(x) - f_3(x) = 0$ if $x \notin A \cup B$, we can therefore compute

$$\int_{A\cup B} f = \int_{E^n} (f_1 + f_2 - f_3) = \int_{E^n} f_1 + \int_{E^n} f_2 - \int_{E^n} f_3 = \int_A f + \int_B f.$$

In the special case $f = 1$, the proposition says that if A and B are subsets of E^n with volume whose intersection has volume zero, then

$$\text{vol}\,(A \cup B) = \text{vol}\,(A) + \text{vol}\,(B).$$

We now have the main existence theorem.

Theorem. *Let $A \subset E^n$ be a set with volume and let $f\colon A \to \mathbf{R}$ be a bounded function that is continuous except on a subset of A of volume zero. Then $\int_A f$ exists.*

Let us first prove this in the special case where A is a closed interval I and f is continuous on I. Here the proof is an easy modification of the earlier proof for $n = 1$. Let $M \in \mathbf{R}$ be such that $|f(x)| \leq M$ for all $x \in I$. Given $\epsilon > 0$, the uniform continuity of f on I gives us a $\delta > 0$ such that $|f(x) - f(y)| < \epsilon$ whenever $x, y \in I$ and $d(x, y) < \delta$. Choose a partition of I of width less than $\delta/\sqrt{n}$. Suppose this partition subdivides I into the closed subintervals $I_1, \ldots, I_N$, so that $I = I_1 \cup \cdots \cup I_N$ and no two of the I_j's overlap except possibly at extremities. Thus vol $(I_j \cap I_k) = 0$ if $j, k = 1, \ldots, N$ and $j \neq k$. If $x, y \in I_j$ then $d(x, y) < \delta$, so we have $|f(x) - f(y)| < \epsilon$. We define $f_1, f_2\colon I \to \mathbf{R}$ by setting

$$f_1(x) = \begin{cases} \min \{f(y) : y \in I_j\} & \text{if } x \in I_j, x \notin I_k \text{ for any } k \neq j \\ -M & \text{if } x \text{ is in at least two of the sets } I_1, \ldots, I_N \end{cases}$$

$$f_2(x) = \begin{cases} \max \{f(y) : y \in I_j\} & \text{if } x \in I_j, x \notin I_k \text{ for any } k \neq j \\ M & \text{if } x \text{ is in at least two of the sets } I_1, \ldots, I_N. \end{cases}$$

Then f_1, f_2 are step functions on I, $f_1(x) \leq f(x) \leq f_2(x)$ for each $x \in I$ and $\int_I (f_2 - f_1) = \sum_{j=1}^{N} \int_{I_j} (f_2 - f_1) \leq \sum_{j=1}^{N} \epsilon \text{ vol } (I_j) = \epsilon \text{ vol } (I)$. Since such f_1, f_2 exist for each $\epsilon > 0$, our criterion for integrability on a closed interval implies the existence of $\int_I f$. This proves the special case.

Now consider the general case of the theorem, with $A \subset E^n$ a set with volume and f a bounded real-valued function on A that is continuous except on a subset of volume zero. If $S \subset A$ is the subset where f is not continuous, then $A - S$ has volume (by part (4) of the proposition on sets of volume zero) and $\int_S f = 0$ (by part (5) of the same proposition). If we can prove that $\int_{A-S} f$ exists, the preceding proposition will imply that $\int_A f = \int_{A-S} f + \int_S f = \int_{A-S} f$. We may therefore replace A by $A - S$, if necessary, to obtain the simplifying assumption that f is continuous on A. Fix a closed interval $I \subset E^n$ such that $A \subset I$. Extend the definition of f to I, redefining f on $I - A$ if necessary, so that $f(x) = 0$ if $x \in I - A$. We then have to show that $\int_I f$ exists. Suppose that $|f(x)| \leq M$ for all $x \in I$. Let $g: I \to \mathbf{R}$ be defined by $g(x) = 1$ if $x \in A$, $g(x) = 0$ if $x \in I - A$. Then $\int_I g$ exists, this being just vol (A). Suppose $\epsilon > 0$ is given. We can then find a partition of I such that any two Riemann sums for g corresponding to this partition differ by less than ϵ. Suppose that this partition subdivides I into the closed subintervals $I_1, \ldots, I_N$, so that $I = I_1 \cup \cdots \cup I_N$ and no two of the subintervals I_j overlap, except possibly at extremities. The points of I that are contained in more than one I_j are a set of volume zero. Let P be the number of subintervals $I_1, \ldots, I_N$ that are entirely contained in A, and let Q be the number of these subintervals that have points in common with A. We may suppose $I_1, \ldots, I_N$ so numbered that $I_j \subset A$ if $1 \leq j \leq P$, that I_j contains both points of A and points of $I - A$ if $P < j \leq Q$ and that $I_j \subset I - A$ if $Q < j \leq N$. Then two Riemann sums for g corresponding to the given partition of I are $\sum_{j=1}^{P} \text{vol } (I_j)$ and $\sum_{j=1}^{Q} \text{vol } (I_j)$. Therefore $\sum_{j=P+1}^{Q} \text{vol } (I_j) < \epsilon$. The restriction of f to each closed subinterval $I_1, \ldots, I_P$ is continuous, so that $\int_{I_j} f$ exists for $j = 1, \ldots, P$. Therefore for each $j = 1, \ldots, P$ there are step functions $f_1{}^j, f_2{}^j: I_j \to \mathbf{R}$ such that $f_1{}^j(x) \leq f(x) \leq f_2{}^j(x)$ for all $x \in I_j$ and $\int_{I_j} (f_2{}^j - f_1{}^j) < \epsilon/N$. Now define a pair of functions $f_1, f_2: I \to \mathbf{R}$ in the following fashion:

If $x \in I_j$ for some unique $j = 1, \ldots, N$ we set

$$\begin{array}{lll} f_1(x) = f_1{}^j(x) & \text{and } f_2(x) = f_2{}^j(x) & \text{if } j = 1, \ldots, P \\ f_1(x) = -M & \text{and } f_2(x) = M & \text{if } j = P+1, \ldots, Q \\ f_1(x) = f_2(x) = 0 & & \text{if } j = Q+1, \ldots, N. \end{array}$$

If $x \in I_j$ for more than one $j = 1, \ldots, N$ we set

$$f_1(x) = -M \quad \text{and} \quad f_2(x) = M.$$

Then f_1, f_2 are step functions on I and $f_1(x) \leq f(x) \leq f_2(x)$ for all $x \in I$. Furthermore, making a repeated application of the preceding proposition,

$$\begin{aligned} \int_I (f_2 - f_1) &= \sum_{j=1}^{N} \int_{I_j} (f_2 - f_1) \\ &= \sum_{j=1}^{P} \int_{I_j} (f_2 - f_1) + \sum_{j=P+1}^{Q} \int_{I_j} (f_2 - f_1) + \sum_{j=Q+1}^{N} \int_{I_j} (f_2 - f_1) \\ &= \sum_{j=1}^{P} \int_{I_j} (f_2{}^j - f_1{}^j) + \sum_{j=P+1}^{Q} \int_{I_j} 2M + \sum_{j=Q+1}^{N} \int_{I_j} 0 \\ &\leq \frac{P\epsilon}{N} + 2M \sum_{j=P+1}^{Q} \text{vol}\,(I_j) \leq \epsilon + 2M\epsilon = \epsilon(1 + 2M). \end{aligned}$$

Since ϵ was an arbitrary positive number, our criterion for integrability on a closed interval again implies that $\int_I f$ exists. The proof is now complete.

So far the only subsets of E^n that are known to have volume are closed intervals, sets of volume zero, and sets that may be obtained from these by part (4) of the proposition on sets of volume zero (page 225). The theorem enables us to give other examples of sets with volume.

Example. Let I be a closed interval in E^{n-1} and let φ_1, φ_2 be continuous real-valued functions on I such that $\varphi_1(x) \leq \varphi_2(x)$ for all $x \in I$. Then the set

$$\{(x_1, \ldots, x_n) \in E^n : (x_1, \ldots, x_{n-1}) \in I,\ \varphi_1(x_1, \ldots, x_{n-1}) < x_n < \varphi_2(x_1, \ldots, x_{n-1})\}$$

has volume. For if $M \in \mathbf{R}$ is such that $|\varphi_1(x)|, |\varphi_2(x)| \leq M$ for all $x \in I$ and $J \subset E^n$ is the closed interval

$$\{(x_1, \ldots, x_n) \in E^n : (x_1, \ldots, x_{n-1}) \in I,\ x_n \in [-M, M]\},$$

then the function $f \colon J \to \mathbf{R}$ which has the value one at each point of the set in question and value zero at all other points of J is not continuous only at points of the form $(x_1, \ldots, x_{n-1}, \varphi_1(x_1, \ldots, x_{n-1}))$ or $(x_1, \ldots, x_{n-1}, \varphi_2(x_1, \ldots, x_{n-1}))$. Since these latter are a set of volume zero, $\int_J f$ exists. This is illustrated in Figure 36.

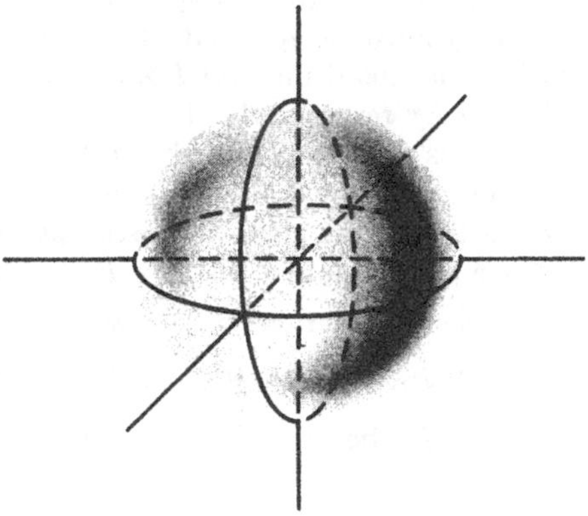

FIGURE 36. The unit ball in E^3 has volume. Except for sets of volume zero it is sandwiched between the graphs of the continuous real-valued functions $\pm\varphi: E^2 \to \mathbf{R}$, where $\varphi(x_1, x_2) = \sqrt{1 - x_1^2 - x_2^2}$ if $x_1^2 + x_2^2 \leq 1$ and $\varphi(x_1, x_2) = 0$ if $x_1^2 + x_2^2 > 1$.

§ 3. ITERATED INTEGRALS.

When the integral of a continuous function of one variable actually has to be computed one usually uses antiderivatives. The main method for computing integrals of functions of several variables is reduction to the one-variable case by means of iterated integrals.

In the following, for any sets $A \subset E^n$ and $B \subset E^m$ we identify $A \times B$ with a subset of E^{n+m} in the obvious way:

$$A \times B = \{(x_1, \ldots, x_n, y_1, \ldots, y_m) \in E^{n+m} : (x_1, \ldots, x_n) \in A,\ (y_1, \ldots, y_m) \in B\}.$$

Theorem. *Let $A \subset E^n$, $B \subset E^m$ and let $f: A \times B \to \mathbf{R}$. Suppose $\int_{A \times B} f$ exists and that for each $x \in A$ the function $f_{(x)}: B \to \mathbf{R}$ given by $f_{(x)}(y) = f(x, y)$ is integrable on B. Then if the function on A whose value at each x is $\int_B f_{(x)}$ is denoted by $\int_B f$ we have*

$$\int_{A \times B} f = \int_A \left(\int_B f \right).$$

(This equation is often written in the slightly more transparent form

$$\int_{A \times B} f = \int_A \left(\int_B f(x, y)\, dy \right) dx.)$$

If we extend f to a function on E^{n+m} by defining it to be zero outside $A \times B$, the theorem becomes equivalent to the analogous statement for the special case $A = E^n$, $B = E^m$. In this special case, the integrability of f on E^{n+m} implies that f must be zero outside some bounded subset of E^{n+m}, therefore outside some closed interval $I \times J$ of E^{n+m}, I and J being closed intervals in E^n and E^m respectively. It follows that the theorem is equivalent to the analogous statement for the restriction of f to $I \times J$. Thus without any loss of generality we may assume that A, B are closed intervals in E^n, E^m respectively. We make this assumption, and first prove the theorem when f is a step function on $A \times B$. In this case, for any $x \in A$ the function $f_{(x)}$ is a step function on B, so that $\int_B f_{(x)}$ exists without any further assumption. If $f_1, f_2, \ldots, f_\nu$ are step functions on $A \times B$ and the theorem holds for each f_j, then the theorem holds for $\sum_{j=1}^{\nu} f_j$, for

$$\int_{A\times B} \sum_{j=1}^{\nu} f_j = \sum_{j=1}^{\nu} \int_{A\times B} f_j = \sum_{j=1}^{\nu} \int_A\left(\int_B f_j\right) = \int_A\left(\sum_{j=1}^{\nu} \int_B f_j\right) = \int_A\left(\int_B \sum_{j=1}^{\nu} f_j\right).$$

But any step function on $A \times B$ is the sum of a finite number of step functions on $A \times B$ each of which is of the following simple type: there are subsets $S_1, \ldots, S_{n+m}$ of $\mathbf{R}$, each S_i being either a single point or an open interval, such that

$$\{(x_1, \ldots, x_{n+m}) \in E^{n+m} : x_1 \in S_1, \ldots, x_{n+m} \in S_{n+m}\}$$

is a subset of $A \times B$ and the step function has a constant value $c \in \mathbf{R}$ on this subset of $A \times B$ and the value zero on the complement of this subset. For a step function on $A \times B$ of the above simple type the theorem can be verified directly, each side of the equality in question reducing immediately to c times the product of the lengths of the sets $S_1, \ldots, S_{n+m}$. Hence the theorem holds whenever f is a step function on $A \times B$. Suppose, finally, that f is any function satisfying the hypotheses of the theorem. For any $\epsilon > 0$ there are step functions f_1, f_2 on $A \times B$ such that $f_1(z) \leq f(z) \leq f_2(z)$ for all $z \in A \times B$ and $\int_{A\times B} (f_2 - f_1) < \epsilon$. We then have $\int_{A\times B} f_1 \leq \int_{A\times B} f \leq \int_{A\times B} f_2$. Furthermore, for each $x \in A$ we have $(f_1)_{(x)}(y) \leq f_{(x)}(y) \leq (f_2)_{(x)}(y)$ for each $y \in B$, so that $\int_B (f_1)_{(x)} \leq \int_B f_{(x)} \leq \int_B (f_2)_{(x)}$. But $\int_B f_1$ and $\int_B f_2$ are step functions on A and

$$\int_A\left(\int_B f_2 - \int_B f_1\right) = \int_A\left(\int_B (f_2 - f_1)\right) = \int_{A\times B} (f_2 - f_1) < \epsilon.$$

Since such f_1, f_2 can be found for any $\epsilon > 0$, it follows that $\int_A \left(\int_B f\right)$ exists. We can therefore write

$$\int_A \left(\int_B f_1\right) \leq \int_A \left(\int_B f\right) \leq \int_A \left(\int_B f_2\right)$$

or

$$\int_{A\times B} f_1 \le \int_A \left(\int_B f\right) \le \int_{A\times B} f_2.$$

Combining this with $\int_{A\times B} f_1 \le \int_{A\times B} f \le \int_{A\times B} f_2$ we get

$$\left| \int_{A\times B} f - \int_A \left(\int_B f\right) \right| \le \int_{A\times B} (f_2 - f_1) < \epsilon.$$

The theorem follows from the fact that the last inequality holds for any $\epsilon > 0$.

We clearly have the symmetric result that if f is integrable on $A \times B$ then $\int_{A\times B} f = \int_B \left(\int_A f\right)$, provided that $\int_A f(x, y)dx$ exists for each $y \in B$.

Corollary 1. *Let f be an integrable real-valued function on the subset A of E^n, let B be a closed interval in E^m, and let $\pi_A\colon A \times B \to A$ be the projection on the first factor, that is, $\pi_A(x, y) = x$ if $x \in A$, $y \in B$. Then*

$$\int_{A\times B} f \circ \pi_A = \left(\int_A f\right) \text{vol } (B).$$

(In another notation, $\int_{A\times B} f(x)\, dx\, dy = \left(\int_A f(x)dx\right)\left(\int_B dy\right)$.*)*

For $x \in A$ and $y \in B$ we have $(f \circ \pi_A)_{(x)}(y) = (f \circ \pi_A)(x, y) = f(x)$ so that $\int_B (f \circ \pi_A)_{(x)} = \int_B f(x) = f(x) \text{ vol } (B)$. If $f \circ \pi_A$ is integrable on $A \times B$ we get $\int_{A\times B} f \circ \pi_A = \int_A (f \text{ vol } (B)) = \left(\int_A f\right) \text{vol } (B)$. Hence we need only show that $\int_{A\times B} f \circ \pi_A$ exists. To do this, we first reduce to the case $A = E^n$ by extending f to a function on E^n that has value zero at each point of $E^n - A$. If we then choose a closed interval $I \subset E^n$ such that f is zero at each point of $E^n - I$, we reduce to the case $A = I$. That is, we may assume that A is a closed interval in E^n. This being so, the existence of $\int_A f$ implies that, given any $\epsilon > 0$, we can find step functions f_1, f_2 on A such that $f_1(x) \le f(x) \le f_2(x)$ for each $x \in A$ and $\int_A (f_2 - f_1) < \epsilon$. Then $f_1 \circ \pi_A, f_2 \circ \pi_A$ are step functions on $A \times B$ such that for each $z \in A \times B$ we have $(f_1 \circ \pi_A)(z) \le (f \circ \pi_A)(z) \le (f_2 \circ \pi_A)(z)$ and $\int_{A\times B} (f_2 \circ \pi_A - f_1 \circ \pi_A) = \int_{A\times B} (f_2 - f_1) \circ \pi_A = \left(\int_A (f_2 - f_1)\right) \text{vol } (B) < \epsilon \text{ vol } (B)$. Hence $\int_{A\times B} f \circ \pi_A$ exists, as was to be shown.

If we apply Corollary 1 to the case $f = 1$, we get the following simple result: if $A \subset E^n$ has volume and $B \subset E^m$ is a closed interval, then $\text{vol } (A \times B) = \text{vol } (A) \text{ vol } (B)$. In particular, if A has zero volume so has $A \times B$.

It is worth remarking at this point that the theorem remains true if we replace the assumption that $f_{(x)}$ is integrable on B for all $x \in A$ by the assumption that $f_{(x)}$ is integrable on B for all $x \in A - S$, where S is a subset of A of volume zero, if we then understand $\int_B f_{(x)}$ to be an arbitrary element of some bounded subset of $\mathbf{R}$ whenever $x \in S$. To see this, note first that B may be assumed bounded, so that vol $(S \times B) = 0$. Therefore $\int_{S\times B} f = 0$. If we define $g\colon A \times B \to \mathbf{R}$ by $g(z) = f(z)$ if $z \in S \times B$, otherwise $g(z) = 0$, then $\int_{A\times B} g = 0$. Therefore $\int_{A\times B} (f - g) = \int_{A\times B} f$. But $(f - g)\colon A \times B \to \mathbf{R}$ has the same restriction as f to $(A - S) \times B$ and is zero on $S \times B$, so that $\int_{(A-S)\times B} f$ exists and equals $\int_{A\times B} (f - g) = \int_{A\times B} f$. Thus

$$\int_{A\times B} f = \int_{(A-S)\times B} f = \int_{A-S} \left(\int_B f\right) = \int_{A-S} \left(\int_B f\right) + \int_S \left(\int_B f\right)$$
$$= \int_A \left(\int_B f\right).$$

Corollary 2. Let the set $A \subset E^{n-1}$ be compact and have volume and let $\varphi_1, \varphi_2\colon A \to \mathbf{R}$ be continuous functions such that $\varphi_1(x) \leq \varphi_2(x)$ for all $x \in A$. Then if f is a continuous real-valued function on the set

$$S = \{(x_1, \ldots, x_n) \in E^n : (x_1, \ldots, x_{n-1}) \in A,$$
$$\varphi_1(x_1, \ldots, x_{n-1}) \leq x_n \leq \varphi_2(x_1, \ldots, x_{n-1})\},$$

we have $\int_S f = \int_A \left(\int_{\varphi_1}^{\varphi_2} f\right)$, *where* $\int_{\varphi_1}^{\varphi_2} f$ *is the function on A whose value at any $(x_1, \ldots, x_{n-1})$ is* $\int_{\varphi_1(x_1,\ldots,x_{n-1})}^{\varphi_2(x_1,\ldots,x_{n-1})} f(x_1, \ldots, x_n)dx_n$.

Let B be a closed interval in $\mathbf{R}$ containing $\varphi_1(A) \cup \varphi_2(A)$, so that $S \subset A \times B$. Extend f to a function on $A \times B$ by setting $f(z) = 0$ if $z \in A \times B - S$. Then f is bounded and is continuous at any point of $A \times B$ that is not of the form

$$(x_1, \ldots, x_{n-1}, \varphi_1(x_1, \ldots, x_{n-1})) \quad \text{or} \quad (x_1, \ldots, x_{n-1}, \varphi_2(x_1, \ldots, x_{n-1})).$$

These later points form a set of volume zero, by part (6) of the proposition on sets of volume zero. Since $A \times B$ has volume (by the comment following Corollary 1), $\int_S f = \int_{A\times B} f$ exists. Also for each $(x_1, \ldots, x_{n-1}) \in A$,

$$\int_B f_{(x_1,\ldots,x_{n-1})} = \int_B f(x_1, \ldots, x_n)dx_n = \int_{\varphi_1(x_1,\ldots,x_{n-1})}^{\varphi_2(x_1,\ldots,x_{n-1})} f(x_1, \ldots, x_n)dx_n.$$

Hence the theorem is applicable to the present A, B and f, giving us the desired result.

In favorable circumstances Corollary 2 may be applied repeatedly to express an integral over a subset of E^n as an n-fold iterated integral.

If we apply Corollary 2 to the case where $n = 2$, $f = 1$ and $A = [a, b]$ (for some $a, b \in \mathbf{R}$, $a < b$) we obtain the well-known fact that the area of the plane set bounded by the lines $x = a$ and $x = b$ and the curves $y = \varphi_1(x)$ and $y = \varphi_2(x)$ is $\int_a^b (\varphi_2(x) - \varphi_1(x))dx$. However we have actually proved something nontrivial because this is now a theorem, not a definition. Similarly, if $n = 3$ and $f = 1$, for any compact subset A of the plane that has area, the volume of the subset of E^3 lying over A and between $z = \varphi_1(x, y)$ and $z = \varphi_2(x, y)$ is $\int_A (\varphi_2(x, y) - \varphi_1(x, y))dx\,dy$.

§ 4. CHANGE OF VARIABLE.

Lemma. *Let D be a compact subset of the open subset U of E^n. Then there exist subsets D', V of U, with D' compact and V open, such that*

$$D \subset V \subset D' \subset U.$$

Each point of D is contained in an open ball of E^n such that the closed ball with the same center and radius is entirely contained in U. Since D is compact, we may find a finite set of such open balls whose union contains D. We may then take V to be the union of this finite set of open balls, D' the union of the corresponding closed balls.

Proposition **(*"Partition of Unity"*).** *Let D be a compact subset of E^n and let $\{U_s\}_{s\in S}$ be a collection of open subsets of E^n whose union contains D. Then there is a finite set of continuous functions $\psi_1, \ldots, \psi_N \colon E^n \to [0, 1]$ such that*

$$\psi_1(x) + \cdots + \psi_N(x) = 1$$

for each $x \in D$ and each ψ_i is zero outside a compact subset of one of the sets $\{U_s\}_{s\in S}$.

Start with any continuous function $h\colon \mathbf{R} \to \mathbf{R}$ such that $h(x) = 0$ if $x \leq 0$, while $h(x) > 0$ if $x > 0$; for example, we may take $h(x) = x$ for $x > 0$. Then the function $g\colon \mathbf{R} \to \mathbf{R}$ given by $g(x) = h(1 - x^2)$ has the properties that $g(x) = 0$ if $|x| \geq 1$ while $g(x) > 0$ if $|x| < 1$. Hence if $\nu > 0$, then $g(\nu x)$ is zero for $|x| \geq 1/\nu$ and positive for $|x| < 1/\nu$. For each point $p \in D$ choose $\nu_p > 0$ such that the closed ball in E^n of center p and radius $1/\nu_p$ is entirely contained in one of the sets $\{U_s\}_{s\in S}$. Let B_p be the open ball in E^n of center p and radius $1/\nu_p$. Since D is compact there is a finite subset $p_1, \ldots, p_N$ of D such that $D \subset U = B_{p_1} \cup \cdots \cup B_{p_N}$.

Set

$$\varphi_i(x) = \frac{g\,(\nu_{p_i}\,d(x, p_i))}{\sum_{j=1}^{N} g\,(\nu_{p_j}\,d(x, p_j))}$$

for $x \in U$ and $i = 1, \ldots, N$. Then each φ_i is a continuous function on the open set $U \supset D$ with values in $[0, 1]$ and $\varphi_1(x) + \cdots + \varphi_N(x) = 1$ for all $x \in U$. In addition, for each $i = 1, \ldots, N$, the points where φ_i is not zero are contained in a compact subset of one of the sets $\{U_s\}_{s \in S}$. Now use the lemma to obtain subsets D' and V of E^n, respectively compact and open, such that

$$D \subset V \subset D' \subset U.$$

Apply what has been proved above to the compact set D' and the collection of open sets $\{V, U - D\}$ (whose union contains D'). We get functions analogous to the above φ_i's and these we group into two batches, according to whether or not the set of points at which the function is not zero is contained in a compact subset of V, then we add the separate batches. We get continuous functions θ_1, θ_2 on an open set $U' \supset D'$ with values in $[0, 1]$ such that $\theta_1(x) + \theta_2(x) = 1$ for each $x \in U'$ while θ_1 is zero at each point of U' outside a compact subset of V and θ_2 is zero at each point of U' outside a compact subset of $U - D$. Since $\theta_1(x) = 0$ for each $x \in U' - V$ and since $V \subset D'$, we have $\theta_1(x) = 0$ if $x \in U' - D'$. Also $\theta_2(x) = 0$ if $x \in D$, so that $\theta_1(x) = 1$ if $x \in D$. For $i = 1, \ldots, N$ we define $\psi_i\colon E^n \to \mathbf{R}$ by $\psi_i(x) = \varphi_i(x)\theta_1(x)$ if $x \in U \cap U'$ and $\psi_i(x) = 0$ if $x \in E^n - U \cap U'$. Since ψ_i is continuous on the open sets $U \cap U'$ and $E^n - D'$ (it is zero on the latter), it is continuous on their union, which is E^n. The other desired properties of $\psi_1, \ldots, \psi_N$ follow immediately from their construction.

Corollary. *Let D be a compact subset of the open subset U of E^n. Then there is a continuous function $\psi\colon E^n \to [0, 1]$ such that $\psi(x) = 1$ for each $x \in D$ and $\psi(x) = 0$ for each x outside some compact subset of U.*

This has essentially appeared in the proof above, but it also follows easily from the statement of the proposition. Let D' and V be subsets of E^n, respectively compact and open, such that

$$D \subset V \subset D' \subset U.$$

Since $D' \subset V \cup (U - D)$, we can apply the proposition to the compact set D' and the collection of open sets $\{V, U - D\}$ to get continuous functions $\psi_1, \psi_2\colon E^n \to [0, 1]$ such that $\psi_1(x) + \psi_2(x) = 1$ for each $x \in D'$ and ψ_1 and ψ_2 are zero outside compact subsets of V and $U - D$ respectively. Since ψ_1 is 1 on D and zero outside D', we may take $\psi = \psi_1$.

We remark that the functions $\psi_1, \ldots, \psi_N$ of the proposition and also the function ψ of the corollary may be chosen so as to possess all partial derivatives of order m, for any given positive integer m, by starting with $h(x) = x^{m+1}$ for $x > 0$. In fact it can be shown that $\psi_1, \ldots, \psi_N, \psi$ possess all partial derivatives of all orders if we take $h(x) = e^{-1/x^2}$ for $x > 0$ (cf. Problem 26, Chapter VI).

Lemma. *The real-valued function f on the closed interval I of E^n is integrable on I if and only if, for each $\epsilon > 0$, there exist continuous real-valued functions f_1, f_2 on I such that*

$$f_1(x) \leq f(x) \leq f_2(x) \quad \textit{for each } x \in I$$

and

$$\int_I (f_2 - f_1) < \epsilon.$$

The proof makes repeated use of the first proposition of § 2. First suppose that f satisfies the given condition. Then given $\epsilon > 0$ there are continuous functions $f_1, f_2\colon I \to \mathbf{R}$ such that $f_1(x) \leq f(x) \leq f_2(x)$ for each $x \in I$ and $\int_I (f_2 - f_1) < \epsilon/3$. Since f_1 and f_2 are integrable on I there are step functions f_1', f_1'', f_2', f_2'' on I such that $f_1'(x) \leq f_1(x) \leq f_1''(x)$ and $f_2'(x) \leq f_2(x) \leq f_2''(x)$ are true for each $x \in I$ and $\int_I (f_1'' - f_1') < \epsilon/3$, $\int_I (f_2'' - f_2') < \epsilon/3$. The step functions f_1', f_2'' are such that $f_1'(x) \leq f(x) \leq f_2''(x)$ for each $x \in I$ and

$$\begin{aligned}\int_I (f_2'' - f_1') &= \int_I (f_2'' - f_2) + \int_I (f_2 - f_1) + \int_I (f_1 - f_1') \\ &\leq \int_I (f_2'' - f_2') + \int_I (f_2 - f_1) + \int_I (f_1'' - f_1') < \frac{\epsilon}{3} + \frac{\epsilon}{3} + \frac{\epsilon}{3} = \epsilon.\end{aligned}$$

Thus f is integrable on I. To prove the converse, suppose first that any step function on I satisfies the given condition. We reason in the same way as above. If f is integrable on I then for any $\epsilon > 0$ there are step functions f_1, f_2 on I such that $f_1(x) \leq f(x) \leq f_2(x)$ for each $x \in I$ and $\int_I (f_2 - f_1) < \epsilon/3$. Since step functions are assumed to satisfy the given condition, there are continuous functions $f_1', f_1'', f_2', f_2''\colon I \to \mathbf{R}$ such that we will have $f_1'(x) \leq f_1(x) \leq f_1''(x)$ and $f_2'(x) \leq f_2(x) \leq f_2''(x)$ for all $x \in I$ and $\int_I (f_1'' - f_1') < \epsilon/3$, $\int_I (f_2'' - f_2') < \epsilon/3$. Thus the continuous functions f_1', f_2'' are such that $f_1'(x) \leq f(x) \leq f_2''(x)$ for all $x \in I$ and

$$\begin{aligned}\int_I (f_2'' - f_1') &= \int_I (f_2'' - f_2) + \int_I (f_2 - f_1) + \int_I (f_1 - f_1') \\ &\leq \int_I (f_2'' - f_2') + \int_I (f_2 - f_1) + \int_I (f_1'' - f_1') < \frac{\epsilon}{3} + \frac{\epsilon}{3} + \frac{\epsilon}{3} = \epsilon,\end{aligned}$$

so that f satisfies the given condition. Therefore it remains only to show that for any step function f on I and any $\epsilon > 0$ there exist continuous functions f_1, f_2 on I such that $f_1(x) \leq f(x) \leq f_2(x)$ for all $x \in I$ and $\int_I (f_2 - f_1) < \epsilon$. But any step function on I is the sum of a finite number of step functions on I each of which is of the following simple type: there are subsets $S_1, \ldots, S_n \subset \mathbf{R}$, each S_i being either a single point or an open interval, such that the set $\{(x_1, \ldots, x_n) \in E^n : x_1 \in S_1, \ldots, x_n \in S_n\}$ is a subset of I and the step function has a constant value $c \in \mathbf{R}$ on this subset and the value 0 on the complement of this subset. Therefore we need only prove that a step function on I of the above simple type satisfies the indicated condition. That is, if for $i = 1, \ldots, n$ the subset $S_i \subset \mathbf{R}$ is a single point or an open interval such that $\{(x_1, \ldots, x_n) \in E^n : x_1 \in S_1, \ldots, x_n \in S_n\}$ is a subset of I and if $f: I \to \mathbf{R}$ has the constant value $c \in \mathbf{R}$ on this subset and the value zero on its complement, we must show that for any $\epsilon > 0$ there are continuous functions $f_1, f_2: I \to \mathbf{R}$ such that $f_1(x) \leq f(x) \leq f_2(x)$ for all $x \in I$ and $\int_I (f_2 - f_1) < \epsilon$. It clearly suffices to prove this for $c = 1$. Assume that I happens to be the closed interval in E^n determined by $a_1, \ldots, a_n, b_1, \ldots, b_n$. First suppose that some S_i is a single point, say $S_1 = \alpha_1 \in [a_1, b_1]$. For any $\delta > 0$ choose a continuous function $\varphi: \mathbf{R} \to [0, 1]$ such that $\varphi(\alpha_1) = 1$ and $\varphi(x_1) = 0$ if $|x_1 - \alpha_1| > \delta$. Define $\psi: E^n \to [0, 1]$ by $\psi(x_1, \ldots, x_n) = \varphi(x_1)$. Then $0 \leq f(x) \leq \psi(x)$ for all $x \in I$ and

$$\int_I (\psi - 0) \leq 2\delta(b_2 - a_2) \cdots (b_n - a_n),$$

which can be made less than ϵ by taking δ small enough. It remains to consider the case where each $S_i = (\alpha_i, \beta_i)$, where $a_i \leq \alpha_i < \beta_i \leq b_i$. For any $\delta > 0$ such that $2\delta < \beta_1 - \alpha_1, \ldots, \beta_n - \alpha_n$, choose a continuous function $f_1: E^n \to [0, 1]$ such that f_1 is 1 on the closed interval determined by $\alpha_1 + \delta, \ldots, \alpha_n + \delta, \beta_1 - \delta, \ldots, \beta_n - \delta$ and 0 outside the open interval determined by $\alpha_1, \ldots, \alpha_n, \beta_1, \ldots, \beta_n$, and choose a continuous function $f_2: E^n \to [0, 1]$ that is 1 on the closed interval determined by $\alpha_1, \ldots, \alpha_n, \beta_1, \ldots, \beta_n$ and 0 outside the open interval determined by $\alpha_1 - \delta, \ldots, \alpha_n - \delta, \beta_1 + \delta, \ldots, \beta_n + \delta$. Then $f_1(x) \leq f(x) \leq f_2(x)$ for all $x \in I$ and

$$\int_I (f_2 - f_1) \leq \int_{E^n} (f_2 - f_1)$$
$$\leq (\beta_1 - \alpha_1 + 2\delta) \cdots (\beta_n - \alpha_n + 2\delta) - (\beta_1 - \alpha_1 - 2\delta) \cdots (\beta_n - \alpha_n - 2\delta).$$

Since polynomial functions are continuous, this latter expression can be made less than ϵ by taking δ sufficiently near zero, and this completes the proof.

Theorem. *Let A be an open subset of E^n, $\varphi: A \to E^n$ a one-one continuously differentiable map whose jacobian J_φ is nowhere zero on A. Suppose that the function $f: \varphi(A) \to \mathbf{R}$ is zero outside a compact subset of $\varphi(A)$ and that $\int_{\varphi(A)} f$ exists. Then*

$$\int_{\varphi(A)} f = \int_A (f \circ \varphi) |J_\varphi|.$$

Since the proof is quite complicated it will be given in a number of steps. We first make a few preliminary remarks to be borne in mind below. The inverse function theorem implies that $\varphi(A)$ is an open subset of E^n and that the map $\varphi^{-1}: \varphi(A) \to A$ is also continuously differentiable. Any compact subset of A (or $\varphi(A)$) is mapped by φ (or φ^{-1}) onto a compact subset of $\varphi(A)$ (or A), since the image of a compact set under a continuous map is compact. Similarly, since the inverse image of an open set under a continuous map is open (by the first proposition of Chapter IV), φ induces a one-one correspondence between the open subsets of A and those of $\varphi(A)$. If f is continuous the assumption that $\int_{\varphi(A)} f$ exists is superfluous, for this fact follows automatically from the assumption that f is zero outside a compact subset of the open set $\varphi(A)$. The reason f is assumed to be zero outside a compact subset of $\varphi(A)$ is that one must allow for the eventuality of A, $\varphi(A)$ or J_φ being unbounded. As usual, the component functions of φ will be denoted by $\varphi_1, \ldots, \varphi_n$, so that $\varphi(x) = (\varphi_1(x), \ldots, \varphi_n(x))$ for all $x \in A$ and $J_\varphi = \det\left(\frac{\partial \varphi_i}{\partial x_j}\right)$.

(1) The theorem is true if $\varphi_1(x_1, \ldots, x_n), \ldots, \varphi_n(x_1, \ldots, x_n)$ are a permutation of $x_1, \ldots, x_n$. For if $\varphi_1(x), \ldots, \varphi_n(x)$ are just $x_1, \ldots, x_n$, but possibly in a different order, then J_φ is the determinant of an $n \times n$ square array that has precisely one 1 in each row and each column, with all the other elements zero, so that $J_\varphi = \pm 1$. Thus the statement is a direct consequence of the definition of the integral, which does not depend on the order in which the coordinates are taken.

(2) We may assume that the theorem is true for $n - 1$ in place of n, if $n > 1$. For suppose we prove the theorem under this assumption. Then if we prove the theorem for $n = 1$ it will be true for $n = 2$, since true for $n = 2$ it will be true for $n = 3$, since true for $n = 3$ it will be true for $n = 4$, etc., so the theorem will hold for all n.

(3) It is sufficient to prove the theorem when f is continuous. For suppose it is known in this special case. Then given an arbitrary $f: \varphi(A) \to \mathbf{R}$ which is zero outside a compact subset of $\varphi(A)$ and integrable on $\varphi(A)$ we must show that $\int_A (f \circ \varphi) |J_\varphi|$ exists and is equal to $\int_{\varphi(A)} f$. Let $D \subset \varphi(A)$ be a compact set such that f is zero outside D. Apply the previous corollary to D and $\varphi(A)$ to get a continuous function $\psi: E^n \to [0, 1]$ that is 1 on D

and 0 outside a compact subset D' of $\varphi(A)$. Let I be a closed interval in E^n that contains the compact set $D' \cup \varphi^{-1}(D')$. For convenience, if F is any function on a subset of E^n we shall denote by $\bar{F}$ the function on E^n which agrees with F where the latter is defined and is zero elsewhere. Thus $\int_{\varphi(A)} f = \int_I \bar{f}$. Now suppose we are given some $\epsilon > 0$. Since $\int_I \bar{f}$ exists, the lemma enables us to find continuous functions $g_1, g_2\colon I \to \mathbf{R}$ such that $g_1(x) \leq \bar{f}(x) \leq g_2(x)$ for each $x \in I$ and $\int_I (g_2 - g_1) < \epsilon$. Then $\psi(x)g_1(x) \leq \bar{f}(x) \leq \psi(x)g_2(x)$ for each $x \in I$ and $\int_I (\psi g_2 - \psi g_1) = \int_I \psi(g_2 - g_1) \leq \int_I (g_2 - g_1) < \epsilon$. If we let f_1, f_2 be the restrictions to $\varphi(A)$ of $\overline{\psi g_1}, \overline{\psi g_2}$ respectively, then f_1, f_2 are continuous real-valued functions on $\varphi(A)$ which are zero outside D', $f_1(x) \leq f(x) \leq f_2(x)$ for each $x \in \varphi(A)$, and $\int_{\varphi(A)} (f_2 - f_1) < \epsilon$. Now consider the real-valued functions on A given by $(f_1 \circ \varphi)|J_\varphi|$, $(f_2 \circ \varphi)|J_\varphi|$ and $(f \circ \varphi)|J_\varphi|$; the first two are continuous, they are all zero outside $\varphi^{-1}(D')$, and they satisfy

$$((f_1 \circ \varphi)|J_\varphi|)(x) \leq ((f \circ \varphi)|J_\varphi|)(x) \leq ((f_2 \circ \varphi)|J_\varphi|)(x)$$

for all $x \in A$. Thus $\overline{(f_1 \circ \varphi)|J_\varphi|}$ and $\overline{(f_2 \circ \varphi)|J_\varphi|}$ are continuous on E^n and

$$\overline{(f_1 \circ \varphi)|J_\varphi|}(x) \leq \overline{(f \circ \varphi)|J_\varphi|}(x) \leq \overline{(f_2 \circ \varphi)|J_\varphi|}(x)$$

for all $x \in I$. By assumption our theorem holds for f_1 and f_2, so that

$$\begin{aligned}\int_I (\overline{(f_2 \circ \varphi)|J_\varphi|} - \overline{(f_1 \circ \varphi)|J_\varphi|}) &= \int_A ((f_2 \circ \varphi)|J_\varphi| - (f_1 \circ \varphi)|J_\varphi|) \\ &= \int_{\varphi(A)} (f_2 - f_1) < \epsilon.\end{aligned}$$

Since ϵ was an arbitrary positive number, the lemma implies that $\overline{(f \circ \varphi)|J_\varphi|}$ is integrable on I. Thus $(f \circ \varphi)|J_\varphi|$ is integrable on A. Furthermore, from the inequalities

$$\int_{\varphi(A)} f_1 = \int_A (f_1 \circ \varphi)|J_\varphi| \leq \int_A (f \circ \varphi)|J_\varphi| \leq \int_A (f_2 \circ \varphi)|J_\varphi| = \int_{\varphi(A)} f_2$$

and

$$\int_{\varphi(A)} f_1 \leq \int_{\varphi(A)} f \leq \int_{\varphi(A)} f_2$$

we deduce that

$$\left| \int_A (f \circ \varphi)|J_\varphi| - \int_{\varphi(A)} f \right| \leq \int_{\varphi(A)} (f_2 - f_1) < \epsilon.$$

This being true for all $\epsilon > 0$, we have

$$\int_A (f \circ \varphi)|J_\varphi| = \int_{\varphi(A)} f,$$

proving the contention of this section of the proof. Therefore from now on we may assume f to be continuous.

(4) If $\{A_s\}_{s \in S}$ is a collection of open subsets of E^n such that $A = \bigcup_{s \in S} A_s$ and for each $s \in S$ the theorem is true for A_s and the restriction of φ to A_s, then the theorem is true for A and φ. For let $f: \varphi(A) \to \mathbf{R}$ be a continuous function that is zero outside the compact subset D of $\varphi(A)$. $\{\varphi(A_s)\}_{s \in S}$ is a collection of open subsets of $\varphi(A)$ whose union is $\varphi(A)$. By the proposition continuous functions $\psi_1, \ldots, \psi_N: E^n \to [0, 1]$ may be found such that $\psi_1(x) + \cdots + \psi_N(x) = 1$ for each $x \in D$ and each ψ_i is zero outside a compact subset of $\varphi(A_{s(i)})$, for some $s(i) \in S$. For each $i = 1, \ldots, N$ we have

$$\int_{\varphi(A)} \psi_i f = \int_{\varphi(A_{s(i)})} \psi_i f = \int_{A_{s(i)}} ((\psi_i f) \circ \varphi) |J_\varphi| = \int_A ((\psi_i f) \circ \varphi) |J_\varphi|.$$

Since $f = \sum_{j=1}^{N} \psi_i f$, we deduce

$$\int_{\varphi(A)} f = \int_{\varphi(A)} \sum_{i=1}^{N} \psi_i f = \sum_{i=1}^{N} \int_{\varphi(A)} \psi_i f = \sum_{i=1}^{N} \int_A ((\psi_i f) \circ \varphi) |J_\varphi|$$

$$= \int_A \sum_{i=1}^{N} ((\psi_i f) \circ \varphi) |J_\varphi| = \int_A (f \circ \varphi) |J_\varphi|,$$

which was to be shown.

(5) The theorem is true for $n = 1$. To prove this we may suppose A to be an open interval, by (4). Let f be a continuous real-valued function on $\varphi(A)$ that is zero outside some compact subset of $\varphi(A)$. The function $f \circ \varphi$ is zero outside some compact subset of A, so we can find $a, b \in A$, $a < b$, such that $f \circ \varphi$ is zero on $A - [a, b]$. Then f is zero on $\varphi(A) - \varphi([a, b])$. In the present case $J_\varphi = \varphi'$, and since this is nowhere zero φ must be either an increasing function on A or a decreasing function. In either case we go back to the change of variable theorem for one variable (Corollary 3 of the fundamental theorem of calculus). If φ is increasing, then

$$\int_{\varphi(A)} f = \int_{\varphi([a,b])} f = \int_{\varphi(a)}^{\varphi(b)} f = \int_a^b (f \circ \varphi)\varphi' = \int_A (f \circ \varphi) |J_\varphi|.$$

If φ is decreasing the computation is

$$\int_{\varphi(A)} f = \int_{\varphi([a,b])} f = \int_{\varphi(b)}^{\varphi(a)} f = \int_b^a (f \circ \varphi)\varphi' = \int_a^b (f \circ \varphi)(-\varphi')$$

$$= \int_A (f \circ \varphi) |J_\varphi|.$$

This proves the theorem for $n = 1$. Therefore from now on we may assume that $n > 1$.

(6) The theorem is true if for certain $i, j = 1, \ldots, n$ we have $\varphi_i(x_1, \ldots, x_n) = x_j$. By virtue of (1), it suffices to prove this in the special case where $i = j = 1$, that is

$$\varphi(x_1, \ldots, x_n) = (x_1, \varphi_2(x_1, \ldots, x_n), \ldots, \varphi_n(x_1, \ldots, x_n)).$$

A is a union of open intervals of E^n, for any point of A is the center of an open ball that is entirely contained in A, and an open ball of radius r contains the open interval having the same center and sides $2r/\sqrt{n}$. Thus by (4) we may assume that A is itself an open interval. Identifying E^n with $\mathbf{R} \times E^{n-1}$, we have $A = B \times C$, where B and C are open intervals in $\mathbf{R}$ and E^{n-1} respectively. Let $f: \varphi(A) \to \mathbf{R}$ be a continuous function that is zero outside a compact subset of $\varphi(A) \subset B \times E^{n-1}$. The function $\bar{f}: B \times E^{n-1} \to \mathbf{R}$ which agrees with f on $\varphi(A)$ and otherwise is zero is continuous. For each $x \in B$, the function $\bar{f}_{(x)}: E^{n-1} \to \mathbf{R}$ given by $\bar{f}_{(x)}(y) = \bar{f}(x, y)$ is continuous and is zero outside a compact subset of E^{n-1}, hence integrable on E^{n-1}, so we have

$$\int_{\varphi(A)} f = \int_{B \times E^{n-1}} \bar{f} = \int_B \left(\int_{E^{n-1}} \bar{f} \right),$$

where $\int_{E^{n-1}} \bar{f}$ denotes the function on B whose value at x is $\int_{E^{n-1}} \bar{f}_{(x)}$.

For any $x \in B$ consider the function $\varphi_{(x)}: C \to E^{n-1}$ which is defined by

$$\varphi_{(x)}(x_2, \ldots, x_n) = (\varphi_2(x, x_2, \ldots, x_n), \ldots, \varphi_n(x, x_2, \ldots, x_n))$$

for all $(x_2, \ldots, x_n) \in C$. $\varphi_{(x)}$ is a one-one continuously differentiable map whose jacobian $J_{\varphi_{(x)}}$ is $(J_\varphi)_{(x)}$, that is for each $(x_2, \ldots, x_n) \in C$ we have $J_{\varphi_{(x)}}(x_2, \ldots, x_n) = J_\varphi(x, x_2, \ldots, x_n)$. Since our theorem holds for $n - 1$ (by (2)), we can compute

$$\int_{E^{n-1}} \bar{f}_{(x)} = \int_{\varphi_{(x)}(C)} f_{(x)} = \int_C (f_{(x)} \circ \varphi_{(x)}) \left| J_{\varphi_{(x)}} \right| = \int_C ((f \circ \varphi) |J_\varphi|)_{(x)},$$

so that

$$\int_{E^{n-1}} \bar{f} = \int_C (f \circ \varphi) |J_\varphi|.$$

Therefore

$$\int_{\varphi(A)} f = \int_B \left(\int_C (f \circ \varphi) |J_\varphi| \right).$$

Now $(f \circ \varphi)|J_\varphi|$ is a continuous real-valued function on $B \times C$ that is zero outside a compact subset of $B \times C$, hence integrable on $B \times C$; also, for each $x \in B$ the function $((f \circ \varphi)|J_\varphi|)_{(x)}$ is a continuous real-valued function on C that is zero outside a compact subset of C, hence integrable on C. Therefore the last iterated integral equals $\int_{B \times C} (f \circ \varphi)|J_\varphi|$. Thus we indeed have

$$\int_{\varphi(A)} f = \int_A (f \circ \varphi) |J_\varphi|$$

in our special case.

(7) We now prove the theorem. For any point $a \in A$ we have $J_\varphi(a) \neq 0$, so that $\dfrac{\partial \varphi_n}{\partial x_i}(a) \neq 0$ for at least one $i = 1, \ldots, n$. For given i,

the subset of A where $\partial\varphi_n/\partial x_i$ is not zero is open and the union of these n subsets is A, so by (4) it suffices to prove the theorem for each one of these subsets. Therefore we may assume that $\partial\varphi_n/\partial x_i$ is never zero on A. By (1), we may assume that $i = n$, that is $\partial\varphi_n/\partial x_n$ is never zero on A. Now consider the map $\sigma: A \to E^n$ defined by

$$\sigma(x_1, \ldots, x_n) = (x_1, \ldots, x_{n-1}, \varphi_n(x_1, \ldots, x_n)).$$

The jacobian of σ is $\partial\varphi_n/\partial x_n$, which is never zero, so by the inverse function theorem each point $a \in A$ is contained in an open subset A_a of A such that the restriction of σ to A_a is a one-one map from A_a onto an open subset $\sigma(A_a)$ of E^n and such that the map $\sigma^{-1}: \sigma(A_a) \to A_a$ is also continuously differentiable. Again by (4), it suffices to prove the theorem for each A_a. Thus we may assume that the map $\sigma: A \to E^n$ is a one-one map from A onto the open subset $\sigma(A)$ of E^n and that $\sigma^{-1}: \sigma(A) \to A$ is also continuously differentiable. The map $\tau = \varphi \circ \sigma^{-1}$ is therefore a one-one continuously differentiable map from $\sigma(A)$ onto $\varphi(A)$, and $\varphi = \tau \circ \sigma$. The maps σ and τ are such that if $x = (x_1, \ldots, x_n) \in A$, then $\sigma(x) = (x_1, \ldots, x_{n-1}, \varphi_n(x))$ and $\tau(x_1, \ldots, x_{n-1}, \varphi_n(x)) = (\varphi_1(x), \ldots, \varphi_n(x))$. By (6), the theorem holds for the map σ of A, provided J_σ is nowhere zero on A, and also for the map τ of $\sigma(A)$, provided J_τ is nowhere zero on $\sigma(A)$. Therefore for any continuous function $f: \varphi(A) \to E^n$ that is zero outside a compact subset of $\varphi(A)$ we have, provided J_σ and J_τ are nowhere zero,

$$\int_{\varphi(A)} f = \int_{\tau(\sigma(A))} f = \int_{\sigma(A)} (f \circ \tau)|J_\tau| = \int_A \big(((f \circ \tau)|J_\tau|) \circ \sigma\big)|J_\sigma|$$
$$= \int_A (f \circ \varphi)|(J_\tau \circ \sigma)J_\sigma|.$$

The theorem will therefore be proved if we can show that

$$J_\varphi = (J_\tau \circ \sigma)J_\sigma$$

at each point of A. For those who know linear algebra this equality is an immediate consequence of the last paragraph of the first section of Chapter IX (since the determinant of the product of two linear transformations is the product of the determinants), but it is possible to give a more "elementary" proof, as follows. Since $\varphi = \tau \circ \sigma$, for any $i, j = 1, \ldots, n$ the chain rule gives

$$(\varphi_i)'_j = ((\tau \circ \sigma)_i)'_j = (\tau_i \circ \sigma)'_j = \sum_{k=1}^{n} ((\tau_i)'_k \circ \sigma)(\sigma_k)'_j$$
$$= \begin{cases} (\tau_i)'_j \circ \sigma + ((\tau_i)'_n \circ \sigma)(\varphi_n)'_j & \text{if } j < n \\ ((\tau_i)'_n \circ \sigma)(\varphi_n)'_n & \text{if } j = n. \end{cases}$$

Thus the $n \times n$ square array $((\varphi_i)'_j)$ is obtained from the $n \times n$ square array $((\tau_i)'_j \circ \sigma)$ as follows: if $j < n$ then each element of the jth column of the former equals the corresponding element of the jth column of the latter plus $(\varphi_n)'_j$ times the corresponding element of the nth column of the latter, while each element of the nth column of the former equals $(\varphi_n)'_n$ times the corresponding element of the nth column of the latter. By the elementary properties of determinants we have

$$\det ((\varphi_i)'_j) = \det ((\tau_i)'_j \circ \sigma) \cdot (\varphi_n)'_n,$$

that is,

$$J_\varphi = (J_\tau \circ \sigma) J_\sigma,$$

which is precisely what remained to be shown.

PROBLEMS

1. Let $I_1, \ldots, I_N$ be disjoint open intervals in E^n. Show that if $J_1, \ldots, J_M$ are open intervals in E^n such that

$$I_1 \cup \cdots \cup I_N \subset J_1 \cup \cdots \cup J_M$$

then

$$\operatorname{vol}(I_1) + \cdots + \operatorname{vol}(I_N) \leq \operatorname{vol}(J_1) + \cdots + \operatorname{vol}(J_M).$$

2. Can you give a less computational argument for Example 3 of § 1?
3. Prove that a continuous real-valued function on a closed interval in E^n is integrable, using only Lemma 1 of § 2 and uniform continuity.
4. Do the n-dimensional generalization of Problem 6, Chapter VI, with $[a, b]$ replaced by a closed interval I of E^n and $\int_a^b f(x)dx$ by $\int_I f$.
5. Write down in all detail the proof of the first proposition of § 2.
6. Let f be a real-valued function on a subset A of E^n. Show that if $\int_A f$ exists, then so does $\int_A |f|$, and $\left|\int_A f\right| < \int_A |f|$. (*Hint:* First assume that A is a closed interval.)
7. (a) Let f be a real-valued function on a closed interval I of E^n. Show that if f is integrable on I then so is f^2.

   (b) Let f, g be real-valued functions on a closed interval I of E^n. Show that if f and g are integrable on I then so is fg.

   (c) Let f, g be real-valued functions on a subset A of E^n. Show that if $\int_A f$ and $\int_A g$ exist, then $\int_A fg$ exists.

   (d) Let f be a real-valued function on a subset A of E^n and let $B \subset A$. Show that if $\int_A f$ exists and B has volume, then $\int_B f$ exists.

(e) Show that if the subsets A and B of E^n have volume, then so do the sets $A \cap B$, $A \cup B$ and $A - B$.

8. Show that if a subset $A \subset E^n$ has volume, then the interior of A (cf. Prob. 15, Chap. III) has the same volume.

9. Show that a bounded subset A of E^n has volume if and only if the boundary of A (cf. Prob. 17, Chap. III) has volume zero.

10. Let f be a bounded real-valued function on a closed interval I of E^n. Prove that f is integrable on I if and only if, for any $\epsilon, \delta > 0$, I is the union of a finite set of closed subintervals such that the sum of the volumes of those subintervals on which f varies by at least ϵ is less than δ.

11. Let f be a bounded real-valued function on a closed interval I of E^n. Prove that f is integrable on I if and only if, for each $\epsilon > 0$, the set of points of I at which the oscillation of f (cf. Prob. 5, Chap. IV) is at least ϵ has volume zero.

12. Use the preceding problem to show that a bounded real-valued function f on a closed interval I of E^n is integrable on I if and only if the set of points of I at which f is not continuous is the union of a sequence of subsets of I of volume zero.

13. Show that the nonempty subset of $[0, 1]$ consisting of those numbers which have decimal expansions none of whose digits is 5 is the set of its own cluster points. Show that this set is of volume zero.

14. For each integer $n > 1$ let S_n be the union of the open balls in $\mathbf{R}$ of centers $1/n, 2/n, \ldots, (n-1)/n$ and radii $1/n2^{n+1}$. Prove that $\bigcup_{n=1,2,3,\ldots} S_n$ is an open subset of $[0, 1]$ without volume. (*Hint:* If this set had volume, the volume would be 1. But the union of any finite number of S_n's has volume less than 1/2.)

15. Let $A \subset E^n$ and let $f: A \to E^m$. Consider the condition that there exist some $M \in \mathbf{R}$ such that $d(f(x), f(y)) \leq Md(x, y)$ for all $x, y \in A$.
(a) Show that the condition is satisfied if f is the restriction to A of some differentiable map into E^m of some open subset of E^n containing A, if the partial derivatives of the component functions of f are bounded on A and A contains the entire line segment between any two of its points.
(b) Show that if the condition is satisfied, if $m = n$, and $\text{vol}\,(A) = 0$, then $\text{vol}\,(f(A)) = 0$. (*Hint:* A is contained in the union of a finite number of cubes of total volume less than any prescribed positive number.)
(c) Show that if the condition is satisfied, if $m > n$, and A is bounded then $\text{vol}\,(f(A)) = 0$. (Contrast with Prob. 31, Chap. IV.)

16. Prove that if $A \subset E^n$ has positive volume and f is a positive-valued function on A such that $\int_A f$ exists, then $\int_A f > 0$. (*Hint:* Reduce to the case where A is a closed interval and for any positive ϵ we have $\text{vol}\,(\{x \in A : f(x) \geq \epsilon\}) = 0$, then try to use compactness.)

17. Let $A \subset E^n$ be a set with volume and $f: A \to \mathbf{R}$ a continuous function. Show that if the set $\{x \in A : f(x) = 0\}$ has volume zero, then the set $\{x \in A : f(x) > 0\}$ has volume.

18. Let A be a bounded subset of E^n and $f_1, f_2, f_3, \ldots$ a sequence of real-valued functions on A that converges uniformly to the limit function f. Show that if $\int_A f_m$ exists for all m, then $\int_A f$ exists, and $\int_A f = \lim_{m\to\infty} \int_A f_m$. Is this true if A is not bounded?

19. Let $A \subset E^n$ be compact and have volume, let $U \subset \mathbf{R}$ be open, and let f be a continuous real-valued function on the set

$$\{(x_1, \ldots, x_n, y) \in E^{n+1} : (x_1, \ldots, x_n) \in A, y \in U\}.$$

Prove that if $\partial f/\partial y$ exists and is continuous on the latter set, then

$$\frac{d}{dy}\int_A f(x, y)\, dx = \int_A \frac{\partial f}{\partial y}(x, y)\, dx.$$

20. Let $V \subset E^n$ be compact and have volume and let A and K be continuous real-valued functions on V and $V \times V$ respectively. Show that if

$$|\text{vol}\,(V)\, K(x, y)| < 1$$

for all $x, y \in V$ then there is a unique continuous real-valued function φ on V such that

$$\varphi(x) = A(x) + \int_V K(x, y)\, \varphi(y)\, dy$$

for all $x \in V$.

21. Let $F: [0, 1] \times [0, 1] \to \mathbf{R}$ be defined by $f(x, y) = 0$ if x and y are not both rational, while $f(x, y) = 1/q$ if x and y are rational and q is the smallest positive integer such that qx is an integer. Show that

$$\int_{[0,1]\times[0,1]} f(x, y)\, dx\, dy = 0 \quad \text{but} \quad \int_{[0,1]}\left(\int_{[0,1]} f(x, y)\, dy\right) dx$$

is not defined. What about $\int_{[0,1]}\left(\int_{[0,1]} f(x, y)\, dx\right) dy$?

22. Compute

$$\iiint_{0\le x,y,z \text{ and } x+y+z\le 1} x\, y^2\, z^3\, dx\, dy\, dz.$$

23. Change the order of integration in

$$\int_0^1\left(\int_0^{(8-8x^2)^{1/3}}\left(\int_0^{(16-16x^2-2y^3)^{1/4}} f(x, y, z)\, dz\right) dy\right) dx$$

(five answers).

24. Show that if $[a, b]$ is a closed interval in $\mathbf{R}$ and $f: [a, b] \times [a, b] \to \mathbf{R}$ is continuous, then

$$\int_a^b\left(\int_a^x f(x, y)\, dy\right) dx = \int_a^b\left(\int_y^b f(x, y)\, dx\right) dy.$$

25. Compute

$$\text{vol}\,(\{(x_1, \ldots, x_n) \in E^n : 0 < x_1, \ldots, x_n \text{ and } x_1 + \cdots + x_n \le 1\}).$$

26. Let V_n be the volume of the closed unit ball in E^n, that is, the set

$$\{(x_1, \ldots, x_n) \in E^n : x_1^2 + \cdots + x_n^2 \le 1\}.$$

Show that if $n > 1$ then $V_n = 2\, V_{n-1} \int_0^1 (1 - t^2)^{(n-1)/2}\, dt$, and hence (applying Prob. 39, Chap. VII) that $V_n = (2\pi/n) V_{n-2}$ if $n > 2$.

27. Let $A \subset E^n$, $B \subset E^m$, let f and g be integrable real-valued functions on A and B respectively, and let π_A and π_B be the projections of $A \times B$ onto its factors, that is $\pi_A(x, y) = x$ and $\pi_B(x, y) = y$ if $x \in A$ and $y \in B$. Show that

$$\int_{A\times B} (f \circ \pi_A)(g \circ \pi_B) = \left(\int_A f\right)\left(\int_B g\right).$$

(*Hint:* Problem 7 can simplify the proof.)

28. Let $A \subset E^n$ and $B \subset E^m$. Show that
(a) if A and B have volume, then (by Problem 27)

$$\text{vol}\,(A \times B) = \text{vol}\,(A) \cdot \text{vol}\,(B)$$

(b) if vol $(A \times B)$ exists and is nonzero, then A and B have volume
(c) if vol $(A \times B) = 0$ then A or B has volume zero.

29. Prove that under the conditions of the change of variable theorem, φ maps any subset of A that is contained in a compact subset of A and has volume onto a subset of $\varphi(A)$ that has volume.

30. Prove that if f is a real-valued function on E^2 such that $\int_{E^2} f$ exists, then

$$\int_{E^2} f(x, y)\, dx\, dy = \int_{0\le r, 0\le\theta\le 2\pi} f(r\cos\theta, r\sin\theta)\, r\, dr\, d\theta.$$

(*Hint:* First prove this if f is zero on some open subset of E^2 containing the positive x-axis. Problem 7(d) can help in passing to the general case.)

31. (a) Use Problem 30 to show that for any $k > 0$ we have

$$\int_{\substack{0\le r\le k\\ 0\le\theta\le\pi/2}} e^{-r^2}\, r\, dr\, d\theta = \int_{\substack{0\le x,y\\ x^2+y^2\le k^2}} e^{-x^2-y^2}\, dx\, dy < \int_{x,y\in[0,k]} e^{-x^2-y^2}\, dx\, dy$$

$$< \int_{\substack{0\le x,y\\ x^2+y^2\le 2k^2}} e^{-x^2-y^2}\, dx\, dy = \int_{\substack{0\le r\le k\sqrt{2}\\ 0\le\theta\le\pi/2}} e^{-r^2}\, r\, dr\, d\theta.$$

(b) Deduce that

$$\frac{\pi}{4}(1 - e^{-k^2}) < \left(\int_0^k e^{-x^2}\, dx\right)^2 < \frac{\pi}{4}\left(1 - e^{-2k^2}\right).$$

(c) Prove that $\int_0^{+\infty} e^{-x^2}\, dx = \frac{\sqrt{\pi}}{2}$ (cf. Prob. 28, Chap. VI).

32. Let A be an open subset of E^n and f a real-valued function on A. Let $\mathfrak{D}$ be the set of compact subsets of A that have volume. Call f *absolutely integrable* on A if $\int_D f$ exists for each $D \in \mathfrak{D}$ and there is a number $L \in \mathbf{R}$ such that for any $\epsilon > 0$ there exists some $D \in \mathfrak{D}$ such that if $D' \in \mathfrak{D}$ and $D' \supset D$ then

$$\left|\int_{D'} f - L\right| < \epsilon.$$

(a) Show that if L exists, it is unique. (Hence we may write

$$L = \int_A^{\text{abs}} f.)$$

(b) Show that if $\int_A f$ exists, then $\int_A^{\text{abs}} f = \int_A f$. (*Hint:* Problem 7(d) implies that $\int_D f$ exists for all $D \in \mathfrak{D}$.)

(c) Show that if $\int_D f$ exists for all $D \in \mathfrak{D}$, then f is absolutely integrable on A if and only if for any $\epsilon > 0$ there exists some $D \in \mathfrak{D}$ such that if $D' \in \mathfrak{D}$ and $D' \subset A - D$ then $\left| \int_{D'} f \right| < \epsilon$, which is true if and only if the set $\{ \int_D f : D \in \mathfrak{D} \}$ is bounded, and this in turn is true if and only if the set $\{ \int_D |f| : D \in \mathfrak{D} \}$ is bounded. (Note that $\int_D |f|$ exists for all $D \in \mathfrak{D}$, by Problem 6.)

(d) Show that if f is continuous and f and A are bounded then $\int_A^{\text{abs}} f$ exists.

(e) Let (a, b) be an open interval in $\mathbf{R}$ and let f be a continuous real-valued function on $\{x \in \mathbf{R} : a < x \leq b\}$. Show that if $\int_{(a,b)}^{\text{abs}} f$ exists then $\int_{(a,b)}^{\text{abs}} f = \int_{a+}^{b} f(x)\, dx$ (cf. Prob. 27, Chap. VI), and that if f takes on only nonnegative values and $\int_{a+}^{b} f(x)\, dx$ exists then $\int_{(a,b)}^{\text{abs}} f$ exists.

(f) Let $a \in \mathbf{R}$, $A = \{x \in \mathbf{R} : x > a\}$, and let f be a continuous real-valued function on $\{x \in \mathbf{R} : x \geq a\}$. Show that if $\int_A^{\text{abs}} f$ exists then $\int_A^{\text{abs}} f = \int_a^{+\infty} f(x)\, dx$ (cf. Prob. 28, Chap. VI), and that if f takes on only nonnegative values and $\int_a^{+\infty} f(x)\, dx$ exists then $\int_A^{\text{abs}} f$ exists.

(g) Show that $\int_0^{+\infty} \frac{\sin x}{x}\, dx$ exists, but $\int_{x>0}^{\text{abs}} \frac{\sin x}{x}\, dx$ does not.

Suggestions for Further Reading

Ahlfors, Lars V. *Complex Analysis.* New York: McGraw-Hill Book Company, 1953; Second Edition, 1966.

Apostol, Tom M. *Mathematical Analysis: a modern approach to advanced calculus.* Reading, Mass.: Addison-Wesley Publishing Company, Inc., 1957.

Bartle, Robert G. *The Elements of Real Analysis.* New York: John Wiley & Sons, Inc., 1964.

Boas, Ralph P., Jr. *A Primer of Real Functions.* New York: John Wiley & Sons, Inc., 1960.

Buck, R. Creighton. *Advanced Calculus.* New York: McGraw-Hill Book Company, 1956; Second Edition, 1965.

Dieudonné, Jean. *Foundations of Modern Analysis.* New York: Academic Press, 1960.

Gleason, Andrew M. *Fundamentals of Abstract Analysis.* Reading, Mass.: Addison-Wesley Publishing Company, Inc., 1966.

Goldberg, Richard R. *Methods of Real Analysis.* New York: Blaisdell Publishing Company, 1964.

Halmos, Paul R. *Naïve Set Theory.* Princeton, N.J.: D. Van Nostrand Company, Inc., 1960.

Landau, Edmund. *Foundations of Analysis.* New York: Chelsea Publishing Company, 1951.

Royden, H. L. *Real Analysis.* New York: The Macmillan Company, 1963.

Rudin, Walter. *Principles of Mathematical Analysis.* New York: McGraw-Hill Book Company, 1953; Second Edition, 1964.

Spivak, Michael. *Calculus on Manifolds: a modern approach to classical theorems of advanced calculus.* New York: W. A. Benjamin, Inc., 1965.

Index